CW01501285

Lecture Notes in Energy

Volume 73

Lecture Notes in Energy (LNE) is a series that reports on new developments in the study of energy: from science and engineering to the analysis of energy policy. The series' scope includes but is not limited to, renewable and green energy, nuclear, fossil fuels and carbon capture, energy systems, energy storage and harvesting, batteries and fuel cells, power systems, energy efficiency, energy in buildings, energy policy, as well as energy-related topics in economics, management and transportation. Books published in LNE are original and timely and bridge between advanced textbooks and the forefront of research. Readers of LNE include postgraduate students and non-specialist researchers wishing to gain an accessible introduction to a field of research as well as professionals and researchers with a need for an up-to-date reference book on a well-defined topic. The series publishes single- and multi-authored volumes as well as advanced textbooks. **Indexed in Scopus and EI Compendex** The Springer Energy board welcomes your book proposal. Please get in touch with the series via Anthony Doyle, Executive Editor, Springer (anthony.doyle@springer.com)

More information about this series at http://www.springer.com/series/8874

Manfred Hafner · Simone Tagliapietra

Editors

The Geopolitics
of the Global
Energy Transition

Springer Open

Editors
Manfred Hafner
Fondazione Eni Enrico Mattei
Milan, Italy

Simone Tagliapietra
Fondazione Eni Enrico Mattei
Milan, Italy

ISSN 2195-1284   ISSN 2195-1292   (electronic)
Lecture Notes in Energy
ISBN 978-3-030-39065-5   ISBN 978-3-030-39066-2   (eBook)
https://doi.org/10.1007/978-3-030-39066-2

This Springer imprint is published by the registered company Springer Nature Switzerland AG
The registered company address is: Gewerbestrasse 11, 6330 Cham, Switzerland

Foreword

In physics, power is the rate of work per unit time. In geopolitics, it is the ability of one nation to influence the behaviour of other nations. It's not surprising that the same word, with its root in the Latin posse (to be able), refers to both concepts. A nation's command of physical power, notably through its control over primary energy resources such as oil, not only shapes its economic development but also its national security and military strength. As such, the international relations of nations are profoundly influenced by the distributions of energy resources and the technologies for their utilization.

Every great transition in energy technology entails a shift in geopolitics as well. Our generation's energy transition to zero-carbon energy, or decarbonization, will reshape geopolitics of the twenty-first century. This superb volume offers a deeply informed tour d'horizon of the geopolitics of global energy decarbonization, and the ways that geopolitics may stymie or support the transition to climate safety.

Energy transitions have defined several key epochs of human history. Early man's harnessing of fire changed the genetic and cultural trajectory of humanity itself. The harnessing of wind power for sailing ships enabled sea-based trade and migrations over vast distances. The harnessing of horse power gave rise to empires. And without doubt, it was James Watt's coal-fired steam engine (building on precursors of Savery and Newcomen) that gave rise to industrialization, and with it, to Britain's remarkable global hegemony in the nineteenth and early twentieth centuries.

Watt's invention, commercialized around 1776, ushered in the Fossil-Fuel Age, a period of more than two centuries in which global economic growth has been powered by solar energy of tens of millions of years ago stored in fossilized remains of plant and animal life. By gaining the ability to harness the fossil fuels—coal, oil and natural gas—humanity tapped into a seemingly limitless reserve of power. Watt's coal-fired steam engine was followed by Daimler and Benz's internal combustion engine for automobiles, and Parsons' and Curtis's gas turbines for transport and mechanical power.

The global distribution of fossil fuel is highly unequal, with some countries blessed with massive reserves and others bereft of fossil fuels that are exploitable on an economical basis. Ownership of plentiful fossil fuels, not surprisingly, has tended to give a huge boost to economic development, military might and geopolitical influence. Britain's highly accessible coal reserves were essential to turn Watt's new steam engine from a mere curiosity to the source of Britain's imperial might. Other countries with accessible coal reserves generally found an earlier path to industrialization in the nineteenth and twentieth centuries.

By the early twentieth century, oil joined coal as a new key currency of geopolitics, especially after Winston Churchill ordered Britain's navy to convert from coal-burning steam engines to oil-burning steam engines, to be followed later by the navy's conversion from steam power to diesel engines. The control of oil became key to military and geopolitical power. Like coal, oil was very unequally distributed around the world. Oil-rich regions like the US and Russia gained vast geopolitical and economic advantages, or alternatively fell victim to military conquests by Britain, the US, Russia and other major powers that acted militarily to secure their oil supplies.

But for the climate crisis, the fossil-fuel era would surely be continuing today and into the future with the exploitation of new fossil-fuel reserves through fracking, deep-sea drilling and other advances in exploration, development and utilization. Despite claims to the contrary, the world has enough coal, oil and natural gas to last for centuries. The fossil-fuel era is ending not by running out of fossil fuels but for a wholly different reason, indeed a quirk of quantum physics. As great nineteenth-century scientists including Fourier, Tyndall and Arrhenius came to realize that fossil fuels have a pesky side effect. When they are combusted, they release carbon dioxide (CO_2) into the atmosphere, and CO_2 has the quantum physical property of absorbing electromagnetic radiation at infrared wavelengths. The implication? Atmospheric CO_2 warms Earth by trapping infrared radiation that would otherwise radiate from Earth to outer space. CO_2 is, in modern parlance, a greenhouse gas. To an extent, this is life-saving good news. Without atmospheric CO_2 and other greenhouse gases, Earth would be as cold and lifeless as the moon. Yet with too much atmospheric CO_2, the Earth's temperature will rise to dangerous levels.

Such is our current predicament and the backstory of this important volume. Humanity has already raised the atmospheric concentration of CO_2 from 280 parts per million (ppm) to around 415 ppm in May 2019, mostly through fossil-fuel use but also through deforestation and other economic activities. The result is that the planet is now around 1.2 °C warmer than when Watt unveiled his world-changing steam engine. 1.2 °C might not seem like much, but it is enough to make Earth warmer than at any time since the start of civilization itself some 10,000 years ago. It is enough to be disrupting societies, economies and ecosystems around the world. We are already in an era of intense storms, rising sea levels, droughts, floods, emerging infectious diseases, massive forest fires and other climate-related disasters. We are already suffering worldwide losses amounting to hundreds of billions of dollars per year, and mass displacements of populations, including losses of life.

And the prospects of further warming are even more dangerous. The Earth is already at, or soon will be at, temperatures high enough to result in a multi-metre rise of ocean levels, with the consequence of disrupting lives for hundreds of millions or billions or people around the world. Moreover, the Earth's climate system is characterized by multiple 'positive feedbacks' that amplify the human-induced warming, such as the loss of sea ice (which reduces the Earth's albedo and accelerates warming), the melting of the permafrost (which releases methane and carbon dioxide) and the drying of rainforests (turning them from carbon sinks to carbon sources).

For all of these reasons, every UN member state adopted the Paris Climate Agreement in December 2015, to hold warming to 'well below 2 °C' and to aim to limit warming to 1.5 °C or less. Each country is to put forward a 'nationally determined contribution' of energy transformation and other economic changes (e.g. in land use and diets) so that the sum of the national efforts will be sufficient to achieve the global goal. The Intergovernmental Panel on Climate Change (IPCC) has, in turn, underscored the importance of adhering to the lower bound of 1.5 °C rather than 2 °C in order to improve the prospects of climate safety. Moreover, to have a likely (>66%) outcome of 1.5 °C or less, the world as a whole must achieve net-zero greenhouse gas emissions by 2050. In shorthand, the world must decarbonize by 2050 and shift from net deforestation to net reforestation and afforestation. So far, the world is grievously off-track, hurtling towards warming by 2100 of 3 °C or higher.

The engineers have made clear that decarbonization by 2050 is feasible, by shifting power generation from coal, oil and gas to zero-carbon energy sources (wind, solar, geothermal, hydro, ocean, biomass, nuclear or fossil fuel with carbon capture and storage), and by using that zero-carbon power to provide the energy for downstream uses, notably transport, buildings and industry. To the extent possible, downstream uses should be directly electrified, as with light-duty battery-electric vehicles and electric heat pumps for home heating. And when direct electrification is not feasible, the zero-carbon power should be used to create synthetic (or 'green') fuels such as hydrogen, green methane, green ammonia and other fuel carriers that can be combusted without releasing CO_2 into the atmosphere. The amazing finding by economists and engineers is that most or all of the decarbonization is within reach at very low cost, perhaps 1% or less of annual world output until 2050.

Yet here geopolitics enters the scene. Indeed, geopolitics may prove to be the single biggest factor in the success or failure of the world's quest for decarbonization and energy safety. The limiting factor in decarbonization, it seems, is not low-cost alternatives, nor the shortage of zero-carbon primary energy sources (solar radiation and wind alone could power the planet). The limiting factor, it seems, is global political cooperation to get the job done through a variety of policy instruments such as carbon pricing, public investments, public–private R&D and government regulation. And what is the real source of this political resistance? One answer is the usual problem of free-riding, as each nation tries to get the others do the hard work and bear the incremental costs. But something else is at play: geopolitics.

For two centuries, national control over fossil fuels has been the key to the global realm. National security, military strength and political power have been deeply intertwined with the nation's control over fossil fuels, especially in the case of countries that are rich in fossil-fuel reserves. And when global superpowers have lacked sufficient domestic supplies of key energy resources (such as the British Empire and oil during 1900–1950, and the US and oil after 1945 at least until recently), they have deployed their military might to secure effective control over overseas oil, often through wars or coups to install pliant regimes, especially in the hydrocarbon-rich Middle East.

The energy transition to zero-carbon energy is, for these reasons, being heavily shaped—and more specifically, slowed—by today's fossil-fuel-rich countries, especially the US, Russia and Saudi Arabia, but also Canada, Australia, Venezuela, Brazil, Mexico, Iran, Iraq and others. If the world succeeds in decarbonizing, these fossil-fuel-rich countries stand to lose, at least relatively, in three ways. First, they will suffer major capital losses as their fossil-fuel reserves are stranded (that is, left under the ground by the requirements of climate safety). Second, they will suffer ancillary economic losses as their treasuries can no longer finance the public sector through fossil-fuel rents. Third, and perhaps as consequential as the first two, they will lose geopolitical relative advantages as their concentrated holdings of fossil fuels are supplanted by globally diffuse access to wind, solar, hydro and other zero-carbon energy sources.

The two populous giants of the planet, China and India, are perhaps the most complex players in the unfolding geopolitics of the energy transition. On the one hand, both of these giants are densely populated and highly vulnerable to climate change, India especially. China is water scarce, especially in the north, and further global warming could play havoc with China's ecosystems. India is heavily trop-ical, water scarce and already under intense ecological stress and degradation. Parts of India are now reaching 50 °C summer days in the context of global warming. Large swaths of India could be rendered essentially uninhabitable through the combination of poverty, water stress, global warming, air and water pollution, soil degradation and other forms of environmental degradation. Nor are these two countries fossil-fuel-rich. They each have enough coal to do dire and lasting damage to the global environment, but they are also both heavily dependent on imported hydrocarbons.

The EU is another key region in this geopolitical puzzle. As one of the three hubs of the global economy, together with North America and Northeast Asia, Europe's energy transition is a vital part of the global story. Europe is mostly fossil-fuel-poor (especially after two centuries of fossil-fuel extraction and use). As such, the European politics are strongly directed towards decarbonization, except in the few remaining pockets of local coal production (Germany, Eastern Europe and Spain).

The energy transition geopolitics to date, therefore, shapes up as follows. The major fossil-fuel powers are generally resistant to decarbonization, now led by the Trump Administration, which has become an outright foe determined to topple the Paris Climate Agreement. While Russia, Saudi Arabia, Australia, Brazil and others still offer rhetorical support for decarbonization, in practice they tend to line up behind Trump in closed-door negotiations. The European Union is on the other end of the geopolitical spectrum, with much to lose from global warming (as is basically true everywhere) and with little at stake in terms of European fossil-fuel reserves. China and India are the world's swing actors, giants that can shape the planet's climate prospects this century, and with mixed motives. On the one hand, they face potential disasters with runaway climate change and are in any event heavily dependent on hydrocarbon imports. On the other hand, they still have significant domestic deposits of coal, which provide vast employment and which constitute a powerful industrial lobby for the status quo.

Of course, as this volume spells out in considerable and illuminating detail, the geopolitical issues extend beyond fossil-fuel haves and have-nots. The zero-carbon energy technologies will also give rise to new global value chains, for example, for the key minerals used in solar panels, wind turbines and batteries. Copper, lead, molybdenum, nickel, zinc, cobalt, lithium and rare earth minerals are all key inputs into the new zero-carbon technologies. These mineral resources are also unequally distributed around the world. Many are located in unstable and impoverished societies (such as the Democratic Republic of the Congo and cobalt reserves). Already the world's major regions are anticipating an intense rivalry and geopolitical competition over global supply lines for the key minerals. As with the past century's many wars over oil, this century could see wars over the mineral inputs to the new energy economy.

The geopolitics of the energy transition will, therefore, involve an intense and highly disruptive era of international relations, one that could end up sinking the world's hopes for climate safety under a multi-metre sea-level rise—if global delay and disruption prevail over global cooperation. To succeed globally, we will need to chart an era of unprecedented cooperation. Here too, this volume is of inestimable value and insight. Global success will require new forms of industrial policy, global diplomacy, transnational infrastructure, global-scale public–private partnerships for research and development and shared policies to adjust to ongoing climate change as well as to decarbonize the energy system.

In sum, we are in for a tumultuous era in geopolitics that will require our greatest reservoirs of wisdom, tolerance, diplomacy and global-scale cooperation. To reach such an unprecedented degree of cooperation, we will need an unprecedented degree of global understanding. This unique volume offers a vital and timely contribution to that much-needed new era of global understanding.

Jeffrey Sachs
Columbia University
New York, USA

Preface

The world is undergoing an historical energy transition, driven by increasingly strong decarbonization policies and quick low-carbon technology developments.

The Paris Agreement marked a major step forward in global efforts to address global warming. For the first time, developed and developing countries committed to act in order to limit global average temperature increase to well below 2 °C, and to pursue efforts to further limit this to 1.5 °C above pre-industrial levels. This reinforces decarbonization measures already being undertaken in different parts of the world. With its Special Report on Global Warming of 1.5 °C, the Intergovernmental Panel on Climate Change (IPCC) stepped up pressure on decarbonization, outlying that limiting global warming to 1.5 °C would require rapid, far-reaching and unprecedented changes in all aspects of society.

Meanwhile, technological advancements have significantly increased the cost-competitiveness of low-carbon technologies such as solar and wind power generation, power storage technologies and electric vehicles. This has already started to reshape the global energy system, notably by giving a greater role to solar and wind in the power generation mix.

By transforming the global energy architecture, international decarbonization policies and low-carbon technology advancements will also have profound geopolitical implications. The large-scale shift to low-carbon energy is disrupting the global energy system, impacting economies and changing the political dynamics within and between countries.

This book seeks to provide a comprehensive analysis of the geopolitical aspects of the global energy transition, from both regional and thematic perspectives. An introductory chapter will set the scene by presenting the big picture of the challenges and opportunities related to the global energy transition, from a technological, economic and geopolitical perspective. The first part of the book will provide a set of regional insights, aimed at analysing the geopolitical implications of the global energy transition in the world's main energy-producing and energy-consuming regions. The second part of the book will provide in-depth focuses on selected issues, spanning from the geopolitics of renewable energy to the mineral foundations of the global energy transformation, up to the governance issues related to the changing global energy order.

Presenting a comprehensive overview of the issue, this book aims to be accessible to a wide readership of both academics and professionals working in the energy industry, as well as to general readers interested in the ongoing debate about energy and climate change.

This book has been written just before COVID-19 hit our societies and economies. The impacts of the pandemic have not spared the world of energy. Lockdown measures put in place all over the world to contain the contagion indeed represented an unparalleled shock to the economies and to energy markets with dramatic impacts to oil and gas markets. The effects of this situation on the global energy transition remain, at the time of writing, difficult to assess. On the one hand, the economic crises and the historical disruption experienced by the petroleum industry might negatively impact the global energy transition, as public and private investments needed for the energy transition may now be channelled into economic recovery and as low-cost oil and natural gas could temporarily slow-down the deployment of clean energy technologies. On the other hand, the aggressive economic recovery policies being adopted by governments across the world could represent a formidable driver for the global energy transition, if focused on green investments as seems to be the case in several countries/regions. Whatever the impacts of COVID-19 on the geopolitics of the global energy transition will ultimately be, they will inevitably build on the challenges and opportunities described in the different chapters of this book. The present work can thus also be considered as a tool to understand the geopolitical fundamentals of the global energy transition that COVID-19 might contribute to reshape.

Support from the Fondazione Eni Enrico Mattei (FEEM) in realizing this book and financing its open access is gratefully acknowledged. Founded in 1989, FEEM is a non-profit, policy-oriented, international research centre and a think tank producing high-quality, innovative, interdisciplinary and scientifically sound research on sustainable development. It contributes to the quality of decision-making in public and private spheres through analytical studies, policy advice, scientific dissemination and high-level education. Thanks to its international network, FEEM integrates its research and dissemination activities with those of the best academic institutions and think tanks around the world.

Within FEEM, the Future Energy research Program (FEP), where this book has been conceived and elaborated, aims to carry out interdisciplinary, scientifically sound, prospective and policy-oriented applied research, targeted at political and business decision-makers. This aim is achieved through an integrated quantitative and qualitative analysis of energy scenarios and policies. This interdisciplinary approach puts together the major factors driving the change in global energy dynamics (i.e. technological, economic, financial, geopolitical, institutional and sociological aspects). FEP applies this methodology to a wide range of issues such as energy demand and supply, infrastructures, market analyses and socio-economic impacts of energy policies.

The editors are thankful to Paolo Carnevale and Luca Farinola, respectively, Executive Director and Financial Director of FEEM, for their strong support in the realization of this book. They also thank Barbara Racah and Pier Paolo Raimondi for their help throughout the editorial process. Finally, a special acknowledgement to Prof. Jeffrey Sachs for his inspiring foreword and for the precious cooperation established between FEEM and SDSN to lay down a scientific pathway towards total decarbonization by 2050.

Milan, Italy   Manfred Hafner
June 2020   Simone Tagliapietra

Introduction

Energy has long shaped global geopolitics, determining great powers, alliances and outcomes of wars. Every international order in modern history has been based on an energy resource: coal was the backdrop for the British Empire in the nineteenth century, oil has been at the core of the subsequent 'American Century', and today many expect China to become the twenty-first century's world renewable energy superpower.

Since World War I, oil has undoubtfully represented the cornerstone of global energy geopolitics. The decision of then-First Lord of the Admiralty Winston Churchill to shift the power source of the Royal Navy's ships from coal to oil in order to make the fleet faster than its German counterpart truly signed the opening of a new era. The switch from the reliable coal supplies from Wales to the insecure oil supplies from what was then Persia, not only made the oil-rich Middle East a key epicentre of global geopolitics, but also turned oil into a key national security issue.

Since the early twentieth century, control of oil resources played a central role in several wars. This was, for instance, the case of the 1967–1970 Biafran War, the 1980–1988 Iran–Iraq War, the 1990–1991 Gulf War, the 2003–2011 Iraq War and of the conflict in the Niger Delta ongoing since 2004.

The second half of the twentieth century also saw increasing tensions between oil-producing and oil-consuming countries, which in two cases erupted in major oil crises. In September 1960, the Organization of Petroleum Exporting Countries (OPEC) was established in Baghdad, with the participation of five member countries: Saudi Arabia, Iraq, Iran, Kuwait and Venezuela. The original aim of OPEC was to prevent its members from lowering the price of oil, by coordinating their production and export policies. During the 1970s, some of OPEC members also had the aim of nationalizing their petroleum resources to preserve sovereignty.

The geopolitical role of OPECs became clear as the Arab–Israeli War—also known as Yom Kippur War—erupted in October 1973. Arab members of OPEC imposed an embargo against the United States, the Netherlands, Portugal and South Africa in retaliation of their support to Israel. A ban of oil exports to the targeted countries as well as oil production cuts was introduced by OPEC. This resulted in a sharp rise in oil prices, and in severe oil shortages and spiralling inflation across the West. As OPEC kept raising prices in the following years, its geopolitical and economic power grew.

In the aftermath of the 1973 oil crisis, and upon proposal of then-US Secretary of State Henry Kissinger, the IEA was established in November 1974 as a platform for oil-importing countries in the West to coordinate a shared response to major disruptions in the supply of oil. This was also allowed by the introduction of a requirement for all IEA member countries to maintain strategic petroleum reserves equal to at least 90 days of their previous year's net oil imports.

A second oil crisis erupted in 1979, as a result of the Iranian revolution and the following 1980–88 war with Iraq, which brought the region into turmoil. By 1981 the price of oil stabilized at USD 32 per barrel, a level ten times higher than before the 1973 oil crisis.

In the following decades, other oil price shocks occurred, notably in relation to major geopolitical developments in the Middle East. For instance, in 1990, an oil price shock took place in the aftermath of the Iraqi invasion of Kuwait, with a doubling of oil price in a matter of few months that contributed to the early 1990s recession in the United States.

But energy geopolitics is not limited to oil. Natural gas, nuclear energy and even renewable energy sources such as wind and solar do have—more or less critical—geopolitical aspects.

In certain areas of the world, natural gas is even considered to be more geopolitical than oil. This certainly is the case of Europe, where natural gas markets have been developed since the 1960s on the basis of large pipeline infrastructures connecting key suppliers such as Russia and Norway to European consumers. This situation has led to an over-reliance of Europe on few major suppliers. Natural gas imports from Russia indeed continue to provide a third Europe's total natural gas supply mix.

For decades, this situation has not raised energy security concerns in Europe. During the 1970s and the 1980s, in the midst of the Cold War, Europe decisively pursued the construction of the long pipelines connecting the large Siberian natural gas fields and Europe, which still today represent the main avenues of Russian natural gas export. Europe pursued these projects notwithstanding the strong opposition of the Reagan Administration, which even sanctioned German and French companies engaged in the construction of the 'Brotherhood' pipeline, which still today represents the major natural gas supply route to Europe.

The (over-)reliance on Russian natural gas supplies started to be considered as a major geopolitical threat in Europe when, first in January 2006 and then in January 2009, natural gas pricing dispute between Russia and Ukraine led to the halt of Russian natural gas supplies to Europe via Ukraine—its primary transit route. This generated economic damages for Europe, notably in South-Eastern European countries heavily dependent on Russian natural gas for both electricity generation and residential heating. Europe responded to these natural gas crises by adopting an energy security strategy mainly focused on reducing its dependency on Russian natural gas supply. In the midst of the 2014 Ukraine crisis, concerns about a potential politically motivated disruption of all European natural gas supplies from Russia lifted again this issue to the top of the European agenda, leading to renewed efforts to lower the European dependency on Russian natural gas supply under the umbrella of the European Union (EU)'s 'Energy Union' initiative.

On its side, nuclear energy presents both security and geopolitical concerns. Issues like safety of nuclear facilities and nuclear waste management represent serious security concerns. The concerns for nuclear safety have particularly amplified after the Chernobyl accident in 1986 and the Fukushima disaster of 2011. These events sparkled, particularly in Europe and in Japan, broad public debates on nuclear energy. In certain cases, these debates led to radical energy policy shifts. For instance, after the Chernobyl accident Italy holds a referendum on nuclear power, which resulted in the decision to close-down all operating nuclear power plants in the country. More recently, after the Fukushima disaster a surge of anti-nuclear protests in Germany pushed Chancellor Angela Merkel to announce the closure of around half of the operating reactors in the country and the complete phase out of nuclear by 2022. These concerns have been most recently accompanied by the emergence of new risks concerning potential terrorist attacks at nuclear power plants.

From a geopolitical perspective, proliferation is the main risk associated to nuclear energy. Elements of the nuclear fuel cycle can, in fact, be used to develop nuclear weapons, either through uranium enrichment or through reprocessing (i.e. the separation of plutonium from the highly radioactive spent fuel). It was precisely the close link between the civil and military use of nuclear energy that led to the establishment in 1957 of the International Atomic Energy Agency (IAEA), a United Nations organization tasked of promoting the peaceful use of nuclear energy. In 1968 (i.e. in the midst of the Cold War), the General Assembly of the United Nations also approved the Nuclear Non-Proliferation Treaty, aimed at the disarmament of countries with nuclear weapons, as well as at the prevention of nuclear weapons adoption by countries still without them.

But if for more than half a century oil, natural gas and nuclear energy have been at the heart of the geopolitics of energy, it is sensible to investigate if and how this will change as a result of the global energy transition, a process driven by decarbonization policies and by quick developments in renewable energy technologies and electric cars.

The Paris Agreement marked an important step forward in global efforts to respond to the challenge of global warming. For the first time, developed and developing countries have committed themselves to act to limit the increase in the average global temperature to well below 2 °C compared to pre-industrial levels. This reinforces the decarbonization measures already in place in several parts of the world, primarily in Europe. Meanwhile, technological advances have increased the competitiveness of solar and wind energy technologies, batteries and electric cars. The convergence of these two elements has already begun to reshape the global energy system. By transforming the global energy architecture, international decarbonization policies and low-carbon technology advancements will also have profound geopolitical implications. The large-scale shift to low-carbon energy is disrupting the global energy system, impacting economies and changing the political dynamics within and between countries. But what will be the consequences of these developments on the geopolitics of energy?

As far as energy-importing countries are concerned, the consequences will certainly be positive. In these cases, as imports of oil and natural gas will decrease, both their 'national energy bill' and the associated geopolitical risks will decrease.

Countries that are able to innovate more in renewables, batteries and electric cars will also be able to reap the industrial and economic benefits of this transition, generating jobs and economic growth.

But, of course, the energy transition will also see the emergence of new geopolitical challenges.

Firstly, the global energy transition represents a challenge for oil- and gas-producing countries, and, in particular, for those with a less diversified economy more dependent on oil revenues. This is the case for many countries in the Middle East and North Africa which, despite the adoption of elaborate strategies for economic diversification, have not really made much progress in this direction. If the global energy transition were to take place more quickly than expected, and if these countries were to remain unprepared, the consequences could be serious from both the socio-economic and geopolitical points of view.

Secondly, the spread of renewable energies will increase electrification and stimulate cross-border trade in electricity. Energy sources such as solar and wind require flexible energy systems that can cope with the variability of weather conditions. Smart electricity grids will, therefore, play an increasingly important role in mitigating this variability and ensuring system stability. The digitization of electricity grids clearly presents security risks, as terrorist groups or hostile countries could seek to either enter the systems to extrapolate information, or to disrupt them to cause economic and social damages.

Thirdly, it is important to stress that the rapid development of wind and solar energy, together with that of electric cars, raises concerns about the security of supply of the minerals needed to manufacture them. These concerns have also developed following events such as those of 2008, when China imposed a limit on the supply of rare earths—of which it holds a large part of the global production—to foreign buyers, leading to panic in the markets and a very rapid increase in prices. Another case was the 'cobalt crisis' of 1978, following the outbreak of a conflict in the province of Katanga—the heart of world mineral extraction—in what was then called Zaire. The crisis caused a global shortage of cobalt, driving the international price of the mineral to the sky. It is clear that if something like this were to happen in the future, the consequences for the production of electric cars would be important. Cobalt is, in fact, a key component of their batteries. These are just examples of how the minerals at the heart of the energy transition will carry their own geopolitical risks, just as oil and natural gas have had theirs.

The global energy transition will not, therefore, lead to the end of the geopolitics of energy, but rather to its transformation. On the one hand, it might strengthen the energy security of most of the countries currently importing oil and natural gas, promoting job creation and economic growth in those who will be able to seize the industrial opportunities of this development. On the other hand, it might create elements of instability in oil- and gas-exporting countries, which might have to reinvent themselves to keep developing in the new energy era, and new security risks linked to electricity grids and minerals.

This book seeks to provide a comprehensive analysis of the geopolitical aspects of the global energy transition, from both regional and thematic perspectives.

The first part of the book provides a set of regional insights, aimed at analysing the geopolitical implications of the global energy transition in the world's main energy-producing and energy-consuming regions.

The second part of the book provides in-depth focuses on selected issues, spanning from the geopolitics of renewable energy to the mineral foundations of the global energy transformation, up to the governance issues related to the changing global energy order.

In chapter "The Global Energy Transition: A Review of the Existing Literature", Hafner and Tagliapietra present an overview of the existing literature in the field, which surprisingly remains fragmented. This should represent for the reader a useful summary to the state of the art of knowledge in the field, and therefore a useful starting point for the book.

In chapter "The European Union and the Energy Transition", El-Mazzega and Mathieu discuss the European Union and the energy transition by looking at the key strategic energy and climate policy issues facing the next 5 years, also elaborating on how the EU energy and climate policies may be shaped, and what are their global implications.

In chapter "US Clean Energy Transition and Implications for Geopolitics", Elkind tackles the U.S. clean energy transition and its geopolitical implications. It argues that in a time of complicated geopolitics, the country's global standing will be materially affected by the way it will manage energy and climate issues, as will its ability to work with international partners on other global challenges.

In chapter "China: The Climate Leader, and Villain", Meidan analyses how China's emergence as a global economic power and energy consumer has shaped global energy production and trade flows. It argues that while China was a technology follower in the fossil-fuel world, in the energy transition it is likely to play a vastly different role, at the forefront of global innovation and projected towards a global clean technology leadership.

In chapter "Implications of the Global Energy Transition on Russia", Henderson and Mitrova discuss the implications of the global energy transition on Russia, arguing that this poses an existential threat for all the key Russian stakeholders, challenging the very sustainability of the economic (and political) system in the country and therefore requiring a new strategy for the energy sector development.

In chapter "A Fine Balance: The Geopolitics of the Global Energy Transition in MENA", Mills analyses the impacts of the global energy transition on the Middle East and North Africa (MENA) region, the cornerstone of global oil and gas production. It argues that while regional countries are—to different degrees—implementing policies to diversify their economies, regional unrest and conflict, climate change and geopolitical competition between the U.S., Russia, China and other local and international powers complicate the diplomacy and energy security challenges of the MENA energy transition.

In chapter "Addressing Africa's Energy Dilemma", Pistelli discusses how the ongoing low-carbon energy transformation could reshape geopolitics within Africa and between the continent and the rest of the world. It analyses the drivers and modalities of Africa's alleged shift to finally explore geopolitical dynamics, questioning whether Africa is still the locus for the global supply of natural resources, introducing patterns of engagement between Africa and international/regional actors, and finally presenting the socio-economic implications of the shift.

In chapter "Technologies for the Global Energy Transition", Hafner and Noussan highlight the main strengths and weaknesses of the current technologies for the global energy transition, to help the reader in understanding which are the main opportunities and challenges related to the development and deployment of each of them. The chapter also provides strategies and policy recommendations from a technology point of view on how to decarbonize the global energy systems by mid-century and of the necessity to take a systems approach.

In chapter "Policy and Regulation of Energy Transition", Daszkiewicz discusses the role of policies and regulations in fostering the energy transition. It looks at the different types of policies that have been effective in delivering these goals and provides examples for the way forward.

In chapter "The Global Energy Transition: A Review of the Existing Literature", Dell'Aquila, Atzori and Raluca Stroe provide a comparative analysis of how China and the U.K. have implemented policies transitioning away from fossil fuels to renewable energy, discussing the commonalities and differences of the two approaches.

In chapter "Financing the Sustainable Energy Transition", Van de Putte, Campbell-Holt and Littlejohn discuss the financing aspects of the global energy transition. They argue that there is also a role for governments in developing countries to develop their capital markets and gradually internalize the direct and indirect subsidies from which the fossil-fuel industry derives an unfair advantage; only when these various change factors come together will it be possible to scale the sustainable energy transition.

In chapter "Minerals and the Metals for the Energy Transition: Exploring the Conflict Implications for Mineral-Rich, Fragile States", Church and Crawford look at the minerals and metals underpinning the energy transition, in view of exploring the extent to which increased demand for the minerals critical to green energy technologies could affect fragility, conflict and violence in producing states, and explore what would be required by the international community to mitigate these local and national threats.

In chapter "The Impacts of the Energy Transition on Growth and Income Distribution", Luciani discusses the impacts of the energy transition on growth and income distribution, claiming that if we want to make progress with the energy transition, it is necessary to acknowledge its cost and seek agreements on the division of the burden. Agreements are needed at the international level, between rich and poor countries, but also at the national level between rich and poor citizens.

In chapter "The Global Energy Transition and the Global South", Goldthau, Eicke and Weko provide a 'Global South' perspective on the energy transition, by shedding light on the specific circumstances pertaining to countries of this part of the world.

In chapter "Governing the Global Energy Transition", Pastukhova and Westphal conceptualize the governance of energy transition and argue that the Paris Agreement should be accompanied by governance mechanisms in the energy realm, being the energy sector the key contributor to global emissions.

In chapter "Setting up a Global System for Sustainable Energy Governance", Zuev discusses the potential ways to set up a global system for sustainable energy governance, arguing that energy governance institutions are key to a global sustainable transformation.

Manfred Hafner
Simone Tagliapietra

Contents

About the Editors

Manfred Hafner is Professor of International Energy Economics and Geopolitics teaching, among others, at The Johns Hopkins University School of Advanced International Studies (SAIS Europe) and at SciencesPo Paris School of International Affairs (PSIA). He is also the Coordinator of the Future Energy research Program of the Fondazione Eni Enrico Mattei. During his over 30-year working experience, he has extensively consulted for governments, international organizations and the energy industry all over the world.

Simone Tagliapietra is a Research Fellow at the Università Cattolica del Sacro Cuore and Adjunct Professor of Global Energy and Environment Fundamentals at The Johns Hopkins University School of Advanced International Studies (SAIS Europe). He is also Senior Fellow at the Fondazione Eni Enrico Mattei and Research Fellow at Bruegel. He is the author of *Global Energy Fundamentals* (Cambridge University Press, 2020).

The Global Energy Transition: A Review of the Existing Literature

Manfred Hafner and Simone Tagliapietra

This chapter presents an overview of the existing literature on the geopolitics of the global energy transition. Notwithstanding its potentially re-defining role for international relations, this issue has, so far, not been analysed in a comprehensive manner but in a rather fragmented way. This chapter represents a useful summary to the state-of-the-art of knowledge in the field, and therefore a useful starting point for the book.

The first attempt to provide a holistic assessment of the geopolitics of the global energy transition has been made by IRENA (2019). This study maps the geopolitical transformations generated by the rise of renewables and the decline of fossil fuels. The study argues that the rise of renewables will reshape relations between states (i.e., oil and gas exporters versus oil and gas importers) and will lead to fundamental structural changes in economies and society. The report affirms that the world of the renewable energy transition will be very different from the one based on fossil fuels. The report claims that to some extent, the global energy transformation may generate a peace dividend, since the world is driving away from fossil fuels, which are often an aggravating factor in armed conflicts within states. However, the growth of digitalization in the energy sector due to the energy transition, can raise security and privacy risks in the absence of an international rules-based framework. IRENA affirms that global power structures and arrangements will change in many ways and the dynamics of relationships within states will also be transformed. Power will become more decentralized and diffused. Those countries that have invested in renewable technologies will increase their influence in the global context; while, by contrast, those states that rely heavily on revenues from fossil fuels will face substantial challenges to their economic and social models.

Following the IRENA report, Goldthau et al. (2019) published a first academic analysis outlining some geopolitical scenarios for the transition by 2030. The authors present four different scenarios for the energy transition and its effects on global

M. Hafner (✉) · S. Tagliapietra
Fondazione Eni Enrico Mattei (FEEM), Milan, Italy
e-mail: manfred.hafner@feem.it

© The Author(s) 2020
M. Hafner and S. Tagliapietra (eds.), *The Geopolitics of the Global Energy Transition*,
Lecture Notes in Energy 73, https://doi.org/10.1007/978-3-030-39066-2_1

geopolitics. The first scenario ('Big green deal') assumes full cooperation due to a global consensus for action on climate change leads to a concerted international policy. In this scenario, a wave of green globalization allows all countries to share in the benefits of decarbonization; petrostates are compensated to transition smoothly to a sustainable economy. The result is a win-win for climate and security, geopolitical friction is low. The second scenario ('Technology breakthrough') assumes that a major technological advance steers the world along a different path. The US and China take the lead in scaling up the technology, but competition between nations also spikes. Indeed, the world fractures into two camps in a cleantech cold war, where technology leaders hold the power and others gravitate towards one of the leaders, reinforcing regional blocs and increasing rivalry. Fossil-fuel producers have to adapt rapidly to falling demand, with tensions rise in some areas. The third scenario ('Dirty nationalism') assumes that nation-first policies put a premium on self-sufficiency, favouring domestic energy sources over imported ones, which drive the development of fossil fuels as well as renewables. In this context, zero-sum logic returns and power rivalries marginalize the UN and undermine multilateral institutions. The fourth scenario ('Muddling on') is a business as usual one, resulting in a mix of energy clubs, with little cooperation. Fossil fuels remain dominant, despite renewables claim an increasing share of the energy mix by 2030, as unit costs keep declining. The speed of the energy transition is too slow to mitigate climate change, but too fast for the fossil-fuel industry to adapt. Oil-producing countries in the Middle East, Russia and Africa see political turmoil as government coffers empty. Motivated by energy security as much as climate change, countries pursue diverse energy strategies. As some regions have inadequate regulation or fail to benefit from these partnerships, existing economic and geopolitical imbalances are reinforced and energy inequality rises. The authors outline three steps that will help to put geopolitics at the heart of debates about the energy transition: (i) researchers and decision-makers need to shift the attention from targets to pathways; (ii) policymakers need to draw lessons from past and parallel experiences and (iii) abating carbon will create losers, thus now the policy focus needs to switch to the potential conflicts resulting from falling fossil-fuel demand, and the related economic and security risks.

Hafner and Wochner (2020) also provide an outlook of how the global energy transition will play out among the different major global geoeconomic/geopolitical blocks and how it may affect and be affected by global governance. They argue that four main unfolding drivers will lead to major tectonic shifts in the global energy system: i—global energy demand, spurred largely by Asia; ii—"top-down" climate policies that contribute to decarbonization of the global economy; iii—bottom-up technology and market-driven digitalization that favour new energy approaches and also a more decentralized energy system; iv—technological innovation that drives the energy industry both in the fields of renewable energy and low-carbon vehicles, but also in unconventional oil and gas production. The authors then present the strategies presently being developed by the different major global countries/block and they argue for instance that Europe being a major energy importer has a much higher incentive to push for decarbonization policies which bring the co-benefit of improved energy security compared to the United States, which is presently experiencing an economic boom thanks to its unconventional oil&gas revolution allowing it to not only to have access to cheap energy but to even become a major gas exporter and possibly a net oil exporter. At the same time, the US thanks to its formidable entrepreneurial spirit, China with its state-driven policies, but also Europe and Japan are investing heavily in developing new low-carbon technologies which should provide them with a technological and economic advantage in a decarbonizing world. Russia and hydrocarbon exporting Middle East countries, on the other hand, face challenges in the energy transformation of their economies due to multiple system inertia. The authors then test the strategies of the different major global blocks under three possible future scenarios: i—Weak Climate Governance; ii—Global Efforts for Climate Acton; iii—Muddling Through. They argue that with globally Weak Climate policies, energy-exporting countries (the Gulf States, Russia and the US) would remain in a strong position while Europe may end up paying a high price in the short term as its investments in low-carbon technologies may not pay off as quickly as planned, though the medium-long term will provide it with an increased independence from increasingly volatile global energy prices. In a Global Efforts for Climate Action scenario, those countries who are at the forefront of the energy transition will be the clear winners, these include Europe but also China and the US, while traditional fossil fuel producers and exporters (Gulf countries and Russia) will need to quickly diversify their economies if they want to continue to have a role on the international scene (they could for instance convert from fossil fuel exports to hydrogen exports). The authors conclude that the present weak global network spanning the energy field does not provide effective governance mechanisms. The most effective way to govern a global energy transition is to create increased ownership of climate policies both among countries (developed, emerging, developing) and inside societies (rich, middle-class and poor) thanks to a more inclusive and just redistribution of burdens (everybody needs to see a win-win solution for itself), and at the same time to strengthen the review mechanisms of the Paris Accord to oblige States to bring their national energy and climate strategies in line with the goal of preserving the planet Earth.

A comprehensive review of the existing literature on individual aspects of the geopolitics of the global energy transition is now presented, to provide the reader with a clear picture of the current status of knowledge in the field. To facilitate the reader, this literature review follows the structure of the book.

From a regional point of view, there is a heterogeneous spectrum about the quantity and quality of the existing literature: profusion on Europe and MENA, scant/meagre on Russia and the US, while there is almost no study on geopolitical analysis of the energy transition in Africa and China.

1 Europe

Bressand (2012) affirmed that the world energy system is undergoing a far-reaching transition in which three agendas collide: an *economic agenda* of supply and demand and of national competitiveness; a *security agenda* reflecting strategic dependence on trade in oil and gas and a *sustainability agenda* now centred on the search for a low-carbon energy mix. The paper seeks to identify the key players and their strategic postures in this new era of energy geopolitics, with a view to drawing implications for the European Union and the US. The paper identifies the energy seven countries with the highest influence on energy and climate relations across energy sources. The author evaluates Saudi Arabia, Iraq, China and Japan as 'status quo' countries in terms of their core policy stance regarding the global energy system and markets. Then, he shows how the Russian Federation and the EU are the two players intent, for very different, reasons, on changing the game. However, the author affirms that the thrust of the conclusions regarding Europe is far less optimistic than is the case in prevailing views that tend to define success. Gains tend to be assessed with reference to the world as Europeans would like it to be rather than as it is. A geopolitical perspective and the less complacent cost-benefit analysis suggests lead to a sharper and more realistic assessment of energy and climate policy options.

Lombardi and Gruenig (2016) consider low-carbon energy security and energy geopolitics in Europe focusing on four thematic clusters: challenging the energy security paradigm; climate change and energy security objectives (the components of a secure and low-carbon energy system); energy security in a geopolitical perspective, as it relates to economics, resource competition, and availability; and the influence of large-scale renewable energy projects on energy security and shifting geopolitical alliances. The book is developed around three themes: energy security in a geopolitical perspective; reshaping equilibria: renewable energy mega-projects and distributed generation; developing policy strategies towards a low-carbon and secure energy system.

Eyl-Mazzega and Mathieu (2019) highlight that the geopolitical and geoeconomic issues related to energy and climate policies are becoming more complex. They affirmed that to the old and existing energy rivalries, there is the emergence of new rivalries related to the energy transition, especially regarding the control of the value chains of low-carbon technologies, which are crucial for competitiveness, economic development, energy sovereignty and security. In this race, China and the US have taken a certain lead. The authors pointed out that the EU has scientific and industrial strengths, but public policies have favoured the breaking up of industrial entities to foster competition and open markets in order to lower prices for consumers, sometimes at the cost of technological leadership objectives. The authors also outlined the necessary steps that the EU, France and Germany should take in order to benefit from the energy and technological transition.

2   United States

Pascual (2015) provides a framework to understand the relationships between energy geopolitics and energy markets, with an underlying premise that neither energy markets nor foreign policy is static. The paper explained that America's new oil and natural gas abundance will not ultimately lead to a more isolated position in global energy markets, because it will not serve its national security interests within a global energy market. The author affirms that the US, like other major energy producers in the past, has used its newly tapped energy resources to support its international objectives. However, interfering with markets can come with unintended consequences that can ultimately undermine the interests of the US and its international partners. Pascual states that if stopping climate change remains a key foreign policy and national security concern, then the financial and technical factors driving these investment trends must become a priority at the intersection of energy markets and geopolitical interests.

3   Russia

Makarov et al. (2017) begin from the fact that Russian budget relies heavily on exports of fossil fuels and they assess the impacts of the §Paris Agreement on the Russian economy and find that climate-related actions outside of Russia lower Russia's GDP growth rate by about a half of a percentage point. Through a number of scenarios, the article shows how the future landscape post-Paris Agreement might affect the Russian economy, which is highly dependent on production and exports of fossil fuels. For Russia, it is critically important to get ready to mitigate the risks associated with the Paris Agreement by adjusting itself to the new energy landscape. They argue that the objective of Russian strategy should be broader than just planning low-carbon development. In addition to the plans to support low-carbon technologies that are most relevant to the Russian market and to introduce new regulations and legislative incentives promoting low-carbon development (including emissions disclosure requirements and a carbon pricing scheme), the strategy should find ways to address three types of risks: of reducing energy exports, of additional market barriers to Russian exporters of energy-intensive goods, and of relying on outdated energy technologies. In conclusion, the authors affirm that the post-Paris energy landscape poses a challenge for Russia and its energy and economic model. The current way of fossil export-based development will be difficult to sustain in the coming decades, regardless of Russia's own climate policy choices.

4 Mena

Despite the MENA region is commonly linked with fossil fuels, only in the last years some studies have been made on the effects of the global energy transformation for the region. One of the most recent study is the one proposed by Goldthau and Westphal (2019), in which the authors challenge the assumption of the end of the petrostates due to the global energy transition. They affirm the global energy transition might even throw some petrostates an additional lifeline, for examples, those petrostates that have already started to move up the energy value chain by building up refining capacity and developing a viable petrochemical industry, namely, Saudi Arabia, the United Arab Emirates or Kuwait. The authors affirm that the global energy transformation does not mean the end of the petrostate. The low-carbon transition may in fact well facilitate new oligopolies, and a higher market concentration among fewer crude suppliers. They state that fast decarbonizers commitment to making the bright clean energy future work will therefore need to prepare for a twofold challenge: managing a rapidly changing energy system in order to secure the transformation dividends it will bring for human security and economic welfare and balancing the (geo) political order after pains of the incumbent fuels leaving the system.

Weatherby et al. (2018) published a report on UAE energy diplomacy and its role in exporting renewable energy to the global south. This analysis identifies three arenas, where the UAE can strategically expand its clean energy diplomacy in order to help mitigate carbon emissions in developing countries: capacity building, strategically targeted foreign aid and increased commercial ties in the renewable energy sector in developing countries. One of the main messages of the analysis is that there is an opportunity for the UAE to build its soft power and reap commercial benefits by helping countries throughout the Global South implement renewable energy projects. The authors affirm that the UAE can play a leading role in Southeast Asia's energy transition, which should be a priority target for UAE clean energy diplomacy.

Luomi (2018) frames the overall challenge for the UAE and other Gulf Arb ab energy exporters which, due to structural similarities, will be facing many of the same external challenges. The paper identifies three interests the UAE has vis-à-vis the transition: remaining a global energy supplier; ensuring that domestic energy targets can be met and ensuring economic prosperity through a diversified economy. She identifies the main economic challenges to the UAE in the context of the transition that is related to: maintaining or increasing oil exports at competitive prices long enough, while increasing the share of higher value oil-based exports to enable a stable transition in terms of government revenue and the broader economy; and meeting domestic energy demand growth without compromising on environmental sustainability. Finally, the author argues for the need for governments in the region to develop outward-oriented and comprehensive 'energy diplomacy strategies' that build on domestic energy agendas, address these opportunities and challenges and proactively engage with a world that is moving away from hydrocarbons.

Griffiths (2017) provides an assessment and outlook for energy policy in the MENA region within the context of the myriad factors impacting policy design and implementation. The author affirms that although the MENA region will continue to be a major oil and gas producer, consumer and exporter for years to come, a global transition to new sources of energy supply is underway and this will continue to impact all MENA countries. The assessment of the current MENA region context suggests that driving forces for the evolution of energy policy are energy security and energy cost minimization. Griffiths underlines that although renewable energy is a central topic of energy system diversification globally, there is significant interest regionally in nuclear energy and coal to complement greater use of natural gas in the power sector. Regional energy cooperation is essential but must be approached in a thoughtful and realistic manner, according to the author. In conclusion, he states that evolution and transition of the MENA energy system will be challenging but will progress out of necessity. Advances for now will mainly be based on transactional exploitation of easier opportunities.

Luomi (2015) examines how the resource-rich GCC countries are positioning themselves in the international relations of the green economy, focusing in particular on how the United Arab Emirates is seeking to acquire the means of implementation for a national green energy economy transition. She affirms that while not unique in a global perspective, the case of the UAE is unique in the GCC context: unlike its neighbours, the UAE has actively embraced the 'win-win' aspects of the green economy agenda, initiating numerous partnerships and programmes. The case study of the UAE provides important lessons for the other GCC states and to other resource-rich developing countries as well. She affirms that the case of the UAE shows how a benefits-oriented approach to the global governance of environmental problems has so far brought its benefits, through its participation in multiple international partnerships that provide invaluable political and technical support and foster new economic partnerships 'free to charge'. Another lesson is that successes of the UAE's sustainable energy drive have resulted from support at the highest levels of decision-making.

Tagliapietra (2019) illustrates the persistent over-reliance of MENA hydrocarbon producers on the hydrocarbon rent. His article presents the ambitious economic reform programmes adopted by MENA hydrocarbon producers since the drop-in oil prices began in 2014, suggesting a positive view on their implementation prospects. The author highlights that two additional arguments have emerged for economic diversification, besides the risk of oil market volatility: the uncertainty regarding the speed of the global energy transition, and the pressing need to create job opportunities for a large and youthful population. In conclusion, the global energy transition might then turn out to be a positive input for MENA hydrocarbons, a stimulus to consider economic diversification as an unavoidable pathway, to be pursued in order to guarantee future economic prosperity in any future scenario—even in a low-carbon world scenario.

We now provide a review of the existing literature related to the in-depth focuses developed in the second part of the book (i.e. the impacts of the transition on economic growth and income distribution, the role of the global South, the relevance of minerals and metals for low-carbon technologies as well as governance and financing of the global energy transition).

5   The Impacts of the Energy Transition on Economic Growth and Income Distribution

IRENA (2016) published a report on the economic benefits generated by the renewable energy, providing the quantification of the macroeconomic impact of doubling the global share of renewables in the energy mix by 2030. The study demonstrates that the benefits of scaling up renewable energy surpass cost competitiveness. It claims that doubling the share of renewables in the global energy mix by 2030 would increase global GDP between 0.6 and 1.1% or between around US$700 billion and US$1.3 trillion. Additionally, according to IRENA's report doubling the share of renewables increases direct and indirect employment in the sector to 24.4 million by 2030. The report explains also the impacts on fuel importers and exporters generated by the increase of the renewable energy share in the global energy system.

Santos Pereira et al. (2019) propose a research, aiming to empirically assess and discuss: (i) whether different types of household have suffered dissimilar effects from the promotion of renewables; (ii) the consequences of promoting renewables on household income; and (iii) if the promotion of renewables has reduced the risk of poverty and social exclusion. The research found that the income of different households has differing effects on RES promotion, benefiting hydropower and solar PV. Secondly, the authors found that the installed capacities of both wind power and hydropower, and the overall share of RES have dissimilar impacts on different households, but they have increased the income of some. However, the unexpected finding was the negative effect of solar PV deployment on household income. Thirdly, the capacity of wind and hydropower have reduced the risk of poverty for some households, but have increased the risk for others.

Concerning the impacts of climate policies on households of different income levels, a relevant study was conducted by Zachmann et al. (2018). The authors present three different kinds of policy: (i) progressive, policies that make low-income households better off relative to high-income households; (ii) regressive, policies that have the opposite effect; and (iii) proportionate, policies that equally affect high- and low-income households. They identify four factors why households with low incomes are affected differently by individual climate policies; the factors are: budget constraints that lead households with low incomes to prefer different consumption baskets; have higher discount rates/feature borrowing constraints that prevent them from procuring more efficient durables; have different skill endowments and hence wages; and, earn less income from capital and land. They find that key climate policy tools such as carbon taxes for different fuels, certain mandatory standards, subsidies and regulatory tools, can be regressive. They affirm that while several 'pure' climate policies can be regressive, the costs and impacts of climate change are also likely to fall disproportionately on low-income households. They suggest to invest more in research on the distributional effects of climate policies; improve policies and making them less regressive; develop climate policies that benefit low-income households, such as support for energy-efficiency investment targeted at low-income households.

Islam and Winkel (2017) offer a unifying conceptual framework for understanding the relationship between climate change and social inequality. The authors affirm that available evidence indicates that this relationship is characterized by a vicious cycle, whereby initial inequality causes the disadvantaged groups to suffer disproportionately from the adverse effects of climate change, resulting in greater subsequent inequality. The paper identifies three main channels through which the inequality-aggravating effect of climate change materializes: (i) increase in the exposure of the disadvantaged groups to the adverse effects of climate change; (ii) increase in their susceptibility to damage caused by climate change and (iii) decrease in their ability to cope and recover from the damage suffered.

OECD (2017) argues that boosting economic growth, improving productivity and reducing inequalities need not come at the expense of locking the world into a high-emissions future; stating that it is the quality of growth that matters. The report affirms that with a climate-compatible policy package, countries can increase long-run GDP by up 2.8% on average across the G20 in 2050. In order to foster a sustainable economic growth, investment in modern, smart and clean infrastructure in the next decade is a critical factor. The report estimates that $6.3 trillion of investment in infrastructure is required annually on average between 2016 and 2030 to meet development needs globally. An additional $0.6 trillion a year over the same period will make these investments climate compatible. Finally, the report affirms that finance and fiscal activities are essential in fostering a sustainable economic growth.

Hallegatte et al. (2016) published a report that emphasizes how climate change could set back poverty eradication efforts. The report underlines that the future is not set in stone; therefore, it affirms that there is a window of opportunity to achieve the poverty objectives in spite of climate change by pursuing both rapid, inclusive and climate-informed development, combined with targeted adaptation interventions, to cope with the short-term impacts of climate change and, secondly, immediate pro-poor mitigation policies to limit long-term impacts and create an environment that allows for global prosperity and the sustainable eradication of poverty.

Cludius (2015) analyses two energy and climate policy instruments, namely, renewable energy policy and the European Union Emissions Trading Scheme (EU ETS), in the context of the distribution of costs and benefits arising from these policies. The author shows that, contrary to public perception, the distributional outcomes are not an inherent feature of the energy or climate policy instrument at hand, but are largely determined by the way in which it is designed. Indeed, industry exemptions from contributing to the cost of both renewable energy policy and emissions trading have been generous and allowed for considerable profits to the companies covered by those schemes. This has led to a situation where households bear the majority of the direct costs associated with those policies. The analysis indicates that low-income households are particularly affected by the associated price increases, because they spend a large fraction of their income on electricity and other emissions-intensive goods. There is scope for governments to improve the situation of (low-income) households through three mechanisms: a reduction of price; household income via financial assistance to households; and energy efficiency measures needed to reduce household consumption of electricity. Finally, the author suggests that there is no basis for pitching climate against equity concerns, but rather there are many opportunities for policymakers to formulate an integrated approach that addresses both issues concurrently.

Dercon (2014) underlines that due to their initial poverty and their relatively high dependence on environmental capital for their livelihoods, the poor are likely to suffer most due to their low resources for mitigation and investment in adaptation. The paper focuses on three elements of green growth policies: pricing and regulation to internalize environmental capital costs; low-carbon and other environmentally sensitive public investments and 'green' adaptation and other resilience-enhancing investments. The report argues that green growth could potentially have important negative consequences for the poor that may even outweigh the benefits for the poor from growth. The author states that environmental pricing and regulation may have considerable negative consequences for the poor as consumers, and would require specific social protection measures to compensate for price rises. Therefore, the authors warn that promises that green growth will offer a rapid route out of poverty are not very plausible; there may well be less rapid an exit than with more conventional growth strategies.

Grösche and Schröder (2014) assess the redistributive effects of a key element of German climate policy, the promotion of renewables in the electricity generation mix through the provision of a feed-in tariff. The authors show that the tariff shapes the distribution of households' disposable incomes by charging a levy that is proportional to household electricity consumption and by transferring financial resources to households who are feeding green electricity into the public grid. The paper analyses the question whether the feed-in tariff scheme increases income inequality in the society and thereby conflicts with the general social goal to reduce disparities in peoples' disposable incomes. They state that the share of renewable fuels in the electricity generation mix increased under the regime of the feed-in tariff from 7% in 2000 to about 20% in 2011, but also imposed substantial cost to the electricity consumer by subsidizing renewables.

6 The Global Energy Transition and the Global South

Hirsch et al. (2017) provide an overview of the different just transition, energy transformation and climate justice discourses of the previous years and how they are ultimately reflected in the Paris Agreement. The authors show how these discourses overlap in terms of transition narratives and policy demands, and they affirm that the shared value base could serve as a starting point for building alliances, which are necessary to make just transition become a reality. The report outlines eight principles related to the just energy transition designed to make justice applicable to energy transition processes in developing countries, which go beyond an abstract call for justice. These principles cover the climate, socioeconomic and political dimensions in a balanced way to reflect the legitimate justice claims of a broad range of potential allies for a just energy transition alliance. These principles are then applied as a reference framework for 12 countries of the Global South (China, India, Vietnam, Philippines, Nepal, Fiji, Morocco, South Africa, Tanzania, Mexico, Costa Rica and Jamaica). The just energy transition country assessments have shown that neither are countries who internationally claim to be climate champions in terms of energy transition necessarily performing well in terms of the social and political dimension of a just transition, nor are those who claim to pioneer justice automatically in the lead in transforming their energy systems in a way that is consistent with a 1.5/2 °C pathway. The scoring indicates that country performance is generally strongest in terms of the climate and energy dimension of the transition, while countries are doing less well in terms of addressing the socioeconomic dimension of a just energy transition.

7   The Geopolitics of Renewable Energy

One of the most analysed topics related to the shift from geopolitics of oil and gas to geopolitics of renewable energy. The most relevant work is the book edited by Scholten (2018), which is the first volume to explore the geopolitical implications of a transition to renewable energy. The book tackles a wide variety of topics, namely, winners and losers of the new emerging global energy scenario, the change in regional and bilateral energy relations between established and rising powers and the governance responses to the transition as well as infrastructure developments. The authors affirm that the future geopolitical world of energy will be a mixed between the one of renewables and the one of conventional energy. It will be different because it will be a more decentralized system; it will be similar to the conventional energy because bigger projects in renewable suffer from very similar security issues, for example, where and who will control certain pivotal power lines. The book states that the geopolitics of conventional energy and that of renewable energy will exist next to each other for a period of several decades.

Overland (2019) addresses four emerging myths about the geopolitics of renewable energy, seeking to stop them developing further. The four emerging myths are: competition over critical materials; new resource curses; electricity disruption as a geopolitical weapon; cybersecurity as a geopolitical risk. Regarding the first myth, Overland affirms that energy transition is about mainly technology and innovation; therefore, she believes that it is highly probable that there will be technological improvements and cost reductions in some areas. About the second myth, she affirms that renewable energy for export could potentially require more long-term maintenance of infrastructure, generate more local jobs and produce more stable revenues than oil and gas have done, especially compared to oil exporters with oil and gas production located offshore and dominated by international oil companies and workers, such as Angola. The third myth claims that interstate electricity cut-offs could become an important foreign policy tool. However, the author explains that net-importer countries will still have the option of developing their own renewables potential and thus face long-term make-or-buy choices. Lastly, the author affirms that the fourth myth—cybersecurity as a geopolitical risk—is overstated sometimes.

She points out that it is probable that increased use of renewable energy will lead to greater decentralization and this may actually make the system more resilient. In conclusion, Overland affirms that renewable energy resources are abundant but diffuse, technologies for capturing, storing and transporting them will instead become more important. International energy competition may, therefore, shift from control over physical resources and their locations and transportation routes to technology and intellectual property rights.

Gielen et al. (2019) explore the technical and economic characteristics of an accelerated energy transition to 2050, affirming that energy efficiency and renewable energy technologies are the core elements of that transition. They notice that countries around the world are in the midst of an energy transition that appears to favour electricity as the preferred final energy carrier, which is favourable from the perspective of both renewables and energy efficiency. Indeed, electricity is an efficient energy carrier and it becomes a clean source of energy when it is sourced from renewables. Their analysis shows that the decarbonization of the energy system is affordable, since the additional cumulative investments over the 2015–2050 period would be $27 trillion, equivalent to $0.77 trillion per year on average in the same period. Also, the energy transition would produce new jobs (around 19 million additional direct and indirect jobs in 2050) offsetting the loss of old jobs (around 7 million); therefore, the global energy transition results in 11.6 million additional direct and indirect jobs in the energy sector.

Hache (2018) aims to analyse the geopolitical consequences of the spread of renewable energies worldwide. Despite it would be tempting to conclude that the energy transition to renewables will gradually end today's geopolitics of fossil fuels, he believes that new challenges induced by energy transition policies could paradoxically turn out being as complex as today's geopolitics of energy. Local and decentralized relations could indeed add a new geopolitical layer to current traditional actors, while technical, economic, sociological, behavioural, spatial and legal dimensions could also complicate the emerging puzzle. A substantial increase of renewables into the wold's energy mix could lead to new, unexpected interdependencies such as dependencies to critical materials, a new geopolitics of patents and the implementation of renewable diplomacy.

Stratfor (2018) publishes an assessment on how renewable energy will change geopolitics. Since the increasing relevance of the renewable sources, there will be a significant geopolitical shift from the current energy geopolitical scenario. It affirms that some countries will fare better than others in the course of the transition, for instance, Germany, the US and China. Indeed, China has raced ahead to become the world's unrivalled leader in the manufacture of clean energy products, in the past decade, as well as the world's biggest miner and supplier of rare earth materials, biggest deployer of renewable energy capacity and biggest market for electric vehicles. It affirms that smaller countries such as Sweden, Denmark, Uruguay, Morocco and Kenya could gain outsize regional influence as a result of the transition, thanks to their potential for exporting renewable energy and technology. The main losers will be traditional oil exporters, such as Venezuela, Kazakhstan and the GCC states. The article argues that renewable energy probably will not have the same power to spark large-scale military confrontations, especially in the Middle East, inspiring increased cooperation among states by encouraging regional grid integration. However, it highlights some risks, such as cyberattacks and interruptions in supply of clean tech minerals. Finally, it affirms that trade will continue to be a source of conflict among states, for example, on intellectual property theft, dumping and domestic content requirements, undermining the global trading regime if they become heated enough.

O'Sullivan et al. (2017) provide the reader with a discussion about the shift of energy geopolitics from the one related to fossil fuels to the one of renewable energy sources. The paper discusses seven mechanisms through which renewables could shape geopolitics. First, the authors affirm that cartels could develop around materials critical to renewable energy technologies. Second, they assess that in a world in which renewables are the dominant source of energy, capital for investment and technology may increasingly become sources of international cooperation or rivalry. Conflicts might be developed over the transfer of technology between developing and developed countries as well as over renewable energy infrastructure. Third, they affirm that the prevalence of the resource curse could be affected by a rise of renewable energy, in different ways, namely: petrostates will lose access to the high rents associated with the curse; whether countries that produce large amounts of renewable energy are likely to become subject to the curse; and potential for a new resource curse in countries rich in rare earth elements. Fourth, the geopolitical complexity of greater electric interconnections between nations, which could create both more vulnerabilities for electricity importers and more interdependence, reducing risks of conflict. Five, the reduction of oil and gas consumption could lead political reform and economic diversification in the fossil fuel producers; but it might lead also to political instability. Six, renewable sources will reduce the risk of conflict and instability that climate change would otherwise generate; Africa is identified as the region where large-scale deployment of renewables may have significant geopolitical consequences. And seven, renewables could help to foster sustainable energy access, contributing to lasting solutions to instability and conflict.

Scholten and Bosman (2016) explore the potential political implications of the geographic and technical characteristics of renewable energy systems. The authors do so through a thought experiment that imagines a purely renewable-based energy system, keeping all else equal. The major implications for renewable energy base markets found by the authors are two: first, countries face a make-or-buy decision, meaning that they have a choice to produce or import energy; second, electricity is the dominant energy carrier, implying a more physically integrated infrastructure with stringent managerial requirements. They illustrate the strategic concerns arising from these implications with two scenarios: continental, following a buy decision and more centralized network; and National, following a make decision and more decentralized network. Three observations stand out compared to the geopolitics of an energy system based on fossil fuels. First, a shift in considerations from getting access to resources to strategic positioning in infrastructure management; second, a shift in strategic leverage from producers to consumers and those countries being able to render balancing and storage services; and third, the possibility for most countries to become a 'prosumer country' may greatly reduce any form of geopolitical concern.

Johansson (2013) explores the security aspects of future renewable energy systems, affirming that renewable energy systems can improve some aspects of security, but they will not automatically lead to the removal of all types of security problems and new problems will most certainly arise. He outlines that the approach to the energy systems as a subject generating or enhancing insecurity can be divided into three different types of risk areas: economic-political, technological and environmental. Regarding the first type of risk area (economic and political risks) he pointed the argument for assuming a reduced risk of single countries being able to exert pressure or influence on individual countries is that renewable energy sources are less concentrated and available in all countries. He affirms the renewable energy source with the greatest technological risks is probably hydropower, where dam safety is a significant issue, while about the third type (environmental risks) he declares that renewable energy will generally lead to a reduced impact in terms of climate change compared with fossil fuels as long as it is sustainably produced. In conclusions, he affirms that the main advantage of renewable energy from a long-term energy security perspective is the fact that it is based on flows instead of exhaustible stocks.

8 Minerals and Metals for Low-Carbon Technologies

Another relevant topic that will be analysed in the book is the minerals and metals for low-carbon technologies, which are often believed as new geopolitical leverages in the global energy transition. IISD (2018) published a report in which identifies 23 key minerals that will be critical to the development and deployment of renewable technologies, such as solar panels, wind turbines, electric vehicles and energy storage technologies. It affirms that a substantial percentage of the minerals required for green energy technologies are located in states with high measures of fragility and corruption; cobalt, graphite, copper and rare earths are of particular concern. Analysing the degree of state fragility and corruption, the report shows a picture where potential hotspots emerge, particularly in South America, sub-Saharan Africa and Southeast Asia. The report states that the increased extraction of many of the identified minerals has, in the past and at resent, been linked with local grievances, tensions and violence, in the worst cases. It examines five case studies: cobalt in the DRC, rare earths in China, Nickel in Guatemala, bauxite and alumina in guinea, lithium in Zimbabwe. Therefore, it highlights the need to ensure the responsible sourcing of the minerals required for green energy technologies and it recognizes some progress, such as strong guidance on responsible supply chains.

Also, the World Bank (2017) took into considerations, the role of minerals and metals for a low-carbon future. The report examines which metals will likely rise in demand to be able to deliver on a carbon-constrained future, particularly aluminium, cobalt, copper, iron ore, lead, lithium, nickel, manganese and rare earths. The report focuses on wind, solar and energy storage batteries as they are commonly recognized as key elements in delivering future energy needs at low/zero GHG emission levels. The study addresses what materials are required in the scaled-up production of these technologies and to what degree will that demand be driven by a range of the global climate scenarios. The report shows that the technologies assumed to populate the clean energy shift are in fact significantly more material intensive in their composition that current traditional fossil-fuel-based energy supply systems. It provides precise estimates on the actual demand for metals which is predicated by at least two independent variables: the extent to which the global community of nations actually succeeds in meeting its long-term Paris climate goals and the nature of intra-technology choices. Finally, the report examines how resource-rich developing countries might best position themselves to take advantage of the evolving commodities market responding to a low-carbon energy transition. The shift to low-carbon energy will produce global opportunities with respect to a number of minerals. The Latin America region (Chile, Brazil, Peru, Argentina and potentially Bolivia) is in an excellent position to supply the global energy transition. Africa, with its reserves in platinum, manganese, bauxite should also serve as a burgeoning market for these resources. With respect to Asia, the most notable finding is the global dominance China enjoys on metals, both production and reserve levels. India is dominant in iron and steel, while Indonesia has opportunities with bauxite and nickel, as does Malaysia and Philippines to a lesser extent. Lastly, the report affirms that, in Oceania, the massive reserves of nickel to be found in New Caledonia should not be overlooked.

Bazilian (2018) considers the implications of the critical role of minerals and metals in the current 'energy transition'. He affirms that the location of the critical minerals and metals shows the clear need to focus on issues around environmental, social, trade and other governance-related issues. Indeed, governance issues can have a major impact on the reliability of the supply of these materials. The author points out that the largest potential reserves exist in developing countries, which are especially dependent on the revenues from mining and this typically serves to exacerbate governance challenges.

In a study, de Ridder (2013) focuses on the geopolitics of minerals for renewable energy technologies. He pointed out that minerals are not scarce because there are not enough minerals to be found in the Earth's crust; the total availability of minerals in the earth's crust in itself is irrelevant for the geopolitics of minerals for renewable energy. Mineral supply depends on whether known mineral deposits are profitable for extraction with current or future technology and under current or future market conditions. The degree of reliance depends on what services and products countries produce and on their economies' position along the supply chain; countries that produce renewable energy technologies sit closer to the refining stage in the supply chain than other countries. He underlines the fact that the global energy transition is taking place within the transition of international system towards a multipolar world, while some state capitalist tendencies are becoming more prominent. The paper looks at how both import-dependent and mineral producing countries are responding to these developments and what the implications are for the balance of cooperation and conflict.

Neil and Speed (2012) published a report about the strategic implications of China's dominance of the global rare earth elements (REE) market. The authors highlight that China's de facto monopoly on rare earth mining and processing and its growing control over rare earth manufacturing enable Beijing to powerfully influence global supply. They affirm that China's near-total domination of the rare earths market is likely to continue over the near term as Beijing works to consolidate its position as the principal global REE supplier. This situation poses a threat to some military capabilities of the US and its principal allies. Nevertheless, due to this dominance, even if the US and its allies take steps to launch, subsidize and protect domestic rare earth mining, processing and manufacturing industries, such measures will take time to become productive, and are unlikely to prevent near-term shortages of these elements. Over the longer term, China's domination of the rare earths market is likely to wane as its reserves are drawn down; as new sources of supply are developed; as recycling becomes increasingly cost-effective; as new technologies replace rare earth-dependent technologies; and as the governments of the advanced, industrialized states look at alternative means to implement 'green' policies and practices.

9 Governing the Global Energy Transition

Goldthau et al. (2018) affirm that the energy transformation essentially implies a systemic shift; from a global perspective, the low-carbon transformation is likely to render the energy system more sustainable, but also much more heterogeneous. The energy transformation will successively reduce dependence on imports, promising a 'security dividend'. If more energy is produced locally, this as an impact on the relations between producer, transit and consumer countries. The authors explain also new risks and challenges, notably in the area of grid stability and cybersecurity. Geopolitically, the restructuring of the energy system will not threaten the major oil and gas producers as quickly and existentially as is generally assumed. A heterogeneous, fragmented energy system would neatly fit an increasingly multipolar world order underpinned by a more protectionist stance towards trade. Yet, mercantilist energy policies present the threat of spiralling rivalries between 'energy block'. Therefore, the authors affirm that this condition makes multilateral cooperation an indispensable policy goal in order to radically and rapidly restructure the energy system worldwide.

WEF (2018) assesses the effective energy transition of 114 countries' energy systems, thanks to an index, the Energy Transition Index (ETI). Through the ETI, the report highlights that over the last 5 years, more than 80% of countries improved their energy systems, but further effort is needed to resolve the world's energy-related challenges. Secondly, countries can foster progress in three ways by: establishing favourable conditions for energy system stakeholders, targeting improvement across all three triangle dimensions, and by pursuing improvement levers with synergistic impact across the system. Thirdly, countries follow different transition paths and need to develop country-specific roadmaps.

Goldthau (2017) explains that, in economic terms, energy assets will move further up the value chain, from commodities to technologies. The author states that there will be winners and losers: technology leaders (OECD nations and China) will benefit most, while countries lacking technology and capital, mainly in the global south, will lose out as well as countries that are rich in fossil fuels could become unable to sell oil or coal, with their economies deeply damaged. Goldthau states that to avert this, the low-carbon transition needs to be governed globally and three factors are key: credible and legitimate leadership; information about climate-related risks to guide investment and global partnerships to advance low-carbon technology. He suggests the G20 as the coalition of nations well placed to take the helm in this transformation. Indeed, about being a credible and legitimate leadership, the G20 includes nations that lead in technology and those that lag behind; industrialized economies; rising powers; resource-rich nations and resource-poor ones. Second, G20 could become global mechanism needed to share information about climate-related investment risk, especially through its Task Force on Climate-Related Financial Disclosures. Third, it would establish global partnerships between technology leaders and laggards to advance the take-up of low-carbon technologies.

In a paper, Andrews-Speed (2016) aims to demonstrate the benefits of applying a wider set of institutional theories to the study of the low-carbon energy transition. He draws principally on rational choice and historical institutionalism with selective reference being made to key concepts within social and organizational institutionalism as well as discursive institutionalism. The paper has sought to show that a broader institutional perspective provides useful insights into the wider context of the organizational field or socio-technical regime. In particular, it has drawn attention to how the general features of the political and economic system and of the national culture may shape the nature, pace and direction of the low-carbon energy transition.

Roehrkasten et al. (2016) publish a study about the G20 and sustainable energy within the global energy transition. The study analyses the G20's potential for advancing a global transition to sustainable energy. It comprises short studies on the energy trends and the domestic and international policy priorities of 13 G20 countries—Argentina, Brazil, China, France, Germany, India, Indonesia, Japan, Russia, Saudi Arabia, South Africa, Turkey and the US—plus the EU. Despite all of the G20 members covered in this study remain highly dependent on fossil fuels, all of them adopted the Paris Agreement. The authors state that a concerted action by G20 countries can offer an important boost to building a sustainable, low-carbon energy system. Therefore, it is essential that the G20 deepens its engagement in key areas, such as renewable energy, energy efficiency, access to energy. Global energy governance faces significant challenges, because governments have been hesitant to engage in global cooperation on energy, mainly due to sovereignty concerns. The study identifies a promising approach is that the G20 partners cooperate with other international institutions, including IEA and IRENA. The G20 can complement and add coherence to the global energy institutional landscape by entitling existing institutions to carry on its initiatives.

Agreeing with the argument that the global energy transition could occur much faster than research on historic transitions suggests, Kern and Srogge (2016) argue that at the heart of the pace of low-carbon energy transitions is firm political commitment at all levels of governance. They highlight three main aspects—agency, international dynamics and the Paris Agreement—which make them optimistic about an acceleration of the global energy transition towards low carbon. They affirm that accelerating the decarbonization of the global energy system is by no means a straightforward exercise but requires hard political work as well as strong political commitment to fighting climate change. They argue that the Paris Agreement has ushered in a new era in which decarbonization and a focus on energy demand reduction and increasing energy efficiency will become the 'new normal', thereby leading to a new paradigm in thinking governing energy transitions.

Instead, Van de Graaf (2013) analyses fragmentation in global energy governance. He explains why the creation of additional institutions is highly unlikely according to conventional institutionalist thinking. The paper proposes an explanation for it, based on the capture of institutions by particular states or interest groups. The capture of an institution can spur the creation of a countervailing organization if there exists a sufficiently strong coalition of dissatisfied states in which the incumbent institution has lost domestic support. He argues that the push to create IRENA can be viewed partly as a symbolic action, taken for internal political reasons of some countries, and therefore challenges the strict functionalist understanding of institutions, revealing that not all institutions are created with the sole purpose of reducing transition and information costs. The author uses the case study as a reminder that international organizations are not neutral vehicles but embody certain interests and principled beliefs. He states that the creation of a specialized renewable energy agency raises the spectre of further institutional fragmentation in global energy governance along sectoral lines with each sector having its own international institution.

Goldthau (2011) lays out the main challenges that need to be addressed during the looming energy transition process. He declares that global governance arrangements need to be inclusive, commit involved actors to achieving a low-carbon future, allow for feedback mechanisms within various levels of the process and be adaptive to previously unperceived challenges along the way. In general terms, the author affirms that future research on global energy governance will need to link the still de-coupled research areas of energy security, energy access and climate change, and address them in the context of a looming energy transition. He underlines that current research bias towards the 'who?' in global energy governance needs to give way to asking more of 'what needs to be governed, and how?'.

10   Financing the Global Energy Transition

The last topic analysed is related to need to finance the global energy transition. Christianson et al. (2017) affirms that finance provided and catalyzed by multilateral development banks (MDBs) will help pay for implementation of the UN Sustainable

Development Goals and the Paris Climate Agreement in many developing countries. The authors notice that it is less known about how investments across their entire energy supply portfolios relate to achieving sustainable development and climate change objectives. The report underlines that the different investment patterns seem to reflect the different mandates of the MDBs, the World Bank and ADB work mainly with public counterparts, while the IFC works with the private sector.

Hall et al. (2018) point out that up to $61 trillion of power systems is needed to fulfil the Paris Agreement, affirming that the mobilization of so much capital is a huge challenge for the world. Indeed, it is unlikely that this kind of amount would be sourced from one form of finance alone. Therefore, it is more likely that various mixes of state, commercial and 'alternative' money capital will be required for low-carbon energy transitions. The paper uses a comparative analysis of two developed economies (Germany and UK) to explore how 'alternative' forms of finance operate in each nation's energy investment landscape. The authors find that alternative finance is often set in opposition to commercial capital. Alternative finance in both nations is motivated by financial justice outcomes that are poorly understood in current energy policy. They identify the six categories of justice most relevant to financing energy transition, which are: affordability, good governance, due process, intra-generational equity, spatial equity and financial resilience. Energy policy that seeks to mobilize capital, should take account of all six principles. The analysis shows that taking account of the variety of capitalism in each nation, and its attendant financial institutions can illuminate several ways in which these principles can be operationalized, from pursuing financial innovation through alternative platforms to expanding public or mutual banking provisions.

IRENA (2018) affirms that the landscape of renewable energy finance has evolved rapidly. Investment reached a comparable milestone in 2015, when renewable power technologies for the first time attracted more finance than non-renewable power technologies. Clearly, investment levels are highly responsive to policy changes. The East Asia-Pacific region was the dominant destination for renewable energy investment, with China as the main driver. The report outlines the significant role of the private sector; indeed, private sources provide the bulk of renewable energy investment globally—over 90% in 2016. Private investors overwhelmingly favour domestic renewable energy projects—93% of the private portfolio in 2013–2015—whereas public investment is more balanced between in-country and international financing.

Frankfurt School-UNEP Centre & BNEF (2018) publish a report representing global trends in renewable energy investment in 2018. The report underlines that the leading location by far for renewable energy investment in 2017 was China, accounting for $126.6 billion or 45% of the global total. This figure has been affected by the extraordinary solar boom in China in 2017. Due to policy uncertainties, renewable energy investment in the US was far below China, at $40.5 billion. The report shows that also Europe suffered a big decline of 36% to $40.9 billion, mainly due to a sharp decline in UK investment, due to the end to subsidies for onshore wind.

OECD/IEA & IRENA (2017) publish a report on the perspectives for the energy transition. Their analysis finds that the energy sector transition could bring important co-benefits, such as less air pollution, lower fossil fuel bills for importing countries and lower household energy expenditures. The energy transition is often linked to significant amounts of money in order to achieve it. However, the report shows that while overall energy investment requirements are substantial, the incremental needs associated with the transition to a low-carbon energy sector amount to a small share of world GDP. Indeed, according to IEA, additional investment needs would not exceed 0.3% of global GDP in 2050, while according to IRENA, the additional investment required would be 0.4% of global GDP in the same year with net positive impacts on employment and economic growth.

IISD (2017) articulates how fossil fuel subsidy reform can contribute to a just transition and how this reform can be more successful under a just transition framework. Fossil fuel subsidies are a barrier to just transition and green economies because they are often socially regressive. The report identifies three core elements for a successful reform: getting the prices right; managing impacts and building supports. The report states that the scale of finance required for the transition is expected to be in the order of many billions of dollars; there is no guarantee to provide the scale or targeted supports required to foster the transition; and fossil fuel subsidies total at least $425 billion per year, which, if removed could go a long way to financing just transition. Therefore, reforming fossil fuel subsidies will contribute to the transition to green economies by removing supports for fossil fuel sectors that harm the environment; utilizing the revenue raised from reform can go a long way to supporting the policies required for just transition.

Acknowledgments The authors would like to thank Dr. Kairat Kelimbetov, Governor of the AIFC, for his insights and for allowing us to to use some of the AIFC development strategy ideas and concepts

References

Global Trends

Goldthau A, Westphal K, Bazilian M, Bradshaw M (2019) How the energy transition will reshape geopolitics. Nature 669:29–31
Hafner M, Wochner A (2020) How tectonic shifts in global energy are affecting global governance. In Grigoriev L, Pabst A (eds.) Global governance in transformation. Springer, Berlin, pp 147–162
IRENA (2019) A new world. The geopolitics of the energy transformation. In: IRENA

Europe

Bressand A (2012) The changed geopolitics of energy and climate and the challenge for Europe. A geopolitical and European perspective on the triuple agenda of competition, energy security and sustainability. In: Clingendael international energy programme (CIEP), CIEP paper 04

Eyl-Mazzega M-A, Mathieu C (2019) Strategic dimensions of the energy transition. Challenges and responses for France, Germany and the European Union. Institut Français des Relations Internationales, Etudes de l'IFRI, April 2019

Lombardi P, Gruenig M (eds) (2016) Low-carbon energy security from a European perspective. Academic Press

United States

Pascual C (2015) The new geopolitics of energy. Center on Global Energy Policy and School of International and Public Affairs, Columbia University, September 2015

Russia

Makarov I, Chen Y-H, Paltsev S (2017) Finding itself in the post-paris world: Russia in the new global energy landscape. MIT joint program global change, report 324, December 2017

MENA

Goldthau A, Westphal K (2019) Why the Global Energy Transition does not mean the End of the Petrostate. Global Policy, Policy Insights

Griffiths S (2017) A review and assessment of energy policy in the Middle East and North Africa region. Energy Policy 102(2017):249–269

Luomi M (2015) The international relations of the green economy in the gulf: lessons from the UAE's state-led energy transition. Oxford Institute for Energy Studies (OIES), OIES paper: MEP 12, May 2015

Luomi M (2018) The foreign relations of energy transitions – framing the issue for the UAE. Emirates Diplomatic Academy (EDA), EDA Insight, December 2018

Tagliapietra S (2019) The impact of the global energy transition on MENA oil and gas producers. Energ Strat Rev

Weatherby C, Eyler B, Burchill R (2018) UAE energy diplomacy: exporting renewable energy to the global south. Trends report, Trends Research and Advisory & the Stimson Centre

The Impacts of the Energy Transition on Economic Growth and Income Distribution

Cludius JM (2015) Distributional effects of energy and climate policy. PhD thesis, Doctor of Philosophy, UNSW Business School, UNSW Australia

Dercon S (2014) Is green growth good for the poor? Policy Research working papers no. 6936. World Bank, Washington DC

Grösche P, Schröder C (2014) On the redistributive effects of Germany's feed-in tariff. Empirical Economics. 46(4):1339–1383

Hallegatte S, Bangalore M, Bonzanigo L, Fay M, Kane T, Narloch U, Rozenberg J, Treguer D, Vogt-Schilb A (2016) Shock waves: managing the impacts of climate change on poverty. World Bank, Climate Change and Development, Washington DC

IRENA (2016) Renewable energy benefits: Measuring the economics. IRENA

Islam N, Winkel J (2017) Climate change and social Inequality. Working paper no. 152. UN Department of Economic & Social Affairs (DESA)

OECD (2017) Investing in climate, investing in growth. OECD editions, Paris. https://doi.org/10.1787/9789264273528-en

Santos Pereira D, Cardosomarques A, Fuinhas JA (2019) Are renewables affecting income distribution and increasing the risk of household poverty? Energy 170:791–803

Zachmann G, Fredriksson G, Claeys G (2018) The distribution effects of climate policies. Bruegel blueprint series, vol 28. http://bruegel.org/wp-content/uploads/2018/11/Bruegel_Blueprint_28_final1.pdf

The Global Energy Transition and the Global South

Hirsch T, Matthess M, Fünfgelt J (2017) Guiding principles & lessons learnt for a just energy transition in the global south. Friedrich Erbert Stiftung

The Geopolitics of Renewable Energy

Gielen D, Boshell F, Saygin D, Dbazilian M, Wagner N, Gorini R (2019) The role of renewable energy in the global energy transformation. Energ Strat Rev 24(2019):38–50

Hache E (2018) Do renewable energies improve energy security in the long run? Inter Econ 156(2018):127–135

Johansson B (2013) Security aspects of future renewable energy systems: A short overview. Energy 61(2013):598–605

O'Sullivan M, Overland I, Sandalow D (2017) The geopolitics of renewable energy. Columbia Center on Global Energy Policy & Harvard Belfer Center for Science and International Affairs, Working paper, July 2017

Overland I (2019) The geopolitics of renewable energy: debunking four emerging myths. Energ Res Soc Sci 49(2019):37–40

Scholten D (ed) (2018) The geopolitics of renewables. Lecture Notes in Energy, vol 61. Springer, Berlin

Scholten D, Bosman R (2016) The Geopolitics of Renewables: Exploring the political implications of renewable energy systems. Technol Forecast Soc Chang 103(2016):273–283

Stratfor (2018) How renewable energy will change geopolitics. Strat for assessments, https://worldview.stratfor.com/article/how-renewable-energy-will-change-geopolitics. Accessed 27 June

Minerals and Metals for Low-Carbon Technologies

Bazilian MD (2018) The mineral foundation of the energy transition. Extr Ind Soc 5(2018):93–97
de Ridder M (2013) The geopolitics of mineral resources for renewable energy technologies. THE HAGUE Centre for Strategic Studies, August 2013
IISD (2018) Green conflict minerals: the fuels of conflict in the transition to a low-carbon economy. IISD report, August 2018
Neill DA, Speed E (2012) The strategic implications of China's dominance of the global rare earth elements (REE) market. Defence R&D Canada, Centre for Operational Research and Analysis, September 2012
WORLD BANK (2017) The growing role of minerals and metals for a low-carbon future, June 2017

Governing the Global Energy Transition

Andrews-Speed P (2016) Applying institutional theory to the low-carbon energy transition. Energ Res Soc Sci 13(2016):216–225
Goldthau A (2011) Governing global energy: Existing approaches and discourses. Curr Opininion Environ Sustain 3(4):213–217
Goldthau A (2017) The G20 must govern the shift to low-carbon energy. Nature 546:203–205
Goldthau A, Keim M, Westphal K (2018) The geopolitics of energy transformation: governing the shift – transformation dividends, systemic risks and new uncertainties. Stiftung Wissenschaft und Politik (SWP), SWP Comment No. 42, October 2018
Kern F, Srogge K (2016) The pace of governed energy transitions: agency, international dynamics and the global Paris agreement accelerating decarbonisation processes? Energ Res Soc Sci 22(2016):13–17
Roehrkasten S, Thielges S, Quitzow R (2016) Sustainable energy in the G20 – prospects for a global energy transition. Institute for Advanced Sustainability Studies (IASS)Potsdam, IASS Study, December 2016
Van de Graaf T (2013) Fragmentation in global energy governance: explaining the creation of IRENA. Glob Environ Polit 13(3):14–33
WEF (2018) Fostering effective energy transition. A fact-based framework to support decision-making. World Economic Forum, Insight report, March 2018

Financing the Global Energy Transition

Christianson GI, Lee A, Larsen G Green A (2017) Financing the energy transition: whether World Bank, IFC, and ADB energy supply investments are supporting a low-carbon, sustainable future. World Resources Institute, Working paper, May 2017

Frankfurt School-UNEP Centre & BNEF (2018) Global trends in renewable energy investment 2018

Hall S, Eroelich K, Edavis M, Holstenkamp L (2018) Finance and justice in low-carbon energy transitions. Appl Energy 222(2018):772–780

IISD (2017) Fossil fuel subsidy reform and the just transition: integrating approaches for complementary outcomes. GSI report, December 2017

IRENA (2018) Global landscape of renewable energy finance. IRENA

OECD/IEA & IRENA (2017) Perspectives for the energy transition. Investment needs for a low-carbon energy system. IEA & IRENA

Regional Insights

The European Union and the Energy Transition

Marc-Antoine Eyl-Mazzega and Carole Mathieu

1 Introduction

The low-carbon energy transition in the world is today taking place unevenly and too slowly to preserve the climate and biodiversity. CO_2 emissions have been rising (2016, 2017, and 2018) albeit they should peak rapidly according to Intergovernmental Panel on Climate Change (IPCC) reports. 2019 and 2020 could see a first slow down though due to the global economic situation and greater deployment of renewables and use of gas in replacement of coal power. But way not enough to follow pace with an 8% yearly decline needed toward 2030 for a significant chance to limit global warming to 1.5 °C. The world is experiencing unprecedented heat waves, forest fires, accelerated glacier melt, draughts, and extreme weather events, resulting from climate change and potentially contributing to harmful feedback loops. And governments' commitments are insufficient: in the long term, with the current policies in place, the world is on a +3 °C trajectory (IPCC 2018). While *our house is burning*, President Trump has introduced a process of withdrawing from the December 2015 Paris agreement[1] and has been diluting multilateral efforts such as in the G7 or G20. Brazil is returning to a policy of deforestation, and Australia is putting short-term economic gains before climate action. Massive fires covering two times the size of Belgium have not yet prompted a policy change. At this stage, only a global recession could curb emissions, provided that recovery measures are then aligned with the Paris Agreement. Most countries in the world have not yet kick-started a low-carbon transition process and in the best cases, renewable deployment

---

[1] This withdrawal, effective in November 2020, should not hide three realities: the boom in deploying renewable energy sources in the US is going ahead, albeit more slowly, and the US intends to vie with China for the leadership in the innovation of low-carbon technologies. Moreover, US civil society, certain states like California, cities and companies remain mobilized and influential.

M.-A. Eyl-Mazzega (✉) · C. Mathieu
Institut Français des Relations Internationales (IFRI), Paris, France
e-mail: eylmazzega@ifri.org

© The Author(s) 2020
M. Hafner and S. Tagliapietra (eds.), *The Geopolitics of the Global Energy Transition*,
Lecture Notes in Energy 73, https://doi.org/10.1007/978-3-030-39066-2_2

27

just avoids building additional coal-fired power generation capacity. Electricity supply security and economic development are legitimate priorities of many emerging economies, ranking higher than the reduction of greenhouse gas emissions. At best, their concern is about reducing air pollution in cities, which often but not always comes with co-benefits in terms of climate change.

Despite the sense of urgency, the inertia of energy systems is strong and coal and oil are going to continue dominating the global energy mix for a long time. With rising demand for electricity, a fundamental challenge is to decarbonize the power sectors globally, while also spending important efforts on the heating and cooling, industry and transport sectors, and on other greenhouse gases than CO_2, such as methane. The world needs a minima to double investments into renewable energy sources and energy efficiency, as well as the deployment of a mix of solutions including energy efficiency, reforestation, Carbon Capture, Utilization and Storage (CCUS) technologies, decarbonized hydrogen, biofuels, biomass, electric vehicles, LNG in the maritime and road freight segments, and nuclear. Last but not least, efforts to fund adaptation measures still need to be strengthened considerably.

Since the Lisbon Treaty, the EU has stepped up its environmental policies and has been able to expand and increasingly integrate both energy and climate policies. This process has seen the introduction and reinforcement of longer term decarbonization objectives, and a growing role of the climate dimension in EU's internal and external energy policies. The EU set for itself sustainable energy targets for 2020, which will be achieved (with some concerns though remaining over energy efficiency) and actively participated in the December 2015 Paris Agreement process.

Responsible for 10% of global yearly emissions, the EU has become a global leader in the energy transition. Decarbonization targets for 2030 have been reinforced in 2018 and a debate over longer term (2050) decarbonization options and pathways has been opened in 2019, with a growing consensus on the need to target climate neutrality by 2050. EU's energy policy is now a decarbonization policy with several pillars: decarbonization and competitive, secure, and integrated energy markets. Yet this is a long and complex process and the long journey toward full decarbonization has just been started, while climate and biodiversity preservation call for unprecedented, massive, and urgent action. Despite having remarkable policies and ambitions, the EU is also confronted with inconsistencies and shortcomings. Matching ambition with actions will be the most challenging.

Following the 23–26 May 2019 European elections and the new five-year mandate for the European Parliament and the European Commission starting on November 01, 2019, and against the backdrop of rapid degradation of the global climate and biodiversity situations, the EU energy and climate policies are at a turning point and on the threshold of unprecedented developments. Ursula van der Leyen, President of the European Commission, referred to environmental protection as the "greatest responsibility and opportunity of our times" and committed to proposing a Green Deal in the first 100 days of her mandate. Over the summer of 2019, a swift, broad, and radical reaction has gained political support and a majority of Europeans see climate action as a priority. The next challenge is to clarify decarbonization pathways, define the key priorities, and adopt the right instruments and policies. In terms of industrial and economic cycles, 2050 is practically tomorrow. Lastly, citizens are often not willing to bear a high financial burden for the transition. Cost-efficiency, and fair redistribution of costs and dividends among citizens, Member countries, economic sectors, and territories, will be essential.

This chapter looks at some of the strategic energy and climate policy issues for the next five years, elaborates on how the EU energy and climate policies may be shaped, and what are the global implications.

2   The Status of the European Energy Transition

The EU is almost on track to meet its 2020 climate and energy package targets (the 20–20–20 objectives),[2] a policy that had been enacted in 2009. In 2018, renewables accounted for over 32% of total electricity generation and CO_2 emissions from the power sector saw a slight decrease from 2017 and an overall significant fall from 1990 levels. In most countries, renewable investments are strong, notably in solar power, with falling deployment costs, especially for offshore wind.

The EU has developed a number of useful tools to support the energy transition: the European Investment Bank, the Emission Trading System (ETS) which has been reformed, pushing up carbon prices over 20 EUR/tons for the first time in a lasting manner since summer 2018, and several infrastructure investment funds, as well as innovation funds and programs, such as Horizon 2020, becoming Horizon Europe under the next long-term EU budget.

Most EU governments have introduced plans to phase out coal from power generation, with 70 GW out of 140 GW of installed capacity expected to be phased out by 2030. Germany's coal generation (37%) will be progressively phased out by 2038 at the latest while Poland's high dependence (80% of power generation) will require special considerations and incentives to support structural change.

Further steps in this energy transition process included the objective to establish a European Energy Union developed by the European Commission to foster the integration of energy markets and convergence of policies. The Clean Energy Package for All Europeans, negotiated and finalized over the period 2016–2018, is a fundamental, highly complex tool that has introduced improvements in the functioning of the internal energy market, encouraged innovation and enabled a more active role for energy consumers. The EU is betting on the complementarity of its national energy mixes and the internal energy market to meet the rising flexibility needs as intermittent renewables are deployed. Institutional changes have also happened with the creation of a Vice President for the Energy Union and the adoption of a Regulation on governance, requiring Member States to adopt consistent planning tools and engage in cross-border consultations, while also giving the EU Commission the

---

[2] 20% increase in energy efficiency, 20% reduction of CO_2 emissions from 1990, and 20% renewables in final consumption by 2020.

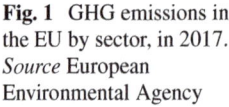

Fig. 1 GHG emissions in
the EU by sector, in 2017.
Source European
Environmental Agency

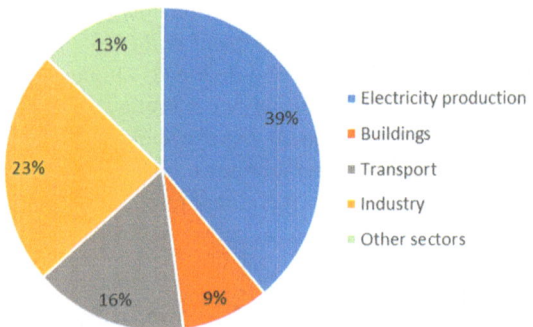

right to review and issue recommendations on the draft plans. The challenge is now to ensure a proper implementation of the package.

With the completion of the 2016 Clean Energy Package, the EU has confirmed the permanence of its climate commitments and its increasing role in policies implemented nationally. This is reflected in the commitment of raising the share of renewables to 32% of final energy consumption (from an initial target of 27%) and improving energy efficiency by at least 32.5% by 2030 (up from 27% compared with 2005). Decisively also, in raising the objective to reduce CO_2 emissions from 30 to 40% by 2030 compared with 1990. Lastly, the package took steps to improve electricity market integration and security (capacity reserves, cross-border flows) and to accelerate the decarbonization of the transport sector (Fig. 1).

These positive results nevertheless hide strong discrepancies among Member countries and important challenges. These include, notably:

- Costs of renewable support schemes grow, not least due to the increasing grid costs; bottlenecks in high transmission lines are not removed, such as in Germany, where four north south lines are planned but will face huge delays; social resistance to renewables and transmission lines is growing, with over 40,000 wind miles installed in Germany and 8000 in France; Overall system costs will further increase with the need to deploy electricity storage technologies, or demand-side management systems. There is a strong case for incentivizing larger renewable projects in suitable locations, where grid connection costs can be minimized. Yet acceptability issues can be an obstacle to cost-effective deployment of renewable energy sources, as recent experience shows in France. It remains to be seen whether reaching 65% of final electricity consumption as is planned in Germany with mainly wind and solar will be possible—that would require at least doubling the number of installed windmills considering that the best spots are often already taken, that interconnection challenges arise, grid costs increase and that social acceptance for mills and grids is increasingly a challenge. Last but not least, the land costs for solar could increase—the declining deployment trends in H1 2019 for renewable energy sources in Germany and France require a careful watch in this context.

- Lignite in most cases is more competitive than natural gas for power generation so that in Germany, for example, the increasing deployment of renewable energy sources has kept coal generation at almost steady level in past years while gas-fired power generation took a hit. The situation is now improving with the fall in gas prices and higher costs for CO_2 allowances, enabling a market-driven switch between low-efficient coal plants and high-efficient gas plants. France has made nuclear power a pillar of its mix and benefits from the legacy of its very low emitting power generation fleet. It has been proposing to complete the EU ETS with a regional floor price for CO_2 in the electricity sector, in order to accelerate and facilitate the phasing out of coal, which Germany has refused so far in order not to give advantages to France, an important industrial competitor and electricity exporter (Matthes 2017). This position could change though as a further reform of the ETS is planned. In line with the 2018 coalition agreement, Germany aims at producing more than 65% of its electricity from renewable energy sources by 2030 (35.2% of output in 2018).
- The EU has no competence in tax policy, nor on the energy mix of national countries, nor on social policy. A question is whether to enlarge the ETS to other sectors currently not covered (aviation, buildings, maritime transport…) or whether to proceed with separate taxation of these sectors. Moreover, the ETS could be fully replaced by a tax system at EU level, which would be politically challenging but worth discussing. The ETS only covers about 45% of total EU CO_2 emissions. Its latest reform was driven by the creation of a capacity reserve mechanism and was an important step forward as prices soar to over 25 EUR/tonne, yet further reforms will be needed to reflect the withdrawal of the United Kingdom following Brexit, but also the closure of coal-fired power plants, the increasing deployment of renewables and higher energy efficiency targets. The carbon market could still concentrate on accelerating low-carbon investments in heavy industries, but the price incentives are limited in this case because of the free allocation of quotas and measures taken nationally to compensate the indirect cost of CO_2 for energy-intensive consumers. These developments once again challenge the relevance of carbon pricing tools in Europe. Examination of the question is today compartmentalized, as unanimous voting by Member States prevails in tax matters. The next ETS reform is expected in 2021.
- Efforts in energy efficiency have slowed down and while low hanging fruits have been tapped, the challenge will be to further deepen efforts in the residential sector and transport sectors. After a gradual decrease over 2007–2014, primary and final energy consumption have increased again over the recent years, in part because of higher economic activity and relatively low fossil fuel prices, but also because of soaring sales of large cars and a slow down in energy efficiency improvements in the building sector. The trend will have to reverse shortly if the EU wants to fulfill its 2020 commitments.
- The decarbonization has so far largely focused on the power sector, with mixed results, but not yet on the transport and industry sectors. The Clean mobility

package adopted in early 2019 is an important milestone in accelerating the decarbonization of the transport sector. The new 2030 targets will accelerate electrification but not remove the thermal engine. They will be hard to achieve and drew very strong resistance from the automotive industry as they were negotiated. Fundamental decisions must also be made as to whether the mobility will be all-electric or whether one should give a greater role to hydrogen, green gases, or LNG/CNG, especially for freight transport. If biofuels, alongside hydrogen, are the only realistic ways in future to decarbonize partly the aviation sector, then a lot of innovation funding should also go in this direction.

– Most of the focus has been on CO_2, but not so much on other polluting gases such as methane—at a time when President Trump is taking steps to ease regulation on US methane emissions and when gas flaring is increasing globally, notably in the US.

– There are important discrepancies among member states' priorities and the recent submission of national climate plans shows that progress toward the common targets is uneven and that more than a last mile is missing to reach the set objectives. The gaps between aggregated contributions and the EU targets could be as big as 1.6% points for renewables, and as big as 6.2% points for primary energy consumption (European Commission 2019a, b, c).

– Set aside long-term contracts with guaranteed prices for renewables projects, there are no price incentives for long-term investments into low-carbon electricity generation capacity, due to the increased volatility on wholesale markets and the expected decline in prices as more renewables with zero marginal production costs are connected, especially during day time. This fundamental market failure is not addressed and will become a growing issue given that the current generation overcapacities will progressively be reduced and that massive investments will be needed to offset the ageing thermal fleet (coal, gas, and nuclear) and expand low-carbon assets. With more renewables, wholesale prices will become increasingly volatile and could well decrease instead of increasing as could be the case in the medium term while carbon prices grow and large thermal capacities are still in the mix. The shift to decentralized production and self-consumption is also triggering new challenges in terms of regulation and distribution of network costs between the different categories of power consumers. To avoid a potential "death spiral" and excessive cross-subsidies, the structure of unitary grid fees may need to be adjusted.

– There are strong divergences as to how far electrification of uses can or should go: while it will have to increase in the transport or residential sectors, the gas industry, for example, claims it needs to continue playing a key role in pointing to the unique value of the gas and storage infrastructure as energy systems will need greater flexibility, especially for the winter, and as the transition needs to happen at the lowest costs possible, which the existing, sunk cost gas infrastructure can facilitate. Yet decarbonizing natural gas is expensive and can only be realized in a system approach. Producing biomethane, for example, costs almost eight times more than natural gas (at 01/2020 prices), but the simple cost comparison with other fuels has limits: when externalities from biomethane production are included (in terms of CO_2 footprint, territorial development, energy security, agro-ecological transition, replacement of fossil fuels in the transport sector) biomethane has clear advantages.. What matters is to properly understand and assess what are the future impacts on electricity demand flexibility, on managing peak loads, and hence on short and longer term storage, what are technological options for electricity storage and their costs, and also to have a clear vision in terms of sequences of the energy transition: natural gas will play a key role for the next decade in providing system flexibility, especially if it is cheap and competitive (which is the case in mid-2019 but could change again), before progressively having to disappear (unless one produces decarbonized hydrogen from gas with CCUS). If climate neutrality is the ultimate objective, remaining emissions that are to be compensated through carbon sinks will not come from natural gas, but from the agricultural or aviation sectors, or specific industries. And the switch to "green gases" is important but will not impede a progressive yet necessarily important reduction in the role of natural gas per se, that could nonetheless boom elsewhere in the world if it is competitive so that global gas players will be able to adapt their strategies. Decarbonized hydrogen will have to play a key role in the refinery sector first and then to decarbonize the industrial sector and probably also, parts of the maritime and freight transport segments. It remains to be seen whether decarbonized hydrogen will play a role in the passenger car segment, residential heating, or electricity storage to meet seasonal fluctuations.

3   The New Political Context from 2019: Pressure for Accelerating and Deepening the Energy Transition

In 2018, discussions have opened on objectives and strategies for 2050, with the exploration of trajectories going as far as carbon neutrality by 2050 in an unprecedented joint undertaking involving many Directorates from the Commission (European Commission 2018). Different trajectories and energy mix options were laid out for different decarbonization objectives in this major exploratory analytical document that aims at informing discussions on the options available.

Systematic transformations in governance and public policies, company strategies, and citizens' behavior are required. These transformations need to be grounded in the broadest consensus possible. Indeed, European policies so far were established in a context and with objectives that do not correspond to deep decarbonization and were largely focusing on market integration, competition, and supply security in order to serve the consumers' interests. The challenge now is to adapt them to this profound transformation.

After the European election that saw climate issues dominate concerns of citizens in many Member States, and in the face of the growing climate urgency, the discussion over the 2050 targets has clearly shifted: while carbon neutrality was opposed notably by Germany, all but one EU member states now decisively support that objective as highlighted at the December 2019 European Council.

The presentation of the European Green Deal on December 03, 2019 marked a tipping point. The Green Deal is targeting all pollutions and environmental degradations. It is a sustainable economic growth strategy which aims at fostering the well-being of Europeans. It aims at decarbonizing all economic sectors, cutting all greenhouse gas emissions, mobilizing more funds and budgets, aligning all Member countries and policies as well as using all available solutions and low-carbon technologies. The Green Deal announcement showcased the new European *raison d'être ensemble*. And a clear commitment to avoid social and territorial inequalities—a Just Transition Fund has been created to help vulnerable regions. This means closing the chapter of marginal improvements and considering profound socioeconomic changes.

At the same time, the EU has agreed on a taxonomy of sustainable investments, setting a global benchmark that will empower finance to foster the energy transition. A number of action plans are to be rolled out across 2020. A Climate Law is to be enacted, precising the objective and the policy framework. Objectives for 2030 are to be revised. An industrial strategy for low-carbon technologies is to be released, focusing notably on batteries and clean hydrogen. The aims should be to secure critical metals, design norms, protect industries, support the scale-up of innovation, create ecosystems and clusters along the value chain, impose carbon low-carbon footprints in tenders and materials (eco-design directive), and foster recycling operations. A new budget for 2021–2027 with higher spending toward climate change mitigation is to be agreed upon. An external strategy will follow. This Hercules task will not be an easy journey and there are already controversies, obstacles, and resistances. But the EU has shown that as one of the world's largest historical emitters, it is taking its responsibilities in bearing higher costs and is aiming for global leadership. If successful, this could reinforce EU's influence globally.

The key challenge now is to deliver as the easy low hanging fruits are no more available. There is no magic solution. The European Green Deal will require new financing (a 1 Trillion package until 2030 has been identified, which is likely to be insufficient though) but also the better use of existing tools and their refocusing on climate concerns. Targeting climate neutrality by 2050 requires choosing the most cost-competitive options and trajectories, fostering regional coordination of policies and investments, avoiding stranded assets, picking the right technologies, and giving them the large deployment scale and perspective needed to reduce costs. It requires giving the industry the rights tools, incentives, and protection to conduct this transformation while remaining competitive. While technological neutrality is a condition for political consensus and efficiency, a holistic approach focusing on the entire system costs is needed, rather than an individual focus on specific technologies or sectors that masks the system-wide implications.

Another policy challenge is to accommodate the new role played by cities and regions, in order to facilitate structural change and strengthen regional cohesion. And last but not least, to ensure social cohesion and a fair distribution of costs and benefits from this transition that will have to rely on a large consensus and joint action from the industry, the citizens and the government. The costs of reaching carbon neutrality will be very high, yet much lower than refusing to take action and facing the most adverse impacts of climate change. The social acceptability of this transformation will depend on two main factors: the ability to provide assistance to those who will be the hardest and most directly affected by the changes, such as low-income households and works in the carbon-intensive sectors; and the ability to make the energy transition a lever for creating quality jobs and added value on the EU territory. Without a robust and steady support from the population, efforts in decarbonizing the economy are at risk of being delayed, diluted or called in question at each election round, which would make the whole process ineffective and costlier.

There is no denying that carbon pricing will need to cover all sectors and increase in a progressive and predictable manner. Acting in a facilitating role, the EU should revise the 2003 Directive on energy taxation to not only ensure that inefficient fossil fuel subsidies are removed and that environmental considerations are included in all national schemes, but also require that fiscal equity is given due attention.

The EU will have to review its state aid policies and the market liberalization and competition dogmas from the 1990s and 2000s as the deep decarbonization of the power system cannot happen in the current market environment. More state intervention is needed to guarantee long-term prices, reduce risks, and borrowing costs. This will have to include nuclear power, which is controversial given the opposition from Austria, for example, but one thing is clear: there will be no effective and deep decarbonization by 2050 without new nuclear power constructions in the EU that at least compensate for the closure of ageing plants, even extended. But new investments in nuclear will be impossible without guarantee schemes like the ones implemented in the United Kingdom (UK) for Hinkley Point C. These triggered an investment decision by state-owned utility EDF, but proved to be insufficient for private actors like Hitachi. Against this backdrop, it is worth paying attention to UK's envisaged introduction of a "Regulated Asset Base" (RAB) model for new nuclear project, as being planned by the Secretary of State for Business, Energy and Industrial Strategy. Its goal is to reduce the cost of raising private finance for new nuclear projects, thereby reducing consumer bills and maximizing value for money for consumers and taxpayers. In the coming months, a discussion will have to start on a new electricity market design to facilitate the large investments needed.

On a similar note, a political agreement is needed to bring down the climate wall between western and eastern European countries. This agreement should include on one side strong commitments for a rapid coal phase-out, and on the other side the mobilization of the dedicated Just Transition Fund complementing existing instruments (Cohesion policy, Modernization and Innovation funds, Invest EU, etc.), and focused efforts by the European Investment Bank to support integrated territorial development strategies in carbon-intensive regions.

4 Strategic Economic Challenges Ahead

a. The industrial dimension of the energy transition

The energy transition process bears new industrial risks and threats related to the control of value chains of low-carbon technologies and the operation of systems:

- technologies, innovations/intellectual property and low-carbon technology value chains (autonomous mobility, nuclear power, decentralized production, renewable energies)[3];
- access to markets and tenders (in public transport, nuclear energy, wind and solar power, hydroelectricity infrastructures, sustainable cities).
- assets (investments and shareholdings in companies operating in electricity, gas, digital technologies, data-processing and data);
- standards (electricity, batteries, electric mobility, interconnectors, networks, data protection);
- cyber-security;
- information and image.

The EU will have to take further actions and steps to address these challenges which it has started to recognize. The EU regulatory framework needs to foster both demand and competitive supply of low-carbon solutions. That will require to further confront policies and actions from emerging and established powers such as China (and its *Made in China 2025* strategy) or the USA.

China has defined a *Made in China 2025* strategy which includes an empowerment dimension and the mastery of energy technologies. The country has already taken, or is seeking to take, dominant positions in the whole value chain of the main technologies involved in the low-carbon energy transition. This is the result of a proactive strategy which combines internal support for innovation (one-third of patents in low-carbon technologies are Chinese) (IRENA 2019), an industrial policy (large state groups receive financing, demand-side support, a capacity to take risks and cooperate throughout the value chain), and technological looting or the transfer of technology as a condition for Foreign Direct Investment. China moreover benefits from its huge domestic market which provides economies of scale while competition between state groups is weak. The country also benefits from the errors and mistakes of its competitors, notably the EU and most of its Member States. They have left aside some of these issues and even directly contributed to China's dominance by transferring polluting industries to China, and by accepting forced technology transfers. Moreover, for a long time they only protested weakly against very unequal Chinese

---

[3] Strategic technologies in the energy transition include nuclear power; onshore and offshore wind turbines and their magnets; the next generation of photovoltaic cells and inverters; cars with highly efficient combustion engines; batteries (especially 4th generation) for mobility and stationary storage; decarbonized hydrogen; electricity storage systems using hydrogen and batteries; smart grids and demand response solutions; recycling technologies; digital technologies and technologies for protection against cyber risks.

market access rules, despite China's membership of the World Trade Organization (WTO) since 2001.

China has mastery of certain value chains which could give it economic supremacy not only in its large domestic market but also abroad: critical metals and rare earths, their refining, the special alloys of certain metals, innovation, the manufacture and assembly of technologies (90% of solar panels, more than 50% of onshore wind turbines), third-generation nuclear reactors (China's first project is under construction), batteries, personal and public transport vehicles using electricity or hydrogen (Voïta 2018; IEA 2018)), equipment for managing smart grids or for telecommunications networks (5G), and soon technologies related to artificial intelligence.

Lastly, China's state companies have unparalleled investment capacities and are making major acquisitions abroad, especially in Europe. They are looking to invest funds at attractive rates of return, but also to take control of technologies, to understand markets and their functioning better, in order to transform their standards, to sell their technologies and identify new assets to acquire. For example, the Three Gorges Company is operating in more than 40 countries and looking to buy assets in the EU.[4] So is the State Grid Corporation of China, which has earnings of more than $300 billion (in 2017) and is seeking to expand its assets across the globe. The development of the 5G network will play a role in piloting energy systems, and is witnessing the Chinese giant Huawei challenge Western companies like Nokia and Cisco.

This dimension will be particularly accurate with regard to the battery and electric car industries that are currently expanding. The EU will have to further develop its European Battery Alliance (Mathieu 2019) and lay out a mineral strategy to address the critical metals challenge and China's increasing domination of that segment (extraction and refining), implement social responsibility standards, foster innovation to reduce dependency on critical metals, ensure that the low-carbon technologies are built in the EU territory and create local value and jobs and decisively, push forward the circular economy concepts to foster the recycling of metals and plastics, and define and implement the necessary standards. Instruments such as the measure of the carbon footprint of products, or localization norms, will be important.

b. **Focus on the vulnerabilities in critical metals**

Our economies have a growing need for critical metals and rare earths in defense, electronics and communications industries, as well as in low-carbon energy transition technologies (alloys, two-thirds of wind turbines use permanent magnets, LEDs, solar panels, glass, smart grids, and digital technologies and batteries).[5] These so-called critical or strategic metals have exceptional optical, catalytic, chemical, magnetic, and semiconductor properties, for example, neodymium and samarium, allowing super-powerful magnets to be made. Some 30 metals are considered indispensable and difficult to substitute (European Commission 2017).

---

[4]See: www.ctg.com.cn.

[5]Car batteries require 10–20 kg of cobalt and up to 60 kg of lithium and other critical metals and rare earths such as neodymium or dysprosium. Solar panels use indium and silicium.

The geographic distribution of these resources, the issues related to the extraction and refining of these metals, the structure of the mining industry as well as their (un-)availability in markets raise numerous challenges. These include geological, political, environmental, technological, social and economic risks which lead to vulnerabilities in the supply chain and so create risks to the value chains of technologies that use them.

The criticality of these metals and rare earths has been much studied and varies depending on the metals and their routes to specific markets: nobody can predict which battery technologies will emerge in the long term, for example, or how solar panels will be built.

These metals are often by-products of more abundant metals, but are present in minute proportions.[6] A ton of rock needs to be processed just to provide a few grams of platinum. Quantities produced are often tiny compared to other metals: 15 million tons of copper are mined each year and only 600 tons of gallium; 2 billion of iron are produced compared to 200,000 tons of lithium. The quality of deposits varies while the concentration of critical metals may range from 0.5 to 15%, depending on the mines and metals.

Significant environmental issues exist on top of these geological challenges because refining uses lots of water, electricity, and often chemical products for hydrometallic processing with acid. Chile has very large lithium deposits, but needs to ration its production because of water shortages and also competition from copper in particular. Developing infrastructures to transport water is expensive, and this intensifies the criticality of the metal given strong demand growth.

Economic risks are substantial because there are tensions both in supply and demand, as well as market structures which are often oligopolistic, or even dominated by Chinese companies. Strong price volatility of some of these resources is another source of concern as it complicates investment and recycling: this is so notably for cobalt whose price surged before falling in early 2019.

The supply of these metals is concentrated in a small number of countries many of which are not members of the OECD (accepting Canada, Chile, and Australia). They include China, the Democratic Republic of Congo (DRC), Argentina, Bolivia, Russia, South Africa, Kazakhstan, and Brazil.

Mining investment in recent years has concentrated in Latin America and to a lesser extent in Africa. Supply takes a long time to adapt to demand, because the development of mining projects is long to implement. These projects are risky and their profitability is often problematic because of price volatility, and prices were also low for a long time. This situation encouraged the closure of mining activities in Europe and North America, while reinforcing the concentration of such activities in the hands of Chinese companies. The latter do not integrate pollution costs, and they have access to cheap credit, cheap labor, or integrated business structures in which losses in one segment of the value chain are compensated by profits elsewhere.

---

[6]Cobalt is a by-product of nickel and cooper, gallium is a by-product of aluminum; indium and germanium are by-products of zinc for example.

Supplies are often not available in transparent, open, and liquid markets: a share of world production is often allocated outside the market. Rosatom is an important producer of good quality lithium which is used in the Russian nuclear and/or military sector. Only surpluses are sold on markets. Cobalt is mined in the DRC, which accounts for 60% of global output, and is largely bought directly by the networks of integrated Chinese companies without it being possible to know exactly the output figures of small artisanal mines for example.

Production companies often operate oligopolies, and China has increased its investments and shareholdings and often dominates the extraction and refining of critical metals and rare earths. Thus, five companies account for 90% of global lithium production, and apart from Abermal and FMC, three of them are Chinese or have Chinese capital (SQM, Tianqi Lithium, and Jiangxi Ganfeng Lithium). The mining of cobalt worldwide is dominated by a few companies including Glencore or Chinese companies which are extending or developing their operations everywhere: in the DRC, in Madagascar, Greenland, and Bolivia. Refining cobalt and lithium is very polluting, and is concentrated in China because producer countries mainly sell intermediate products.[7] This spectacular strategy by China to expand mining activities and buy up assets has several aims: meeting its needs for metals which are not available in China; pre-empting markets; dealing with growing environmental problems in China; developing more competitive resources; and limiting declines in its own reserves (Seaman 2019). Lastly, apart from these issues, conditions for accessing resources may change: while Argentina and Australia have a very stable investment framework, the DRC has recently adopted a new mining code which increases royalties from 2 to 10% and plans further increases as mining nationalism is developing notably along the lines of the Bafokeng in South Africa. These developments are often perfectly legitimate but constitute risks for investors and favor actors who can protect themselves from them.

The demand for critical metals is expanding rapidly and is concentrated in emerging countries or countries which are technologically advanced, especially the EU, the US, Japan, and China. Demand for lithium is set to triple by 2025, to reach 600,000 tons per year, and increase by 20% for copper, and could rise by 60–100% for cobalt according to various scenarios, requiring at least an increase in output equal to that of the DRC. More generally, the energy transition is likely to be as hungry for other metals and resources which for the moment are not critical, but which could become so. These include copper, iron, or even sand for cement.

Control over the supply chain of critical metals is a strategic asset in developing low-carbon technology value chains and developing advantages over competitors. The EU greatly depends on imports to cover its growing needs because it practically produces none, even though it has non-negligible reserves. The investment framework there, however, is relatively unfavorable and societal opposition is an obstacle despite rising prices and low-interest rates which should allow production to be relaunched. Finland is exceptional in creating a mining cluster[8]: projects have

---

[7]The refining giants are Huayou, its subsidiary CDM, Jinchuan and GEM.
[8]See: www.miningfinland.com.

been launched on the Keliber site especially so that 11,000 tons per year should be produced by 2020, while the country has a significant lithium refining industry which will allow it to become a hub in batteries, just as New Caledonia is for nickel. There are some French and European mining and processing groups, such as Eramet, Solvay, Umicore, Imerys, ThyssenKrupp. But their size and global weight are far behind the Asian, Swiss, Canadian, and American giants. There is significant output potential in Greenland but this has already been partly captured by China. Mining projects are emerging in Portugal, Serbia, Hungary, and Germany, but they only represent about 5% of global annual investment and do not change the overall situation: European dependency on imports will grow.

In a context of heightened economic and technological rivalry, China has a strategic advantage because it can favor its companies at the expense of European customers and so limit the availability of resources or create distortions in competition or use its grip on the chain of critical metals to obtain economic, trade, and technological advantages over European actors. There are important risks in terms of value chains, employment, as well as foreign economic and industrial dependencies. China has, for example, already temporarily reduced its exports of rare earths to Japan following political tensions (Lepesant 2018). Although strategies based on cartel behavior and pressure have not been pursued openly, vulnerabilities remain.

The concentration of resources in a small number of countries outside the OECD, the oligopolistic nature of markets and the fact that these resources are in the hands of powers which are often rivals (especially China and Russia) generate risks for access to resources, and even of emergence of cartels. Both could raise the total costs of the energy transition and block or threaten the development of national industries. This is especially so as competition is strengthening from military technologies, which are also big consumers of critical metals. Faced with trade tensions from the US, China may enhance its strategy of self-sufficiency and reinforce its pre-emption of resources.

Issues related to water, pollution, and the social conditions of mining are also a challenge to corporate social responsibility for European economic actors. There are up to 100 million informal mineworkers in the world who sometimes work in deplorable safety and environmental conditions, while working conditions often do not comply with the standards of the International Labor Organization. The EU should push for higher ESG standards in its imports of critical metals or related materials.

These challenges, risks, and even threats are not new and have been the subject of political and strategic consideration for several years. The EU has a list of 27 critical metals out of 61 that are taken into account (European Commission 2019a, b, c). The US has a substitution strategy, while NATO has formulated goals for reducing dependency of the military industry on China. Yet given the ever greater hegemony of China and the rising challenges of energy transition, a new strategy and evaluation of risks are required.

c. **The battery cells chain**

The demand for Electric Vehicles (EVs) is set to rise strongly as of 2020, due to the combination of: lower costs for electric batteries; restrictions imposed on vehicles with combustion engines (new European emission standards, and traffic restrictions in cities especially); the development of charging infrastructures, and above all the serious commitment of global and European car producers, partly linked to the "dieselgate" scandal. Sales of EVs jumped between 2017 and 2018 and should account for nearly a third of light vehicles sales by 2030. In a favorable scenario, there could be 220 million electric vehicles on road by 2030 compared to 3 million today (IEA 2018).

Factors determining this path include changes in public support measures; the cost of batteries; vehicle autonomy; the availability of fast-charging infrastructures; the environmental and societal footprint of cars; competition and trade strategies; and antipollution regulations in force.

Systems linked to electric vehicles are expanding rapidly but face the following drawbacks:

- the value chain is largely dominated by Asian actors benefiting from subsidies and economies of scale (China, Japan, and Korea), especially in battery cells production. European carmakers need to control the risk of competition moving upstream in the sector;
- load challenges should not be neglected: the development of networks, the management of peaks, and the charging speed: 1 million vehicles generate only about 2 TWh of extra consumption, yet there may be challenges in terms of power demand surges. The partial electrification of France's vehicle fleet could lead to demand peaks of up to 10 GW, whose consequences on the network must be anticipated;
- if the carbon footprint of electric vehicles is really to be lower than for conventional vehicles, then they need to be charged with low-carbon electricity. This is the case in France, but not in Poland, for example, which has strong ambitions for reducing city pollution. European regulation at present does not take into account the electricity mix of vehicles. Likewise, life-cycle assessments show that battery manufacturing conditions require substantive amounts of energy and therefore the location of battery gigafactories should also be decided on the basis of environmental criteria;
- without a significant improvement in the energy density of batteries, the search for greater autonomy will run into technical-economic limits. Indeed, the greater vehicle autonomy is, the heavier batteries are and the more metals they consume (500 kg for a Tesla). The creation of interchangeable batteries could nevertheless facilitate the expansion of electric vehicles used for long distances, and reduce charging times.

A combination of technological options should, therefore, be favored so that the goals of cutting CO_2 emissions are best achieved in three types of usage:

- electric mobility: city buses, city fans, cars, two-wheelers, and off-road vehicles;
- carbon-free hydrogen mobility: professional mobility (trucks, long-distance transport), certain trains, shipping; in the long run, possibly aviation;
- natural gas mobility, based on LNG and NGV, but also renewable gas as much as possible: maritime and river transport, long-distance freight transport, family vehicles (presently 1.3 million in the EU). There will still be other forms of pollution like nitrogen oxides, and innovation needs to take place, for them to be filtered especially;

Given that Asian manufacturers are today best placed to capture the bulk of global battery demand, the recent launch of a "European Battery Alliance" is to be welcomed. It is intended to foster the emergence of a European industrial ecosystem by creating a favorable framework for investment in manufacturing capacity. Initial discussions began in the autumn of 2018 and led to a clear diagnosis: without major contracts with the European car industry, it will not be possible to have European-led projects that aim directly at achieving an annual production capacity of around 30 GWh per year each, based on the model of the Tesla-Panasonic Gigafactory in Nevada. European car producers are indeed in global competition and believe their negotiating capacity with Asian cells suppliers is sufficient to obtain, today at least, the best cost/performance levels. However, the balance of power is likely to evolve and it is important to be fully aware of becoming technologically dependent on Asia, knowing that a global capacity shortage cannot be excluded. It is, therefore, important to support all intermediate projects (producing 8–10 GWh per year) by 2025, in order to enter the industrial race and establish the credibility of European suppliers, and for order books to be expanded in the future as development strategies are pursued.

The European Battery Alliance could foreshadow a revival of a European industrial policy which needs to take into account changes in the international rules of the game and find an equilibrium between a wait-and-see position and dirigisme. The Battery Alliance is open and not prescriptive. It is geared to mobilizing private actors and the search for industrial synergies between European actors. The alliance should also draw on a proactive approach by public authorities. All avenues should be explored to improve cost and non-cost competitiveness of European battery manufacture, including an accompanying differentiation strategy; promoting European supply by introducing standards concerning the environmental footprint of batteries; and introducing criteria for the public procurement of electric buses for example. Investment decisions should also be facilitated by mobilizing public funding instruments (the EIB, R&D programs, IPCEI status authorizing state aid for transnational industrial projects, etc.), or even the design of skill development plans to favor better matching of market needs. Lastly, this new industrial policy should include an external dimension, especially a frank dialogue with China on barriers preventing access to its gigantic electric vehicle market. Success is not guaranteed, but this European Battery Alliance demonstrates a willingness to act without delay, in consultation with European industrial actors and by activating all the available levers of public policy.

From this point of view, it is to be hoped that the approach will be renewed for other technologies considered to be strategic for the future of the energy transition.

d. **The external dimension**

With the expected acceleration and deepening of the European energy transition efforts and policies, the external context will become critical and dedicated and specific policies will have to be developed in order to protect the European transition process. The risk for the EU is to lose its competitiveness, that its economies and companies are weakened, that imports replace carbon-intensive production in the EU ("carbon leakage") and that external actors steel EU's innovations and technologies. It is also that external actors use information technologies to destabilize social and political processes related to this systemic economic and social transformation. Last but not least, it is that the EU is a global island that is pushing for climate neutrality while the rest of the world is on a business as usual trajectory.

The EU will have to develop power instruments, and to make use of these instruments when needed. It could not stop Trump from leaving the Paris Agreement, but it could keep the agreement alive. It could seal a climate pact with China in April 2019. It could not so far prevent Brazil from destroying its Amazonia forest, but is having influence on India, for example, that could strengthen its climate commitments. Overall, instruments that could be mobilized are

- Trade agreements, which so far have no binding climate obligations as highlighted by the Mercosur agreement and follow-up controversy;
- Carbon border adjustment mechanisms at EU's borders for certain goods;
- Public finance: the EBRD, direct contributions such as to the Green Climate Fund, development aid;
- Public diplomacy in bilateral relations and multilateral institutions (OECD, G20, G7);
- Sanctions such as trade restrictions, or limited access to cooperation mechanisms

Priorities for actions or issues that will increasingly be in the focus include

- Raising national climate commitments from leading emerging and OECD economies;
- Stopping direct or indirect investments/financing of coal-fired power generation globally;
- Scaling up sustainable electrification in Africa and revisiting multilateral aid mechanisms and projects;
- Greening suffocating cities in emerging economies;
- Fighting deforestation and reforestation;
- Reducing fossil fuel subsidies;
- Improving energy efficiency efforts, especially in the Eastern and Southern neighborhood;
- Reducing the carbon footprint of maritime transport;
- Reducing fugitive methane emissions such as from the natural gas sector.

5   Conclusion

The year 2019 marks a turning point in the European energy transition process which ultimately culminated with the adoption of the carbon neutrality by 2050. This will require systemic transformation of the EU economies and societies, of its energy policies which will have to include a strong industrial component and decisively, of its external policies. One does not know if the EU will succeed, but this can be achieved. To succeed, the EU will have to reshape all its domestic policies and priorities toward that fundamental goal, while maintaining social and political cohesion. Launching a European Green Deal and mainstreaming the climate neutrality target in all EU policies will have tremendous repercussions for the rest of the world, possibly generating new tensions, but also fueling new alliances and profound geopolitical and geoeconomic transformations.

At the start of this new political cycle for EU institutions, it is essential that the following dimensions are addressed.

Defining carbon neutrality in including imported emissions, determining responsibilities for setting and enforcing climate targets and raising the 2030 targets, reforming the ETS further, notably post-Brexit.

Increasing the capacity for states, regions, and cities to experiment new ways of supporting investment and innovation in low-carbon technologies, while working for enhanced cooperation in the industrial and regulatory fields. These initiatives would begin on a voluntary basis but could be supported and coordinated by a European Energy Transition Agency.

- Implementing a common electricity strategy between France, Belgium, the Netherlands, and Germany within a context of readjusting national electricity mixes and progressive decarbonization. This analysis of regional production equilibriums should also feed the debates on the most relevant interconnection scheme post-Brexit, and on whether new nuclear power stations should be built in the coming decades.
- Given vulnerabilities in critical metals, the EU needs to act and favor new, responsible mining projects on their land, and link their development aid to the implementation of environmental and social standards in the mining sector, while supporting traceability initiatives. Four areas must be pursued simultaneously on the demand side: re-use, recycling, reduction, and reindustrialization.
- Consolidating Europe's industrial policy for low-carbon technologies, by drawing on the initial lessons of the European Battery Alliance. Drawing on a sound diagnosis of present and future technological dependence, as well as on a close dialogue with academia and business, the EU should mobilize all possible public policy tools available (regulations and standards, funding, education, etc.) in order to improve Europe's cost- and non-cost competitiveness. At the same time, the EU should organize a frank dialogue with its trade partners to guarantee fair access to their domestic markets.

- Accelerating work on the taxonomy in order to promote the large-scale development of green and responsible finance and to encourage investments compatible with the Paris Agreement within the EU, but also with emerging countries.
- Developing a strategic, coherent, and coordinated external energy and climate strategy that would aim to protect the European energy transition process and develop and accelerate the energy transition not only globally, but also decisively, in the EU Eastern and Southern neighborhood, where emissions are 20% higher than in the EU and could grow further (Eyl-Mazzega 2020).

The energy and climate dimensions have been increasingly dominating the EU policy agenda. A new, broader definition of energy security is required to frame EU's actions and policies. It would entail

- continuous, unimpeded, and guaranteed energy supplies;
- the competitiveness of energy prices in a context of international economic competition;
- the decarbonization of energy sources and their use;
- the mastery and control of innovation, as well as of economic and technological value chains;
- the smooth operation and reliability of integrated, low-carbon energy systems;
- social and territorial cohesion, which are prerequisites for a sustained low-carbon transition.

References

European Commission (2019a) United in delivering the Energy Union and Climate Action—Setting the foundations for a successful clean energy transition. Communication 285. https://eur-lex.europa.eu/legal-content/EN/TXT/?qid=1565713062913&uri=CELEX: 52019DC0285. Accessed 18 June 2019
European Commission (2019b) Towards a more efficient and democratic decision making in EU tax policy. Communication. https://ec.europa.eu. Accessed 15 Jan 2019
European Commission (2018) In-depth analysis in support of the commission communication: A clean planet for all, a European long-term strategic vision for a prosperous, modern, competitive and climate neutral economy, November 2018. https://ec.europa.eu
European Commission (2017) Communication from the European Commission to the European Parliament, the Council, the European Economic and Social Committee and the Committee of the Regions, on the 2017 list of critical raw materials for the EU. https://ec.europa.eu. Accessed 13 Sept 2017
European Commission (2019) Critical raw materials. http://ec.europa.eu
Eyl-Mazzega M-A (2020) The green deal's external dimension. Re-engaging with neighbors to avoid carbon walls, Édito Énergie de l'Ifri, 3 March 2020
IEA (2018) Global EV outlook 2018, May 2018. www.iea.org
Intergovernmental Panel on Climate Change (2018) Global Warming of 1, 5 °C. Special report, October 2018. www.ipcc.ch
IRENA (2019) Patents evolution of renewable energy. http://resourceirena.irena.org
Lepesant G (2018) La transition énergétique face au défi des métaux critiques. Une domination de la Chine? Études de l'Ifri, Ifri, January 2018

Mathieu C (2019) The European battery alliance is moving up a gear, Edito de l'Ifri, May 2019

Matthes FC (2017) Decarbonising Germany's power sector: ending coal with a carbon floor price, Notes de l'Ifri, Ifri, December 2017

Seaman J (2019) Rare earth and China: a review of changing criticality in the new economy, Études de l'Ifri, Ifri, January 2019

Voïta T (2018) Going green: are Chinese cities planting the seeds for sustainable energy systems?, Études de l'Ifri, Ifri, February 2019

US Clean Energy Transition and Implications for Geopolitics

Jonathan Elkind

1 Introduction

The impacts of climate change pose severe challenges around the globe: rising sea levels, disrupted precipitation patterns, altered agricultural conditions, increasingly frequent droughts, brutal heat waves, and severe storms to name a few. Different regions of the United States will encounter all these impacts, just as is true elsewhere around our globe. Consequently, those Americans whose livelihoods or homes are vulnerable to a changing climate, and especially those who do not share in the country's overall prosperity, are at risk.

In addition, however, there is another form of risk that climate change poses for the United States: geopolitical risk. This country has been central to historical emissions of greenhouse gases and is today the world's second-greatest emitter of greenhouse gases. It is no less central to the global efforts to refine climate science and develop climate solutions, but it is preoccupied by a bitter, protracted debate over current and future climate policy.

So precisely when scientists and citizens around the globe are calling for an accelerated and scaled-up response to global warming, national leaders in the United States are profoundly divided over whether and how to contribute to climate solutions. Some Americans–especially policy-makers in leading states and municipalities and executives in an increasing number of companies–see climate as an existential challenge that requires a local, regional, national, and global response. These leaders intend for the United States to be part of the global solution. Others–especially Federal officials during the Trump Administration and many leaders in one of the two major U.S. political parties–view climate with disdain and are abdicating the country's duty to respond.

Already, this conflicted stance is causing damage to U.S. geopolitical interests. In the coming period, either the United States will return to playing a proactive and

J. Elkind (✉)
Center on Global Energy Policy, Columbia University, New York, NY, USA
e-mail: jhe36@columbia.edu

© The Author(s) 2020
M. Hafner and S. Tagliapietra (eds.), *The Geopolitics of the Global Energy Transition*,
Lecture Notes in Energy 73, https://doi.org/10.1007/978-3-030-39066-2_3

constructive role on climate change mitigation, or its influence and key relationships with partners around the globe will suffer serious harm. This chapter reviews the complicated and internally contradictory U.S. attitudes about the clean energy transition from the perspective of debates, bottlenecks, and progress occurring at the Federal, state, and municipal levels.[1] It then assesses the implications of the climate issue for the geopolitical interests of the United States.

2 U.S. Climate Policy: Struggling for Speed, Scale, and Durability

Many policy analysts have referred to climate change as a "wicked problem"–a challenge of extraordinary complexity that defies easy resolution (APSC 2007). This rubric fits the climate challenge well because responding to climate change requires policies with a highly unusual combination of attributes that exist in tension with one another. This reality has greatly complicated the American response to climate for years, and especially in the last decade.

Success in responding to climate change requires policies that combine (a) speed, (b) scale, and (c) durability. The first part of the puzzle–speed—is easy to grasp for anyone who accepts the findings of the vast majority of the international scientific community: Atmospheric concentrations of greenhouse gases (GHGs) are rising due to human activity, and these concentrations are causing unprecedented changes in the global climate. Climate science is far from simple, and important uncertainties remain, especially in regard to regional climate impacts over the medium and long terms. But as successive analyses from the UN's Intergovernmental Panel on Climate Change and other authoritative sources such as the US National Academies of Science, Engineering, and Medicine have repeatedly documented, human beings are changing the climate by engaging in activities that result in GHG emissions, chiefly through energy-sector emissions of carbon dioxide (IPCC 2018; McNutt et al. 2019).[2] Responding quickly is clearly a necessity for successful climate policy.

---

[1] The Federal government of the United States formulates national-level policy and legislation. Throughout this chapter, the terms Federal and national are often used synonymously.

[2] The IPCC's most recent assessment, which examined the implications of an effort to limit aggregate temperature rise to 1.5 °C, garnered a great deal of public attention around the globe because it stated starkly the necessity of urgent changes, starting immediately, especially in regard to the global energy economy (IPCC 2018). No less noteworthy was a short press statement, issued seemingly out of the blue by the presidents of the US National Academies in mid-2019. It reiterated the conclusions of the Academies' repeated climate assessments. The fact that the Academies' presidents felt it important to issue such a reiteration is itself testimony to the sense of alarm that the U.S. scientific community feels in light of policy-makers' seeming inability to take action on climate change. The statement read in part: "Scientists have known for some time, from multiple lines of evidence, that humans are changing Earth's climate, primarily through greenhouse gas emissions. The evidence on the impacts of climate change is also clear and growing. The atmosphere and the Earth's oceans are warming, the magnitude and frequency of certain extreme events are increasing, and sea level is rising along our coasts." (McNutt et al. 2019).

A second required attribute is scale. No climate solution can be successful alone if it only addresses a narrow slice of total global greenhouse emissions. This is true in the case of sector-specific emissions reductions that are not accompanied by other, parallel measures addressing other sources of emissions. In the United States, as we will see shortly, laudable emissions reductions have occurred in the electric power sector, but they are insufficient because they have not been matched in sectors such as industry and transportation. It is equally insufficient to apply approaches that address emissions in one set of countries without also addressing emissions elsewhere. The risk of simply "exporting" emissions by creating incentives for industrial activity (and thus employment) to move off-shore has historically been a sensitive and often legitimate argument used by certain policy advocates in the United States and elsewhere who worry that poorly designed climate policies could harm national economies without meaningfully reducing emissions. The growth of U.S. oil and natural gas production over the past decade creates new complexity in regard to concerns over causing emissions to move to other countries.

A third necessary attribute is durability. Barring surprises in climate science, the business of building and sustaining climate-friendly economies must become an indefinite part of our global future. We need to become so skilled and effective at climate mitigation that we can maintain our level of effort without flagging. Indeed, over time we may need to respond to climate change with progressively more robust and far-reaching measures. We will not succeed in this effort if citizens feel that their quality of life is suffering unduly, or more than their neighbors. We must find solutions that will be durable over time, solutions that will not fall victim to popular backlash the next time we have an economic downturn.

Here it is worthwhile to flag the fact that some policy prescriptions for mitigating global warming could translate into major socioeconomic impacts. For example, some environmental advocates argue in regard to fossil fuel production that the United States should "leave it in the ground," and should in fact end fossil fuel consumption before the year 2030. But the oil and gas industries currently employ over 1.4 million Americans (NASEO-EFI 2019). This does not mean that the United States should fail to respond to the climate challenge in a timely and effective manner. Nor should one forget that the low- and no-carbon energy economy is resulting in the creation of many new jobs across the country. But the new clean energy jobs do not magically appear just where old jobs are being lost, and they do not always require the same skillsets that oil and gas workers have. These facts underscore the need for policies that adequately account for the impacts on hundreds of thousands of people's livelihoods. Most of those same people are voters; for the chosen climate policies to be durable, they must account for near-term negative socioeconomic impacts; the policies must provide for what many have come to refer to as a "just transition."

Adding to the complexity of this triple challenge is the fact that is not easy to implement policies and measures that *simultaneously* deliver speed, scale, and durability. For example, as will be discussed in greater detail below, twenty-nine of the fifty US states have found it to be *comparatively* straightforward to institute renewable portfolio standards (RPSs) that prescribe greater and greater shares of non-emitting wind and solar power in the electricity resource mix. An RPS can thus be implemented on a relatively speedy timeframe, and RPSs seem relatively durable in that many consumers express support for clean energy. But, important though these standards may be as early wins for avoiding GHG emissions, they will prove insufficient to respond to the totality of climate change because they fall short on scale; currently-available and currently foreseeable renewable power technologies do not address emissions from certain hard-to-abate sectors such as steel, cement, or heavy-duty transport (especially by air and sea).[3]

Other policy approaches might in theory deliver speed and scale but fail to offer durability. After the protracted and unsuccessful attempt to enact a cap-and-trade system during President Obama's first term in office, and after his Republican political opponents made clear that they intended to oppose almost any Obama proposals in order to handicap him politically, the Obama Administration concluded that there was no chance of securing bipartisan legislative compromises to enable progress on climate—one of Obama's highest policy priorities. His Administration chose instead to implement policies that they felt could be based on existing legal authority. Thus, the Obama Clean Power Plan (CPP), which aimed to reduce power-sector emissions (discussed more below), was implemented pursuant to authority in the Clean Air Act. For the Obama team, this choice to opt for comparative speed seemed like the only path forward at the time. Unfortunately, the U.S. Supreme Court later suspended implementation of the CPP because of questions about whether the particular requirements under the Plan exceeded Clean Air Act authorities. And when Obama was succeeded by Donald Trump, the new U.S. Administration set about an extensive effort to dismantle the Clean Power Plan as well as many other Obama climate initiatives (Adler 2019).

Lastly, other policy prescriptions can excel in terms of their scale and perhaps even durability, but they face a significant challenge in terms of the speed with which they can be enacted. Arguably the classic example of this dynamic is the longstanding effort to introduce an economy-wide price on U.S. carbon emissions. If designed well, such an instrument could exert broad and powerful pressure to reduce greenhouse gas emissions with modest energy price impacts.[4] And indeed, more US Members of Congress and policy advisors are currently expressing support for carbon pricing than at any time in recent memory. Nonetheless, the politics surrounding a carbon tax and other pricing mechanisms are especially daunting in the United States because of the country's almost legendary opposition to taxes. Few

---

[3] A possible exception to this statement about the difficulty of using renewables to address hard-to-abate sectors could be solutions like so-called "green hydrogen"—where renewable power sources are employed to electrolyze water and produce hydrogen. Whether green hydrogen can, over the medium term, prove to be a commercially viable options remains to be seen.

[4] Colleagues at the Center on Global Energy Policy, led by Dr. Noah Kaufman, have undertaken important work assessing potential design tradeoffs and implications of a carbon tax at the Federal level in the United States. Their analysis examines the emissions reductions that could be achieved with different levels of a tax, designed in different ways, as well as the macroeconomic, fiscal, and distributional impacts of potential carbon taxes. Their work makes clear that if a well-designed carbon tax could be instituted, it could be a powerful incentive for emissions reductions and could even reduce energy expenditures of many households, including less-affluent households (Kaufman and Gordon 2018).

analysts expect that any meaningful carbon pricing mechanism will be instituted under the Trump Administration. Carbon pricing's potential strengths in terms of scale and durability thus are not matched in terms of speed.

With this understanding of the importance and challenge of implementing climate solutions that integrate speed, scale, and durability, let us now examine the status and main trends of the clean energy transition in the United States.

3   Headline Trends in the U.S. Clean Energy Transition

At present, the United States is the world's second-largest emitter of greenhouse gases. The U.S. energy sector represents more than 80% of total U.S. greenhouse gas emissions. Consequently, the country's success in executing the transition to clean energy has massive implications for the success of the global response to climate change.

Over most of the past decade, the United States has recorded mixed results on the transition to clean energy. The progress that has occurred has largely been in the electric power industry, and largely thanks to energy market developments and technological advances. Policy at the national level and macroeconomics have each played contradictory roles over this same timeframe, with some years when they contributed to clean energy progress and other times when they led to retrenchment. State-level and municipal policies, however, have played a progressive and important role, as we will discuss below.

The electric power sector has achieved the greatest clean energy results. That sector represented nearly 80% of total reductions in U.S. greenhouse gas emissions from 2008 through 2017, with its CO_2 emissions dropping by more than 25%, from 2.4 to 1.8 gigatonnes (EIA 2019a). After decades as the leading sectoral source of greenhouse gas emissions, the power sector fell to second position, with emissions lower than those from the transportation sector, as shown in Fig. 1. The greatest contributing factors for this change have been fuel-switching from coal to natural gas[5] and, more recently, the growth of investments in solar and wind energy, the results of which can be seen in both the precipitous drop-off in coal consumption shown in Fig. 2 and the decrease in power generation from coal in Fig. 3.

The industrial sector, which represents slightly more than one-fifth of total U.S. greenhouse gas emissions, saw sustained improvements in recent years in the clean energy transition, but that trend appears now to have halted. Industrial emissions of carbon dioxide fell by 22% over the two decades from 1997 to 2018, from 1.8 to 1.4

---

[5]The emissions reductions resulting from coal-to-gas fuel-switching are the subject of ongoing debate in the expert community. Some peer-reviewed analysis suggests that standard methods for estimating leakage of methane in the oil and gas supply chain are significantly underestimating actual leakage rates. See, for example, Alvarez et al. (2018). Particularly for United States, where increased production and power-sector consumption of natural gas have been such an important feature of recent years, accurate understanding of methane leakage rates will play a vital part in evaluating true emissions rates and emissions reductions.

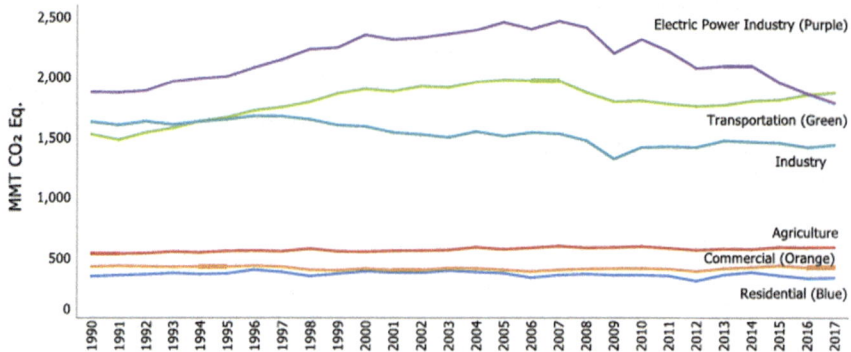

Fig. 1 U.S. greenhouse gas emissions by sector. U.S. Environmental Protection Agency (EPA 2019a)

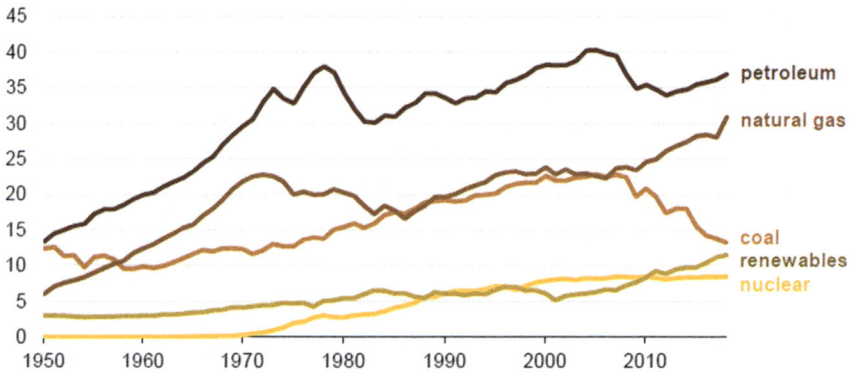

Fig. 2 U.S. total energy consumption by fuel (1950–2018) in quadrillion Btus. Energy Information Administration (EIA 2019b)

gigatonnes. Changes in energy-related industrial emissions over that period reflected, among other factors, overall economic activity, structural features in American industry (e.g., shares of energy-intensive industries in the overall mix), and technologies employed in energy-intensive industries (such as the adoption of higher efficiency technologies for heavy energy consumers like steel-making).

The residential and commercial sectors, which together represent just less than 30% of total greenhouse gas emissions (roughly 15% from residential, 14% from commercial), have held reasonably steady over the past decade. Key factors in these sectors' energy consumption and GHG emissions levels have been weather-related parameters, translating into the number of heating- and cooling-degree-days, and the efficiency of household equipment and appliances, which has improved steadily (EIA 2019b).

The sector that has been most clearly moving backward in terms of greenhouse gas emissions is transportation. This sector is an especially important bellwether for

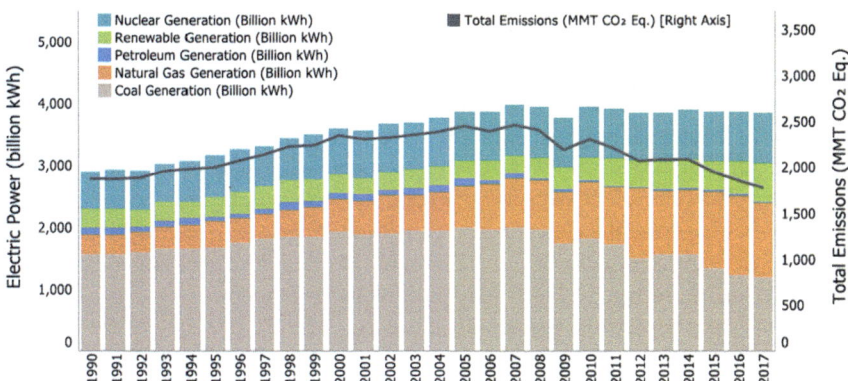

Fig. 3 Electric power generation (Billion kWh) and emissions (MMT CO_2 Eq.). U.S. Environmental Protection Agency (EPA 2019a)

progress in the U.S. clean energy transition, as transportation represents over 70% of U.S. petroleum consumption and nearly 29% of total energy use (EIA 2018). For each year since 2012, transportation emissions have grown (EIA 2018), and transportation has surpassed electric power as the single largest source for greenhouse gas emissions by sector.

Whether transportation demand and emissions will continue to grow is a critical component of the future U.S. clean energy transition. Since 2004, the efficiency of the U.S. road vehicle fleet has increased steadily, with light-duty vehicle efficiency improving by 29% over that period and CO_2 emissions per mile dropping by 23% (EPA 2018). The improvement since 2004 has been undercut, however, by increases in fuel demand for aviation and freight vehicles (Houser et al. 2018) and increases in the amount of vehicle-miles-traveled (VMTs) for light-duty vehicles, which grew by 49% over the period from 1990 to 2017 (FHWA 2018). VMT increases in turn reflect a variety of factors, including population growth, overall economic growth, land-use patterns (especially urban sprawl), and consumer responses to low fuel prices (EPA 2019a). The Trump Administration announced its intention to freeze increases in vehicle efficiency requirements under the Corporate Average Fuel Economy (CAFÉ) standards, a proposal that is still being reviewed as of this writing, discussed in more detail below.

One transportation trend that holds the promise of reduced GHG emissions is the electrification of the light-duty vehicle fleet. After years of only gradual sales growth, and total numbers amounting to less than one percent of the total vehicle fleet on the road, electric vehicle (EV) purchases boomed in 2018. Sales grew by more than an 80% increase to bring the total fleet to over 360,000 EVs (Pyper 2019). EVs are not yet of sufficient scale to materially alter the emissions profile of the transportation sector in the near term, but they may now be gathering momentum with car buyers.

4 Federal Policy for Clean Energy

Having surveyed the major trendlines in the U.S. clean energy transition—trendlines that are far from uniformly encouraging—we now turn to a discussion of policy at the Federal, state, and local levels. One finds a country whose Federal-level leaders have struggled to reach consensus on what constitutes effective policy with regard to climate change and the clean energy transition. Indeed, political leadership in the United States splits into those who are convinced that we need an urgent and broad response, and those who say they see no problem to respond to.

During the past decade and a half, the U.S. Federal government has played a variety of roles in regard to the clean energy transition. The first U.S. president in this century, President George W. Bush, did not prioritize policies intended to spur reductions in greenhouse gas emissions. The Bush Administration did, however, sustain strong investments in scientific understanding of the global climate system, which have been a foundational part of U.S. Federal policy since the administration of his father, George H.W. Bush. Moreover, through both policy and legislation, the George W. Bush Administration gave prominent emphasis to energy research and development, much of it targeting clean energy R&D.

From the very start of President Barack Obama's term in office in January 2009, his Administration signaled that it saw climate change as a top policy priority. President Obama described climate change as one of the defining challenges of the current age, a moral threat imperiling the future of humanity. In his first term, Obama included funding of more than $90 billion for the deployment of clean energy technologies in the American Recovery and Reinvestment Act, which the U.S. Congress passed in response to the 2008 financial crisis (White House 2016a). Obama also tried but failed to pass economy-wide legislation sponsored by Democratic Representatives Henry Waxman of California and Edward Markey of Massachusetts that would have capped emissions of greenhouse gases from the power sector and many other industrial sources and established a trading system for GHG emission reductions (CBO 2009). The failure of the Waxman–Markey cap-and-trade measure was a bitter defeat. In its wake, the Obama Administration sought during its second term to respond to climate and the clean energy transition by taking actions based on its interpretation of existing legal authority.

The centerpiece of this executive action was President Obama's Climate Action Plan (CAP), released in June 2013 (White House 2013). CAP called for a portfolio of actions aimed at reducing U.S. greenhouse gas emissions and increasing the country's preparedness for the impacts of climate change, plus enhanced collaborations with key international partners. The single most significant element of CAP in terms of projected emissions reductions was its instruction to the Environmental Protection Agency (EPA) to promulgate new regulations to limit greenhouse gas emissions from electricity generating plants. The Clean Power Plan (CPP), as this new program of regulations came to be known, called for newly built and existing power plants to emit less than prescribed maximum levels of CO_2 per unit of output. During the CPP's development, the plan was expected to drive retirements of subcritical coal plants and older, inefficient gas-fired plants.

In reality, two considerations emerged that meant that CPP never lived up to the Obama team's expectations. First, many utility companies were opting to retire those high-emitting coal plants anyway, and even faster than the CPP required. Local pollution requirements (such as those limiting emissions of mercury), favorable economics for high-efficiency natural gas plants, and plummeting costs for solar or wind power plants meant that coal plants were increasingly unattractive for utilities. Moreover, twenty-seven states introduced legal challenges to the CPP, and the Supreme Court temporarily "stayed" (or halted) the CPP's legal implementation. When President Trump came into office, his Department of Justice announced that it would not proceed with a legal appeal against the stay. In June 2019, Trump's EPA issued a much weaker proposal, the so-called Affordable Clean Energy (ACE) rule, which called for improvements in the operating efficiency of remaining coal plants, but which was projected by many analysts to have no discernible impact on reducing power-sector GHG emissions. The ACE rule was immediately challenged in court (Friedman 2019).

Another major component of the Obama CAP was a concerted focus on the energy efficiency of vehicles, buildings, appliances, and equipment. CAP called for the promulgation by the U.S. Department of Energy of new appliance and equipment standards that would, by 2030, result in a cumulative total of three billion tonnes of avoided CO_2 emissions. CAP also focused on transportation-related greenhouse gas emissions—through the first-ever efficiency standards for heavy-duty vehicles as a complement to the Corporate Average Fuel Economy (CAFÉ) increases that the Obama Administration had set in place during its first term, and also through research and communications that would facilitate the adoption of electric vehicles.

CAP also set in motion new efforts to assess and reduce the leakage of methane from both energy- and non-energy-related sources. A final energy-related aspect of CAP was its call for extensive international leadership by agencies of the U.S. Federal government, including through mechanisms such as the Clean Energy Ministerial and collaborative bilateral research efforts with China, India, and other key partners.

The Obama Administration pledged through its Nationally Determined Contribution (NDC) under the Paris climate agreement to reduce its greenhouse gas emissions by 26–28% by 2025, using a 2005 baseline. The Obama Administration also recognized that the goal of protecting the global climate would require far greater, sustained reductions in greenhouse gas emissions. For this reason, the Obama Administration prepared, and formally submitted at COP-22 in Marrakech in November 2016, the U.S. Mid-Century Strategy for Deep Decarbonization. The Mid-Century Strategy proposed to extend out to mid-century the downward-sloping emission trajectory started by the 26–28% reductions of the U.S. NDC and indicated the kinds of changes that would be necessary to achieve deep decarbonization by 2050 (White House 2016b).

The Mid-Century Strategy was announced after the 2016 presidential election, and it was already clear from statements of the incoming President that the new Administration was likely to think very differently about climate change and the role of Federal policy in responding to it. President Donald Trump came into office pledging to remove regulatory burdens from American business, even regulations intended to protect the environment or public health. Trump appointed as key members of his initial cabinet and sub-cabinet people who had long histories as climate skeptics—such as Scott Pruitt, Trump's initial Administrator of the U.S. Environmental Protection Agency. Obama's sense of urgency about climate—his call to answer a moral struggle—was long gone. The United States now had a president who, as a candidate, had referred to climate change as a hoax.

In numerous public remarks and Congressional hearings, Administration officials called into question the science of climate change (Gustin 2018). Administration officials and the President himself repeatedly accused his predecessor of engaging in a "war on coal." The Trump team pledged to "end the bureaucratic blockade" that they accused the previous Administration of creating. They promised to put miners back to work across the United States (Perry et al. 2017). In June 2017, the President announced his intention to withdraw the United States from the Paris climate agreement.[6]

Text Box A: Proposed budget cuts for clean energy R&D

The Trump White House proposed deep cuts for fiscal years (FY) 2018, 2019, and 2020 in appropriations for the U.S. Department of Energy in precisely those elements that provide the greatest Federal support for clean energy R&D. Congress refused to comply. Proposed cuts, averaged over those three years, were as follows:

- Advanced Research Projects Agency for Energy (ARPA-E) -- 123% cut
- Advanced Technology Vehicles Manufacturing (ATVM) LOAN program – 91% cut
- Office of Energy Efficiency and Renewable Energy (EERE) – 74% cut
- Office of Electricity Delivery and Energy Reliability (OE) and successor offices – 17% cut
- Office of Fossil Energy (FE) R&D progs – 20% cut
- Office of Nuclear Energy (NE) – 31% cut
- Office of Science – 11% cut (incorporates clean-energy-related basic science such as fusion, basic energy science, high energy physics)

---

[6]It is important to note that President Trump announced his *intention* to withdraw the United States from the Paris climate agreement, but contrary to many assertions, the United States has not yet withdrawn from the Paris agreement. Trump's intention can only be fulfilled, in accordance with the terms of the agreement, after the elapsing of a prescribed amount of time. This means that the actual withdrawal cannot occur until the day after the 2020 U.S. election. Until that time, U.S. negotiators continue participating in meetings occurring under the terms of the Paris agreement.

Trump Administration spokespeople summed up their energy policy as the promotion of "energy dominance." Like many political slogans, the practical meaning of "energy dominance" never came to be crisply defined. What was clear was that the Trump Administration favored a production-oriented, export-oriented approach to energy policies—one where environmental protection was more likely to be treated as an unwelcome hindrance than an essential building block.

As part of the Trump de-regulatory agenda, the new Administration set about dismantling many of the policies that the Obama team had used to create incentives for the clean energy transition. EPA Administrator Pruitt halted the implementation of the Clean Power Plan and set in motion a re-writing of vehicle efficiency standards. Pruitt also announced the intention not to implement new rules on methane leakage, a move that was quickly rejected by the courts.

In reality, removing regulations is not as simple as some in the Trump Administration may have wanted for people to think. The Administrative Procedures Act requires that new regulations be justified by benefit-cost analysis and be enacted only after public notice-and-comment procedures. The same strictures apply when one is removing existing regulations; the action must be justified by benefit-cost analysis and must undergo notice-and-comment. Many of the de-regulatory "victories" trumpeted by the Trump Administration were therefore premature claims (Adler 2019).

Another feature of Federal policy on clean energy under President Trump has been his selective—and climate-unfriendly—budget-cutting agenda. Here, the President persistently sought to cut parts of the Federal budget that would facilitate the clean energy transition—programs aimed at deployment of current technologies and those focused on development of the next generation of technologies. Text Box A presents the average funding reductions (in percentage terms) that the Trump White House requested in its first three budget proposals. Both from the Administration's persistence and from its public comments, the Trump team made clear that these proposed cuts reflected the Administration's broader hostility toward engaging on climate change. For example, in presenting Trump's first budget request to the press in March 2017, then-budget director Mick Mulvaney dismissed the notion of funding for climate-related research: "We're not spending money on that anymore. We consider that to be a waste of your money" (Phillip 2017).

Under the U.S. constitution, however, the Executive Branch does not have unilateral decision-making authority regarding the Federal budget. The White House proposes each fiscal year's budget to Congress; the Legislative Branch then develops and passes appropriation bills. Generally speaking, enacted budgets substantially reflect Congressional priorities—not simply those of the White House. The President can, theoretically, veto Congressional spending bills, but doing so can alienate members of the President's own party. Once the budget is signed and enacted, it must be implemented by the Executive Branch. Congressional decision-making is often motivated in part by funding for institutions in Congressional members' home districts, and consequently most clean energy R&D budgets have actually grown over the years since President Trump came into office (DOE 2017, 2018, 2019).

One particularly positive development for Federal-level clean energy policy during Trump's tenure was the provision, in the Bipartisan Budget Agreement from February 2018, to expand the provisions of the so-called 45Q tax credits. The expanded tax credits were expected to provide meaningful incentives for the development of carbon capture, utilization, and sequestration (CCUS) projects. CCUS is viewed by many as a critical area for technological progress to enable a successful response to climate change.

If, as we have seen above, Federal policy on climate change suffered from pronounced political divisions in policy, an obvious question is whether the 2020 presidential election will bring more pendulum swings or perhaps a new national consensus on climate. It would certainly defy popular expectations to suggest that President Trump might relent in his hostility toward climate in the remaining months of his current term. And the current president can almost certainly be relied upon to veto most climate legislation—if indeed any such legislation could pass the Republican-led Senate and reach his desk. Nonetheless, starting after the 2018 midterm elections, members of both the President's own party and the Democratic opposition began spending an increasing amount of time considering policy options ranging from the mild to the ambitious. At a minimum, these options hint at the possibility of legislation that could pass the Congress and be placed before the next president, whether that is a second-term President Trump or a newly arriving Democrat.

Some of the main policy proposals that are circulating in the U.S. Congress and among other political leaders include these: First, a number of Republican members of Congress are exploring ideas to support clean energy innovation, often with support from Democratic counterparts. Many of these ideas are comparatively non-controversial. Some focus on support for next-generation nuclear energy or CCUS technologies. Another and more hotly debated area of focus has been carbon pricing. A veritable who's-who of former Republican cabinet members and a number of noted Democrats, for example, have formed the Climate Leadership Council and called for the passage of a carbon tax. Notwithstanding vigorous opposition from the Republican Congressional leadership, one Republican representative, Carlos Curbelo of Florida, even proposed legislation for a carbon tax in the summer of 2018, only to be defeated by a Democratic opponent in the November 2018 midterm elections. Over fifty Democrats have signed on as joint sponsors of carbon tax legislation in the 2019 Congressional term, with very cautious Republican support, although the prospects for such proposals is highly uncertain.

Many observers partially attribute the new Congressional attention on climate policy proposals to the fact that the midterms delivered a new Democratic majority in the House of Representatives. With the Democrats' resulting ability to set legislative and hearing agendas, climate ideas that never saw the light of day under Republican control of the House of Representatives are now up for detailed debate.

Another factor that is sometimes credited for the renewed vigor in climate debates is the proposal known as the Green New Deal (GND). The GND is an expansive set of goals, advocated most prominently by first-term Representative Alexandria Ocasio-Cortez of New York. It calls broadly for (a) rapid and deep decarbonization accompanied by (b) employment guarantees in the clean energy sector, especially including minority and economically disadvantaged communities, (c) health care reform, and (d) measures to address local pollution problems that have beset many low-income minority populations. The GND was embodied first in a non-binding resolution that was passed by the Democratic-controlled House of Representatives in early 2019 (Congress 2019).

Despite the fact that the GND was not immediately translated into formal legislative proposals, and despite the fact that it served as a political pinata for President Trump and certain other Republican leaders, many of whom equated it to socialism, the GND seemed to alter the political playing field on climate and clean energy. It seemed to capture a growing sense of public concern about climate change—which is reflected in opinion polling especially among Democratic voters. Each Democratic presidential candidate was asked on the record whether she/he supported the GND. Widespread climate protests in numerous American cities (and in other cities worldwide) only intensified the pressure for accelerated climate action.

Republicans also felt forced to respond to the pressure created by public opinion and the GND. One Republican Representative, Matt Gaetz of Florida, proposed an alternative package entitled the Green Real Deal, which contained certain pro-climate measures but stripped out the GND's broader socioeconomic ambitions. Former Rep. Curbelo summarized the dynamic surrounding the GND by giving public credit to its proponents: If we Republicans do not like the Green New Deal's elements, he said, we will have to say what we as a party do support.[7]

5 State-Level Policies for Clean Energy

If Federal-level policy under the Trump Administration markedly moved away from focusing on the clean energy transition, and if the field of play on climate was confused and uncertain as the next presidential election approached, the same could not be said for the majority of states and a significant number of municipalities. Here one saw a mounting sentiment that responding to climate change is a policy priority.

A plurality of the 50 states—states with Democratic leaders as well as some with Republican leaders—banded together in a variety of formats and initiatives to reject very publicly the climate-hostile policies that have dominated the views of the Trump Administration. Some elected to put in place policies that create direct incentives to build low- or no-carbon energy systems or create indirect incentives to do so. Others pledged to reduce their state-level emissions to levels in keeping with the Obama-era pledge under the Paris climate agreement. Others laid out sweeping programs for deep decarbonization by mid-century.

---

[7]Curbelo's on-the-record comment was made in the course of a program on proposed climate solutions hosted by the Center on Global Energy Policy at Columbia University (CGEP 2019). For an excellent analysis of the Green New Deal and its significance in the debates about Federal climate policy, see Bordoff (2019).

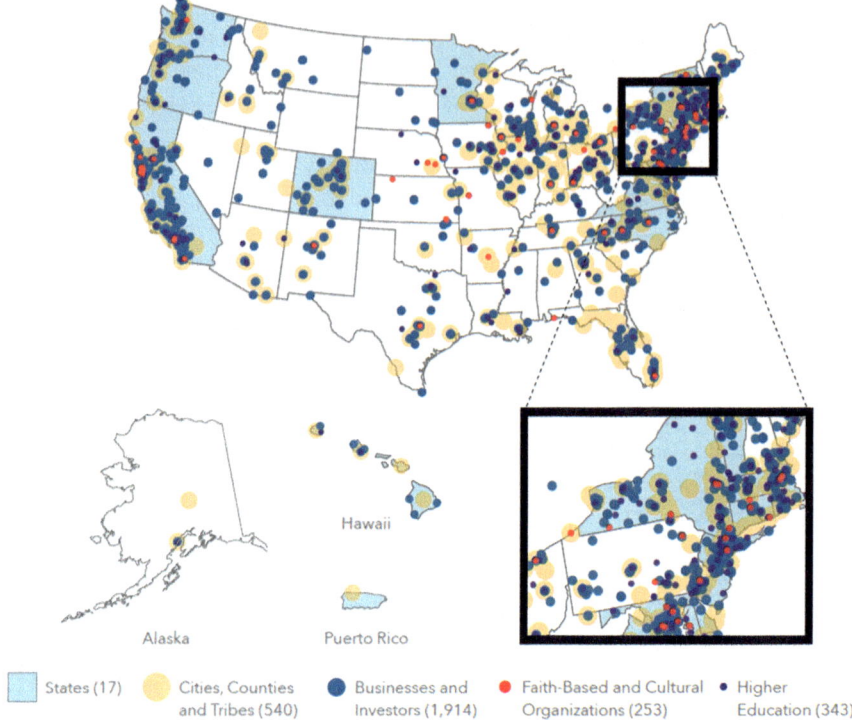

Fig. 4 States, cities, and other entities pledged to fulfill Paris pledge (as of 2018). America's Pledge (AP 2018)

President Trump's Rose Garden announcement in June 2017 that he intended to withdraw the United States from the Paris climate agreement triggered vociferous, public rejection by a number of governors. Most prominently and immediately, Governor Jerry Brown of California, Governor Andrew Cuomo of New York, and Governor Jay Inslee of Washington declared that their three states would adhere to the emissions reductions pledged in the Obama Administration's Paris pledge, and they called for others to join them.

Their grouping came to be referred to as the United States Climate Alliance. As of this writing it includes 25 states and one territory in its membership. The Climate Alliance thus represents 55% of the population of the United States, 40% of U.S. GHG emissions, and 60% of the total U.S. economy, or $11.7 trillion on an annual basis (USCA 2019b). As can be seen in Fig. 4 (which is itself already dated since several additional states subsequently joined the Climate Alliance), the states, cities, and other entities pledging themselves to fulfill the Paris climate pledge are heavily—but not exclusively—clustered along the Atlantic and Pacific coasts with other concentrations in the industrial Midwest and a scattering of participants across the Plains states (AP 2018).

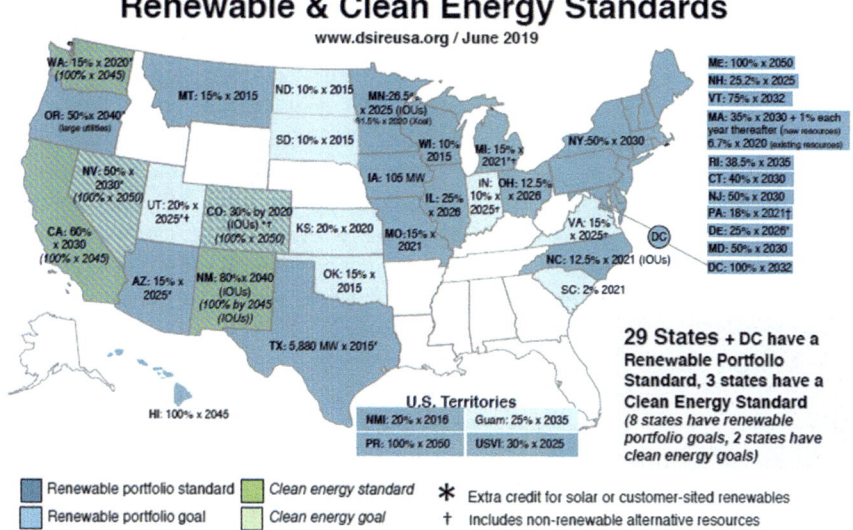

Fig. 5 States employing Renewable Portfolio Standards (RPSs) and Clean Energy Standards (CESs). Database of State Incentives for Renewables and Efficiency (DSIRE 2019)

The states participating in the Climate Alliance committed to reduce their emissions in a manner consistent with the Obama-era Nationally Determined Contribution—namely, by 26–28% below 2005 levels by 2025. They also committed to track and report their progress, to institute new policies to reduce carbon pollution and promote the clean energy transition, and to facilitate knowledge-sharing so that states can learn from each others' experience (USCA 2019b). The participating states represent a considerable mass: If collectively they formed a single national economy, they would rank third in the world, smaller than only the entire United States and China (USCA 2019a; AP 2018).

Arguably the single most visible state-level policy that has been used to incent deployment of clean energy is the renewables portfolio standard (RPS). RPSs apply to 56% of retail electricity sales nationwide (Barbose 2018). RPSs typically prescribe the share of total electricity sales that must come from renewable generation sources in the course of a year. Different states' RPSs are not uniform in scope, application, or rules, however, because they are developed under the authority of individual state governments and therefore respond to state-level politics, resource endowments, and policy priorities. Some apply to all electric utilities in the state; others apply only to investor-owned utilities but not to municipally-owned utilities or cooperatives (or perhaps they apply to munis and coops, but with less stringent requirements). Some have caps on the cost of the RPS. Some mandate the absolute amount of generating capacity rather than percentage share of retail sales coming from renewable energy generation (Barbose 2018).

As is shown in Fig. 5, a total of 29 of the 50 states in the union, plus the District of Columbia and three U.S. territories, now employ legally-binding RPSs. Many of these states have progressively increased the stringency of their standards in recent years as costs for wind and solar power dropped and concerns about climate change rose. In fact, in the decade from 2009 to 2018 alone, states elected to strengthen their RPSs on 65 occasions (some doing so more than once) (Barbose 2018). Seven states now have RPSs requiring utilities to generate 50% or greater shares from renewables (NCSL 2019). In addition, a growing number of states are instituting clean energy standards (CESs)—sometimes as an alternative structure to an RPS and sometimes as a complement thereto. Because CESs generally permit the use of other non-GHG-emitting technologies such as nuclear power or natural gas paired with CCUS, CESs are expected to facilitate greater flexibility and thus greater cost control as states move toward higher levels of decarbonization.

The legal status of these RPS and CES programs varies. Some are simple policy goals; others are binding requirements from state-level utility commissions and still others are enforceable mandates enshrined in state-level law. The unmistakable trend toward progressively greater ambition is clear—including 100% carbon-free electricity in several states and net-zero-carbon entire economies in other states.[8]

Among these is California, which now has an RPS of 50% by 2026, followed by a 60% RPS by 2030, followed by a requirement of 100% net-zero-carbon electricity by 2045. The latter step leaves open the use of clean generating resources other than wind or solar—namely, nuclear, or hydro-electric, or carbon capture and utilization or storage paired with either industrial installations or natural gas power generation. The legally binding timelines are also augmented by non-binding executive orders that call for (a) economy-wide carbon neutrality by 2045 plus (b) an economy-wide reduction of all GHG emissions by 80% below baseline by 2050 (EFI 2019). California's policies bear particular significance because of the sheer size of the state's economy. If it were an independent nation, it would have the fifth-largest economy in the world.

Also noteworthy are the ambitious new clean energy requirements for the state of New York, which became law in June 2019. Under the mandate provided by the Climate Leadership and Community Protection Act, New York will have 100% carbon-free electricity by 2040 and a net-zero carbon economy as a whole by 2050—including sectors such as industry and transportation, which present greater decarbonization challenges than the power sector (Roberts 2019). And other states have also enacted or are considering laws or goals to deeply decarbonize their power sectors or entire economies—including Colorado, Hawaii, Maine, Minnesota, New Jersey, New Mexico, Oregon, and Washington.

---

[8] State-level interest in *clean energy* and particularly clean electricity does not always coincide with state leadership that attaches an explicit priority to pro-*climate* policies. Nor does support for clean energy development align neatly with leadership by the typically more pro-climate Democratic party. The state of Texas, for example, has been a predominantly conservative and petroleum-oriented state for years and has had a succession of Republican governors. Nonetheless, Texas is a hotbed of wind power with over 22 GW of installed capacity. In this regard, it would rank sixth in the world if it were an independent nation (Irfan and Zarracina 2019).

Several U.S. states have also introduced legally binding structures that place a price on carbon emissions that are expected to help accelerate the clean energy transition. The California Greenhouse Gas Emissions Cap-and-Trade Program, which was created as a result of the California Global Warming Solutions Act of 2006 (also known as Assembly Bill 32), took effect in 2012 and now covers 85% of the state's greenhouse gas emissions. It prescribes that 2030 emissions levels should be 40% below 1990 levels and establishes emissions caps and a quarterly trading system for covered facilities. It is also linked to a similar program in the Canadian province of Quebec (CARB 2019).

Nine northeastern states participate in the Regional Greenhouse Gas Initiative (RGGI)—Connecticut, Delaware, Maine, Maryland, Massachusetts, New Hampshire, New York, Rhode Island, and Vermont—with New Jersey slated to join in 2020 (RGGI 2019). RGGI, which was created in 2009, covers emissions from electricity generating plants with a capacity greater than 25 MW. States participating in RGGI have the obligation to create and implement programs to deliver annual power generation emissions reductions of 2.5% to 2020.[9]

In addition to the electric power sector, certain states have also implemented clean energy policies in the transportation sector, some of which policies interact in complementary fashion with Federal transportation policies, and some of which exist in some tension with policies from Washington, D.C. As an example of the latter, California has a long history of seeking more stringent automobile efficiency standards than those enshrined in the national CAFÉ program.

Since the passage of the Clean Air Act in 1970, California has repeatedly sought, and usually secured, the authority to impose more stringent vehicle standards within the state than are applied elsewhere across the nation, based on the rationale that (a) it had efficiency standards in place before the national program was established, motivated by the state's long struggle with smog, and (b) California's stricter standards protected the state's public health and environment better than the weaker national requirements. Over time, a dozen other states adopted California's more exacting fuel economy standards as well. In 2012, the Obama Administration set in place a package of elevated efficiency standards for passenger cars and light trucks produced from 2017 to 2025, in exchange for which California agreed to harmonize its own regulations with the new Federal requirements. In April 2018, however, the Trump Administration halted the implementation of the Obama package. California brought suit against the Administration, charging it with arbitrary and capricious action. Sixteen other states and the District of Columbia joined that lawsuit, which is still pending as of this writing.

Beyond vehicle efficiency standards, state-level policies for clean transportation have included a wide range of other measures. California and nine other states have established percentage sales targets for so-called "zero-emission vehicles," meaning

---

[9]One facet of both RGGI and the California-Quebec cap-and-trade systems is that to date they have resulted in modest prices per allowance—only $5.62 per RGGI allowance in June 2019 (RGGI 2019), $17.45 per California allowance in May 2019 (CARB 2019). Nonetheless, supporters of the two programs claim that they are driving clean energy investments.

battery electric vehicles, hydrogen vehicles, and others with no tailpipe emissions. Numerous states provide preferential treatment on the roads for what have been traditionally referred to as "alternative fueled vehicles," with the benefits including access to high-occupancy lanes on congested roads, or reduced taxes. Some of these benefits have been made available not only to battery electric vehicles or high-efficiency internal combustion vehicles like hybrids, but also to vehicles that simply have the capability to burn part-ethanol fuel, which has given rise to debates over the environmental merits of corn-based ethanol. An increasingly important set of state-level policy interventions has been in the area of charging infrastructure for EVs.

6   Clean Energy Policies in U.S. Cities

Just as has been true with a number of U.S. states, many cities and towns have chosen to institute policies to promote the clean energy transition, even though national policy does not currently prioritize that transition. Action by cities carries potential significance because of several factors. For starters, America's urban areas represent between 70 and 80% of the total population. In addition, many of the decisions that exert significant influence on future emissions occur at local levels.

Land-use patterns reflect local zoning rules, for example, and those rules affect materially considerations such as the degree of urban sprawl, which both spurs individual vehicle use and creates structural challenges for the development of public transit systems. Transportation plans usually are driven by concerns at local levels, although funding usually combines local, state, and Federal resources. On the other hand, many cities lack certain authority, financial power, independent market status, or other attributes needed to implement and sustain effective low-carbon policies. Thus, the role of city-level clean energy policies can be important, but cities do not always have at their disposal all the tools needed to ensure delivery of the desired outcomes.

Municipal-level policies promoting the clean energy transition include some of the same features as state-level policies such as goals for zero-carbon power supply, goals for deep decarbonization by mid-century, and collaborations to signal continuing focus on political commitment. In regard to clean power supply, a growing number of cities and towns have instituted mandates requiring that all electricity be supplied by renewable energy sources. Some of the earliest cities and towns to adopt such policies were modest in size—such as Aspen, Colorado; Burlington, Vermont; and Greensburg, Kansas. More recently, however, many major cities have also adopted municipal clean power supply policies—including New York, Los Angeles, Washington D.C., and others.

American cities are also employing a wide array of other policies to promote the clean energy transition. These range from the development of sustainable transportation plans and increased efficiency standards for municipal vehicle fleets to energy conservation building codes and efficiency-/climate-related procurement processes (such as bulk purchasing arrangements for municipal electric vehicle fleets or for zero-carbon electric power to be used to operate city facilities).

In addition, over 400 U.S. cities and towns representing more than 70 million inhabitants (more than 20% of the population of the United States) have joined the Mayors' National Climate Action Agenda, also known as the Climate Mayors coalition. Under this effort—initiated in 2014 by Eric Garcetti, the mayor of Los Angeles; Annise Parker, then the mayor of Houston; and Michael Nutter, then the mayor of Philadelphia—cities commit to developing and publicizing near- and long-term climate targets, engaging in peer-to-peer learning, and collaborating on policy development and advocacy. For example, the bipartisan Climate Mayors initiative filed formal statements opposing the Trump Administration's repeal of the Clean Power Plan and its review of the CAFÉ standards for vehicles (Climate Mayors 2019). Many cities also participate in major international groupings of cities such as the Carbon Neutral Cities Alliance and the C40 grouping. These groups promote exchanges of best practice and advocate for cities' interests on global climate issues, including at the meetings of the Conference of Parties under the UN Framework Agreement on Climate Change.

7 Geopolitics and the U.S. Clean Energy Transition

The preceding discussion has outlined the tensions and contradictions that constitute the current state of play in the transition to clean energy within the United States. The reality is one that defies simple, overly broad generalizations: The United States has certainly not achieved broad societal agreement on aggressive reductions in greenhouse gas emissions, especially not at the level of national politics. The divisions are stark, and they are powerful as a political "wedge issue"—a cudgel that can be used by partisan actors to attack opposing leaders. Nonetheless, it is inaccurate to assert that the country is failing entirely to address climate.

Progress has occurred in certain areas, even in the face of the national-level political divisions. The U.S. Congress continues to support scientific and technological research and development in regard to both the operation of the climate system and clean energy technologies that can deliver solutions. Numerous states and cities have made clear that they "are still in" and intend to adhere to emissions cuts pledged by the Obama Administration under the 2015 Paris climate agreement. Moreover, the political sands appear to be shifting slightly at the national level. President Trump has given no indication that he intends to soften his opposition to action on climate. But physical calamities that are thought to be occurring with increasing severity and frequency as a result of a changing climate are afflicting voters more and more—including those who have traditionally supported the Republican party. More and more legislators from the President's own party are generating legislative proposals that call for enhanced clean energy research and development, energy efficiency, and even in some rare cases a price on carbon.

Thus, at present the U.S. transition to clean energy at the national level is characterized by tensions: a mounting sense of urgency on one side, a stubborn denialism on another, and halting steps forward on a third. These tensions seem unlikely to disappear completely over the next decade, but based on the increasing prominence of the climate issue in public policy debates, the country is likely to move along one of two broad paths over the coming decade: Under one path, U.S. policy would continue to be riven by political fragmentation, climate rejectionism, and the resurgent isolationism (including in regard to international climate cooperation) that has flourished during President Trump's tenure. Under an alternative path, the United States would shift to a mode of greater national action—to a sense of purpose and possibility to achieve deep decarbonization by mid-century. These alternative scenarios imply in turn distinct geopolitical implications for the United States and others around the globe, the topic to which we now turn.

A two-way relationship links the U.S. clean energy transition and the world of geopolitics. The timing, extent, and nature of the U.S. transition causes impacts in numerous geopolitical dimensions—including international partners' trust of U.S. leadership and relationships with key partners around the globe.

No less important are links that run in the opposite direction—ways in which geopolitics exert influence on the U.S. clean energy transition. Table 1 summarizes this two-way relationship under two alternative scenarios—one in which the U.S. remains divided and ambivalent in its commitment to clean energy, and the other in which it renews its focus on responding to climate and acts with renewed purpose and commitment.

Several main features emerge from a review of the links between the U.S. clean energy transition and the world of geopolitics. One is that U.S. actions on clean energy and climate seem likely to forge strong impressions around the globe about whether the United States is a heedless contributor to a massive problem or a leading force for solutions. International partners in typically well-heeled settings such as the United Nations and its agencies, the G-7, the G-20, and even the Arctic Council have repeatedly experienced petulant behavior from the Trump Administration. U.S. representatives insisted that climate be given reduced prominence if any at all.

To leaders and voters elsewhere around the world, such a refusal to acknowledge what is a matter of broad scientific consensus, and is an issue with life-or-death implications for some, is wholly inconsistent with the actions of a country that wishes to be influential on the world stage. Thus, continued U.S. reluctance on climate, if that scenario plays out, can reasonably be expected to cause damage to American prestige, soft power, and influence. Less clear, however, is whether any such damage to U.S. prestige would recede into memory if the more favorable scenario occurs and the United States returns to acting with a renewed sense of purpose on climate.

A second geopolitical dimension that merits note is that the posture of the United States on the clean energy transition has major implications for the credibility and effectiveness of international institutions such as the United Nations, and international cooperation in general. The Paris climate agreement, which President Trump declared his desire to leave at the earliest permissible moment, was a triumph of international diplomacy. Its strength lies in its flexibility, its emphasis on differentiated

Table 1 A two-way relationship links the U.S. clean energy transition and geopolitics

	U.S. CLEAN ENERGY → GEOPOLITICS Clean energy policy influences geopolitics	GEOPOLITICS → U.S. CLEAN ENERGY Geopolitics influence clean energy transition
Scenario 1: United States continues to be a **divided country** on climate and clean energy	U.S. actions contribute to a *vicious cycle* which undercuts a global response to climate change: • U.S. reluctance to commit to the clean energy transition is justified by some U.S. leaders who claim that other large emitters fail to shoulder a fair share of the burden (often blamed: China and India); the same U.S. reluctance only erodes support on the part of other countries to engage on climate mitigation, which in turn serves as further justification for the U.S. to be less proactive *Soft power* of the United States: • Internal U.S. divisions over climate progressively erode confidence that U.S. can be a constructive partner on the global stage, addressing the biggest challenges of our age; this alienates partners and allies and undercuts the effectiveness of international cooperation Global *stability and security*: • Greater economic divides open between global "haves" and "have nots" • Increased immigration flows occur as climate impacts create unmanageable local stresses; • New risks of conflict as countries seek to manage new migration flows Exacerbation of *tensions with historical allies* and partners (e.g., EU, Canada, Mexico, Japan, ROK, Australia, UK): • Partners and allies see the United States as insufficiently active in responding to what is certainly one of the greatest challenges of our time • Partners press for more active U.S. policy, progressively alienating the U.S. and its partners from each other Ceding markets to *economic competitors:* • Ambiguity of pro-climate Federal policy fosters inattention to new economic and commercial opportunities in clean energy, leaving those opportunities to China and other current or potential competitors	*Trade tensions:* • Disputes with competitors and even historical partners lead to long-lived tariffs that raise costs of energy (and other) infrastructure projects; • Trade disputes retard the dissemination of new, zero-emitting technologies by U.S. companies (and others), in part by disrupting globalized supply chains that have grown over recent decades Global systems of *governance:* • Because of the weakening of global systems of governance (for example, the United Nations system)—weakening exacerbated in part by U.S.—actions the United States itself may lack a system through which to argue for, and implement, a truly global response to climate *Competing China* (contrast to next item) • China under its current leadership is more assertive and proactive than in recent memory; China is mobilizing capital and technical resources, and it seeks to augment its global influence and to diminish that of the United States; • The longer that the United States remains divided regarding climate, and reluctant to mobilize sufficient resources to press ahead with the clean energy transition both at home and with partners abroad, the more that China is able to advance its own economic and international visions without challenge *Essential China* (contrast to previous item) • Numerous sources of tension characterize the U.S.–China relationship at present, but this does not diminish the fact that China is the single largest emitter of GHGs globally, with the United States in second place; to succeed in addressing the global climate and clean energy challenge, *both the United States and China need to be part of the solution*

(continued)

Table 1 (continued)

	U.S. CLEAN ENERGY → GEOPOLITICS Clean energy policy influences geopolitics	GEOPOLITICS → U.S. CLEAN ENERGY Geopolitics influence clean energy transition
Scenario 2: United States acts with **new sense of purpose** on climate and clean energy	Potential for a *virtuous cycle* on climate *mitigation*: • United States is seen as engaging seriously to accelerate its own clean energy transition, including through policies and action undertaken by the U.S. Federal government, thus providing an opportunity to encourage other major emitters to take, or to reinforce, their own policies and actions Opportunity to *avoid geopolitical risks* arising from climate impacts: • An accelerated U.S. transition to clean energy can help reduce to some degree the insecurity and instability that will otherwise arise from climate impacts—such as migration exacerbated by heat stress, reduced agricultural outputs, competition over resources, and the like A model for *success on big international challenges other than* climate: • Countries that are currently locked in non-climate tensions (such as the United States and China) may find reasons for mutual accommodation on those non-climate issues if (a) they find mutual interest in avoiding the worst impacts of climate change, and (b) they then see each other as essential and effective partners Restored U.S. *soft power*: • Serving as a global leader on addressing one of the signal challenges of our era will help restore the view that the United States can be relied upon to seek and contribute to solutions to big global problems—including both those related to climate change (like clean energy R&D collaborations) and those related to other topics like infectious disease, nuclear non-proliferation Domestic *economic power*: • By both funding clean energy R&D and supporting accelerated deployment of clean energy technologies across the U.S. economy, the United States will reinforce the attractiveness of U.S. universities, research laboratories, and firms as global centers for innovation and wealth creation	Changing times for certain *traditional relationships*: • Geopolitical relationships that have grown since the Second World War around the hydrocarbon economy (e.g., the U.S.–Saudi relationship) may experience certain tremors as the clean energy transition proceeds if those partners perceive that the U.S. commitment to the clean energy transition will come at the expense of their own interests • The speed with which traditional relationships change remains to be seen, but one can see evidence of the shifting relations, such as Saudi Arabia's interest in building new ties with East Asian consumers New supply chains, *new relationships*: • New relationships will likely develop around the new supply chains for materials required for the new energy economy—the so-called rare earth minerals and other critical materials needed for batteries, wind turbines, solar panels, high-efficiency lighting, and other applications; already we have seen a concentration of critical material production capacity in the hands of Chinese firms and at least one instance when that market power was applied by Beijing to exert pressure on another country (Japan)

national action, its call for regular assessments of progress, and its ability to marshal momentum for additional action if countries fail to institute sufficiently strong and timely mitigation measures.

Paris embodied critical features for which American negotiators called for years.[10] Indeed, in several regards, the Paris agreement amounted to a conscious departure from the structures that were agreed in the Kyoto protocol of 1997 and the Copenhagen accord of 2009 (Stern 2018). If the U.S. Federal government not only walked away from the Paris climate agreement but also persisted in its hostility to the very idea of a clean energy transition, this behavior would create ample justification for people around the globe to conclude that the United States cannot be relied upon to play a consistent and constructive role in international cooperation and diplomacy. If on the other hand, the U.S. Federal government returned to the business of providing national and international leadership for the clean energy transition, then solving hard problems would seem to be more within our reach.

A third geopolitical dimension worth highlighting is that the current reluctance of the U.S. Federal government to press ahead with the clean energy transition comes against a particular geopolitical backdrop—an assertive China that is simultaneously a major geopolitical challenge for the United States and also an essential partner in managing the global climate problem. Beijing under President Xi has been exceptionally self-confident and proactive on the global stage. The massive Belt and Road Initiative (BRI), for example, makes available massive amounts of capital that are presently not rivaled by the scale of funds available from the Bretton Woods institutions, bilateral aid programs, or private capital flows. To date, despite declared goals of creating a "green" Belt and Road, the Initiative has instead mobilized billions for coal-fired power plants, which will pollute air and water in the recipient countries and will emit decades worth of carbon dioxide affecting the global climate. Regardless, as long as Washington is inclined toward greater isolationism, budget cuts for development assistance and diplomacy, and Federal inaction on the clean energy transition, the United States will be unable to offer a meaningful alternative for decision-makers in emerging economies. Beijing's development model can mobilize capital quickly, and even if it results in more coal plants, that may be fine from the perspective of some political leaders in certain capitals of the developing world (Elkind 2019).

Just to make the U.S.–China relationship a still more complicated part of the present calculation, China is not just a competitor but also an essential partner in dealing with the global climate problem. No country emits more GHGs than China (26.8% of total GHG emissions in 2017), but the United States, which represents a much greater share of historical emissions is number two (13.1% in 2017) (UN Environment 2018). Success therefore requires effective action by both China and the United States to execute the clean energy transition at home, and ideally to

---

[10] A number of observers have criticized the Paris agreement for failing to institute binding, top-down reductions in greenhouse gas emissions and penalties for those countries that fail to meet their targets. Nothing short of such stringent requirements, the argument goes, corresponds to the urgency of the climate problem. This view misses entirely the nature of climate change as a "wicked problem," as discussed earlier in this chapter. The climate challenge requires a more nuanced and flexible diplomatic response (Stern 2018).

promote solutions in other countries as well. In the face of ongoing disputes over tariffs, intellectual property protection, technology theft, market access, exchange rates, the South China Sea, economic sanctions, and many other matters, the U.S.– China bilateral agenda is overloaded. Nonetheless, the two countries need each other on climate. There is no alternative.

The U.S.–China relationship is not the only one that affects, and is affected by, U.S. decisions about clean energy. The United States has numerous longstanding ties that have been framed in part by U.S. support for rules-based international trade. Some of these relationships may, to some degree, actually suffer if the United States shifts back to a more proactive Federal effort to promote the clean energy transition. Leaders in the oil-producing Gulf Arab states, for example, may worry about whether the United States will remain engaged as a constructive partner for them, invested in their countries' long-term security.[11] Concerns of this type appear to be encouraging the growth of new relationships, such as the OPEC–Russia relationship and new engagement between Riyadh and Beijing. Even in a world characterized by reduced U.S. importation of traditional fuels (whether as a result of the clean energy transition or simply the last decade's growth in U.S. domestic hydrocarbon production), the United States will continue to have an interest in international trade in goods and services. This reality may limit the extent of change in U.S. relations with the Gulf Arab partners, but some change is afoot regardless.

A final geopolitical consideration that could arise from U.S. policies on the clean energy transition relates to materials required for production of low-carbon energy systems. Analysts and policy-makers have recognized for some time that clean energy technologies rely on a number of critical materials whose supply chains are not very diversified (DOE 2011). Indeed, concerns have arisen because the predominant supply of a number of those materials is controlled by Chinese companies, and China attempted to exploit that dominance in 2010 after a fishing dispute with Japan (Bradsher 2010). As the United States and other economies proceed with the clean energy transition, their ability to deploy cost-effectively energy-efficient lighting, wind and solar generating plants, batteries, and other technologies will require adequate and assured supply chains for critical materials. This does not mean automatically that problems will arise with the availability of critical materials; new supplies, innovation, and re-use may all help to alleviate any shortages. But the transition to clean energy will nonetheless place in the spotlight a new set of supply relationships that will need to be managed effectively.

---

[11] A further complication: The very same Gulf Arab states, where some leaders worry about a potential reduced long-term U.S. commitment to the region, are among those countries with the most to lose as median temperatures rise. For an excellent discussion of climate change vulnerability in the Gulf, see Krane (2019). In this case, the links between geopolitics and the clean energy transition amount to a situation in which geopolitical strains may arise from either action or inaction on clean energy and climate.

8 Conclusions

Especially since the end of the Second World War, the United States has played a major role on the geopolitical stage. It has sought to influence the rules that governed international trade, global financial systems, security interests, and even environmental standards. The issue of climate change presents an unparalleled challenge to the postwar global order because the roots of global warming run so deep in how we conduct our daily lives and operate our economies. To address this "wicked problem" requires unprecedented ingenuity, dedication, and persistence. It requires policies at all levels of governance that combine speed, scale, and durability.

To date, unfortunately, the very country that has played such an important geopolitical role since the middle of the last century has not demonstrated clear and consistent leadership when it comes to climate. To be sure, especially during the Trump presidency, leaders at the state and municipal levels across the United States have signaled a progressively greater willingness to act on the climate problem. At the national level, however, the political leadership of the United States remains divided. There is a possibility that the United States is approaching a turning point from which there could emerge a restored national focus on solutions for clean energy and climate. An alternative option, however, is that the United States continues to struggle to find a unified voice on these matters.

If climate change is indeed one of the defining issues of our day, then the geopolitical implications of the U.S. clean energy transition are hard to exaggerate. Without American climate leadership, it is hard to imagine global success in responding adequately, and in this scenario American standing on the world stage will be significantly diminished. With American leadership, one can imagine the possibility of success in responding to the threat of climate change. But the hour is late, and indecision by U.S. political leaders is already taking a toll on American interests on the global stage.

References

Adler D (2019) U.S. climate change litigation in the age of trump: year two. Sabin Center for Climate Change Law, Columbia Law School. http://columbiaclimatelaw.com/files/2019/06/Adler-2019-06-US-Climate-Change-Litigation-in-Age-of-Trump-Year-2-Report.pdf?utm_source=newsletter&utm_medium=email&utm_campaign=newsletter_axiosgenerate&stream=top. Accessed 1 July 2019

Alvarez RA, Zavala-Ariaza D, Lyon D, Allen D, Barkley Z, Brandt A, Davis K, Herndon S, Jacob D, Karion A, Kort D, Lamb B, Lauvaux T, Maasakkers J, Marchese A, Omara M, Pacala S, Peischl J, Robinson A, Shepson P, Sweeney C, Townsend-Small A, Wofsy S, Hamburg S (2018) Assessment of methane emissions from the U.S. oil and gas supply chain. Science 361:186–188. https://science.sciencemag.org/content/361/6398/186. Accessed 2 July 2019

America's Pledge Initiative on Climate (2018) Fulfilling America's pledge: how states, cities, and businesses are leading the united states to a low-carbon future. https://www.americaspledgeonclimate.com/fulfilling-americas-pledge/. Accessed 2 July 2019

Australian Public Service Commission (2007) Tackling wicked problems: a public policy perspective. https://www.apsc.gov.au/publications-and-media/archive/publications-archive/tackling-wicked-problems. Accessed 25 June 2019

Barbose G (2018) U.S. renewable portfolio standards: 2018 annual status report. Lawrence Berkeley National Laboratory. http://eta-publications.lbl.gov/sites/default/files/2018_annual_rps_summary_report.pdf. Accessed 2 July 2019

Bordoff J (2019) Getting real about the green new deal. Democracy. J Ideas. https://democracyjournal.org/arguments/getting-real-about-the-green-new-deal/. Accessed 3 July 2019

Bradsher K (2010) Amid tension, China blocks vital exports to Japan. New York Times. https://www.nytimes.com/2010/09/23/business/global/23rare.html. Accessed 5 July 2019

California Air Resources Board (2019) Cap and trade program (website). https://www.arb.ca.gov/cc/capandtrade/capandtrade.htm. Accessed 3 July 2019

Center on Global Energy Policy (2019) Prospects for climate solutions – February 6, 2019 event video. https://energypolicy.columbia.edu/events-calendar/prospects-climate-solutions. Accessed 3 July 2019

Climate Mayors website (2019). http://climatemayors.org. Accessed 3 July 2019

Congressional Budget Office (2009) Cost estimate – HR 2454, American clean energy and security act of 2009. https://www.cbo.gov/sites/default/files/111th-congress-2009-2010/costestimate/hr24541.pdf. Accessed 11 July 2019

Database of State Incentives for Renewables and Efficiency (2019) Renewables & clean energy standards, June 2019. North Carolina Clean Energy Technology Center. https://s3.amazonaws.com/ncsolarcen-prod/wp-content/uploads/2019/06/RPS-CES-June2019.pdf. Accessed 2 July 2019

Elkind J (2019) Toward a real green belt and road. Center on Global Energy Policy, Columbia University. https://energypolicy.columbia.edu/sites/default/files/file-uploads/Real%20Green%20Belt%20and%20Road_CGEP_Report_042219_Final.pdf. Accessed 2 July 2019

Energy Futures Initiative (2019) Optionality, flexibility, and innovation: pathways for deep decarbonization in California. https://static1.squarespace.com/static/58ec123cb3db2bd94e057628/t/5ced7013ee6eb03a466f546d/1559064604282/EFI_CA_Decarbonization_SFPM.pdf. Accessed 2 July 2019

Energy Information Administration (2018) Oil: crude and petroleum products explained – use of oil (updated 28 September 2018). https://www.eia.gov/energyexplained/index.php?page=oil_use. Accessed 1 July 2019

Energy Information Administration (2019a) Annual energy outlook 2019. https://www.eia.gov/outlooks/aeo/. Accessed 26 June 2019

Energy Information Administration (2019b) Today in energy: in 2018, the United States consumed more energy than ever before, 16 April 2019. https://www.eia.gov/todayinenergy/detail.php?id=39092. Accessed 1 July 2019

Federal Highway Administration (2018) Travel monitoring: historical monthly VMT report (modified 26 April 2018). https://www.fhwa.dot.gov/policyinformation/travel_monitoring/historicvmt.cfm. Accessed 1 July 2019

Friedman L (2019) EPA finalizes its plan to replace Obama-Era climate rules. New York Times. https://www.nytimes.com/2019/06/19/climate/epa-coal-emissions.html. Accessed 1 July 2019

Gustin G (2018) Climate denial pervades the trump white house, but it's hitting some limits. Inside Climate News, 8 January 2018. https://insideclimatenews.org/news/08012018/climate-change-denial-trump-hoax-2017-year-review-pruitt-tillerson-endangerment-finding. Accessed 1 July 2018

Houser T, Marsters P, Rhodium Group (2018) Final U.S. emissions numbers for 2017, 29 March 2018. https://rhg.com/research/final-us-emissions-numbers-for-2017/. Accessed 29 June 2019

Intergovernmental Panel on Climate Change (2018) Special report: global warming of 1.5°C – summary for policymakers. https://www.ipcc.ch/2018/10/08/summary-for-policymakers-of-ipcc-special-report-on-global-warming-of-1-5c-approved-by-governments/. Accessed 26 June 2019

Irfan U and Zarracina J (2019) 4 maps that show who's being left behind in America's wind-power boom. Vox, 14 June 2019. https://www.vox.com/energy-and-environment/2018/5/2/17290880/trump-wind-power-renewable-energy-maps. Accessed 2 July 2019

Kaufman N and Gordon K (2018) The energy, economic, and emissions implications of a federal U.S. carbon tax. https://energypolicy.columbia.edu/sites/default/files/pictures/CGEP_SummaryOfCarbonTaxModeling.pdf. Accessed 26 June 2019

Krane J (2019) Energy kingdoms: oil and political survival in the Persian Gulf. Columbia University Press, New York, pp 160–172

McNutt M, Mote CD Jr, and Dzau VJ (2019) National Academies Presidents affirm the scientific evidence of climate change. National Academies of Science, Engineering, and Medicine, 18 June 2018. http://www8.nationalacademies.org/onpinews/newsitem.aspx?RecordID=06182019. Accessed 25 June 2019

National Association of State Energy Officials (NASEO) and Energy Future Initiative (EFI) (2019). The 2019 U.S. energy and employment report. https://www.usenergyjobs.org/2019-report. Accessed 7 October 2019

National Conference of State Legislatures (2019) State renewable portfolio standards and goals, 1 February 2019. http://www.ncsl.org/research/energy/renewable-portfolio-standards.aspx. Accessed 2 July 2019

Perry R, Zinke R, Pruitt S (2017) Paving the path to U.S. energy dominance. Washington Times, 26 June 2017. https://www.washingtontimes.com/news/2017/jun/26/us-energy-dominance-is-achievable/. Accessed 16 May 2018

Phillip A (2017) Mulvaney defends proposed cuts to climate research and poverty programs, deeming them wasteful or ineffective. Washington Post, 16 March 2017. https://www.washingtonpost.com/news/post-politics/wp/2017/03/16/mulvaney-defends-proposed-cuts-to-climate-research-poverty-programs-deeming-them-wasteful-or-ineffective/?utm_term=.35976682f1ac. Accessed 1 July 2019

Pyper J (2019) US electric vehicle sales increased by 81% in 2018. GreenTech Media, 7 January 2019. https://www.greentechmedia.com/articles/read/us-electric-vehicle-sales-increase-by-81-in-2018#gs.ohnerc. Accessed 11 July 2019

RGGI Inc. (2019) RGGI states welcome New Jersey as its CO2 regulation is finalized (press release – 17 June 2019). https://www.rggi.org/sites/default/files/Uploads/Press-Releases/2019_06_17_NJ_Announcement_Release.pdf. Accessed 3 July 2019

Roberts D (2019) New York just passed the most ambitious climate target in the country. Vox, 20 June 2019. https://www.vox.com/energy-and-environment/2019/6/20/18691058/new-york-green-new-deal-climate-change-cuomo. Accessed 2 July 2019

Stern T (2018) The Paris agreement and its future. Brookings Institution. https://www.brookings.edu/wp-content/uploads/2018/10/The-Paris-Agreement-and-Its-Future-Todd-Stern-October-2018.pdf. Accessed 5 July 2019

United Nations Environment Programme (2018) Emissions gap report 2018. https://www.unenvironment.org/resources/emissions-gap-report-2018. Accessed 29 June 2019

United States Climate Alliance (2019a) 2019 fact sheet. https://static1.squarespace.com/static/5a4cfbfe18b27d4da21c9361/t/5ccb5aa56e9a7f542fe4233c/1556830885910/USCA+Factsheet_April+2019.pdf. Accessed 2 July 2019

United States Climate Alliance (2019b) Montana Governor Steve bullock becomes 25th governor to join U.S. climate alliance (press release 1 July 2019). https://www.usclimatealliance.org/publications/2019/7/1/montana-governor-steve-bullock-becomes-25th-governor-to-join-us-climate-alliance. Accessed 2 July 2019

United States House of Representatives (2019) H.Res. 109 – recognizing the duty of the Federal government to create a Green New Deal. https://www.congress.gov/bill/116th-congress/house-resolution/109/text. Accessed 3 July 2019

U.S. Department of Energy (2011) Critical materials strategy, December 2011. https://www.energy.gov/sites/prod/files/2016/12/f34/2011%20Critical%20Materials%20Strategy%20Report.pdf. Accessed 5 July 2019

U.S. Department of Energy (2017) FY 2018 congressional budget request – budget in brief. https://www.energy.gov/sites/prod/files/2017/05/f34/FY2018BudgetinBrief_3.pdf. Accessed 1 July 2019

U.S. Department of Energy (2018) FY 2019 budget justification. https://www.energy.gov/sites/prod/files/2018/04/f50/FY2019ControlTablebyAppropriation.pdf. Accessed 1 July 2019

U.S. Department of Energy (2019) FY 2020 budget justification. https://www.energy.gov/cfo/downloads/fy-2020-budget-justification. Accessed 1 July 2019

White House (2013) The President's climate action plan, June 2013. https://obamawhitehouse.archives.gov/sites/default/files/image/president27sclimateactionplan.pdf. Accessed 1 July 2019

White House (2016a) Fact sheet: the recovery act made the single largest investment in clean energy in history, driving the deployment of clean energy, promoting energy efficiency, and supporting manufacturing, 25 February 2016. https://obamawhitehouse.archives.gov/the-press-office/2016/02/25/fact-sheet-recovery-act-made-largest-single-investment-clean-energy. Accessed 1 July 2019

White House (2016b) United States mid-century strategy for deep decarbonization. https://unfccc.int/files/focus/long-term_strategies/application/pdf/mid_century_strategy_report-final_red.pdf. Accessed 1 July 2019

China: Climate Leader and Villain

Michal Meidan

1 China's Changing Energy Landscape

Since China's reform and opening up in 1978, the country has undergone a profound transformation: The Chinese economy in 1978, as measured in Gross Domestic Production (GDP) stood at $150 billion (current US$ according to the World Bank (2019), and was half the size the Italian economy. Three decades later, China's economy is the second largest in the world and its per capita GDP has grown by nearly 24 times from 1978 to 2017. Even though income inequality has increased dramatically, the country has all but eradicated extreme poverty, with the share of China's population living in extreme poverty (according to the World Bank definition) plummeting from 90% in 1971 to less than 2% by 2013.

Significantly, in 1980, agriculture was a larger part of the Chinese economy than industry and services, but it now accounts for under 10%, while the service sector is approaching 40%, and industry accounts for the lion's share of economic activity. Urbanisation has therefore been a defining feature of China's economic transformation, with the rural population, which accounted for roughly 85% of China's population on the eve of China's reform and opening up, now down to around 40% (World Bank 2019).

1.1 A Voracious Appetite for Fossil Fuels

Fuelling the country's rapid industrialisation and urbanisation process is a voracious appetite for energy, with primary energy consumption increasing sevenfold, from just under 400 million tonnes oil equivalent (toe) in 1978 (BP 2019), to 3.27 billion toe

M. Meidan (✉)
Oxford Institute of Energy Studies (OIES), Oxford, England
e-mail: michal.meidan@oxfordenergy.org

© The Author(s) 2020
M. Hafner and S. Tagliapietra (eds.), *The Geopolitics of the Global Energy Transition*,
Lecture Notes in Energy 73, https://doi.org/10.1007/978-3-030-39066-2_4

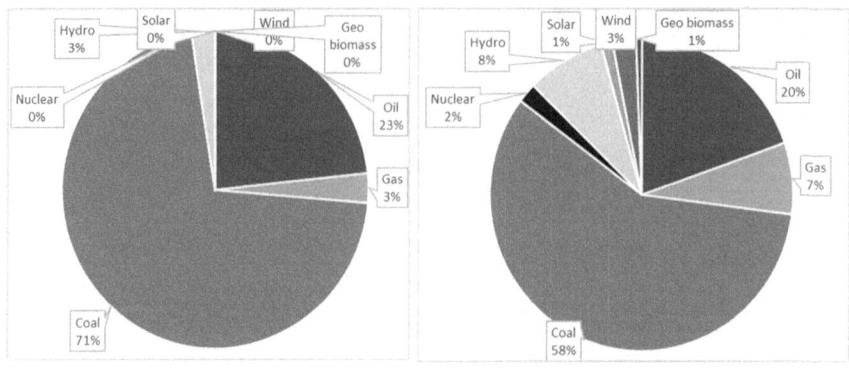

Fig. 1 China's energy mix 1978, 2018. While China's energy mix still relies heavily on coal, it has become much more diverse over the course of the last four decades. *Source* BP

in 2018. Domestically produced coal accounted for 70% of the energy mix in 1978, alongside oil which accounted for another 23%. In the late 1970s, China consumed a mere 17% of global coal (BP 2019), but by 2018, however, China burned 1.9 billion toe of coal, half of the coal used worldwide (see Fig. 1).

In light of China's heavy reliance on coal, the country has, since 2006, become the world's largest emitter of carbon dioxide (CO_2). In 2018, according to the BP Statistical Review (BP 2019), the country accounted for 28% of global CO_2 emissions—more than the US and the EU combined, with coal accounting for an estimated 70% of energy-related CO_2 emissions (Myllyvirta 2019).

China's carbon footprint has also expanded due to the country's oil demand, which has tripled over the past 20 years, from 4.1 million barrels per day (mb/d) in 1998, to 13.5 mb/d in 2018 (BP 2019). Urbanisation and rising incomes have increased the number of passenger vehicles on China's roads. Indeed, in 2010, China's passenger vehicle park was estimated at 55 million vehicles, but in 2018, it counted 199 million. Still, there are currently only 150 vehicles per 1,000 persons in China, compared to around 600 vehicles per 1,000 persons in France and Germany, and as the middle class continues to grow wealthier and buy cars, oil demand is set to rise further. That said, China's ambitious programme to electrify its fleet could displace some oil demand growth in the future, allowing the country also to tackle local air pollution problems. But with a power sector heavily reliant on coal, until China decarbonises the power sector, electric vehicles only displace the problem. At the same time, air travel and demand for consumer goods (and the freight to transport them around the country) have all contributed to the surge in China's oil demand growth and will continue to fuel the country's reliance on crude oil. But with limited domestic reserves, China has become dependent on imported crude, leaving it vulnerable to geopolitical disruptions in producer countries across the globe, and especially in the Middle East, which supplies just under half of China's total imports (see Fig. 2).

This surge in China's external exposure to natural resources has been just as rapid as its economic boom: In 1993, China became a net importer of oil for the first time,

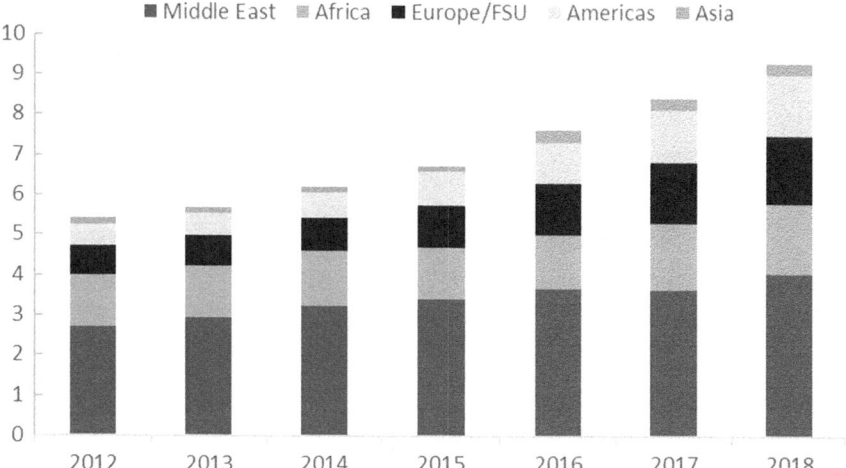

Fig. 2 China's crude oil imports by region, mb/d. *Source* China Customs

and 10 years later, it was importing 5.6 million barrels per day (mb/d), roughly the equivalent of Iranian and Iraqi oil production combined that year. By 2018, China's crude oil imports reached 9.2 mb/d.

1.2 Changing Policy Priorities

Over the course of the past four decades, since the 'reform and opening up' and especially since the country's accession to the World Trade Organisation (WTO) in 2001, to accommodate China's rapid economic development, energy policy was geared first and foremost towards ensuring ample supplies: to keep factories churning, to deliver energy from producer hubs to consumer centres, and to keep the rising numbers of urban homes warm in winter and cool in summer. But as China's appetite for energy continued to grow and outstrip its domestic production, the country's policymakers began fretting about rising dependence on waterborne flows of commodities and the US's ability to curb supplies via maritime routes (Downs 2019; Meidan 2014). Chinese companies, with government support, embarked on costly M&As and overseas infrastructure investments, hoping to secure a footprint across the energy value chain. Producer countries from the Middle East through to the Americas saw new opportunities to produce and sell commodities to China, and also to attract investments in infrastructure, even in conflict-torn countries.

China's concerns about import dependence were gradually also compounded by the rising toll of environmental degradation (Meidan et al. 2009). China is home to around 20% of the world's population but has five to seven percent of freshwater resources and under 10% of the world's arable land (Ely et al. 2019). China's mega deltas are particularly vulnerable to rising sea levels, while hazardous smog in densely populated cities—from industrial activities and road transport—is driving demand for the government to tackle air pollution (Ramaswami et al. 2017). Moreover, health problems due to air pollution are estimated to have led to the loss of about 133 million workdays in China in 2007, or 1.34% of real GDP (ChinaPower 2016).

Yet unlike many developed economies, that began to regulate air pollution after their de-industrialisation was underway, the Chinese economy continues to grow and industrialise, leaving the government to grapple with the need to protect its environment while also ensuring affordable and secure sources of energy. The need to diversify the domestic energy mix and ensure more sustainable fuels for growth has coincided with a broader desire to shift the country's economic structure away from industrial-led growth towards a consumption-driven development path. Environmental protection which was once seen as a costly impediment to growth has become both a social necessity and an industrial opportunity.

This change in priorities was reflected in the 12th Five-Year Plan (12th FYP, spanning 2011–2015), in which the government set out for the first time binding targets for a 16% reduction in energy consumption per unit of GDP, an 8% reduction in sulphur dioxide (SO_2) emissions and a 10% reduction in nitrogen oxide (NOx) emissions by 2015, from 2010 levels. As a result, PM2.5 monitoring efforts intensified with the government setting more stringent targets for heavily polluted regions. The 12th FYP also incorporated a number of specific measures to shut down heavily polluting industrial facilities and expand the use of clean energy, including natural gas. Against this backdrop, China introduced its first 'Airborne Pollution Prevention and Control Action Plan' in 2013 (Action Plan 2013), which recognised coal as a key driver of air pollution and sought to limit its use (ChinaPower 2016).

The Action Plan 2013 established mid- to long-term targets for reducing total coal consumption and cutting its share of the energy mix (Miyamoto and Ishiguro 2018), replacing industrial coal furnaces with natural gas. Gas demand, which has long played second fiddle to both the coal and oil industries, began to surge on the back of the coal-to-gas switch. In the early 2000s, natural gas was largely used as feedstock in industry and only played a marginal role in the power sector, where coal remains the dominant fuel. While this still holds true for the power sector, the coal-to-gas switch as mandated by the Action Plan 2013 led to a strong surge in industrial and commercial gas consumption. As a result, China's natural gas demand went from 177 bcm in 2013 to 280 bcm in 2018 (Fig. 3) but the strong uptick in demand also led to an increased dependence on imported gas, mainly liquefied natural gas (LNG).

But with global oil and gas prices falling on rising US shale production, supply security become a secondary concern in Beijing. Moreover, in the early Trump years, increased oil and gas imports from the US were touted as a potential way of evening out some of the US–China trade deficit, with Chinese traders increasing their purchases of US crude oil and looking to sign long-term LNG contracts.

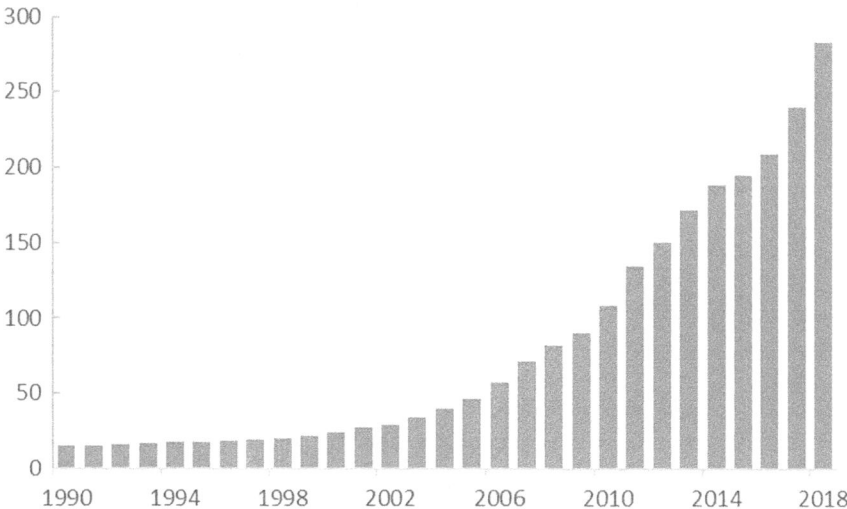

Fig. 3 China's natural gas demand, BCM. *Source* NDRC, BP

1.3 China's 'Energy Revolution': The Nexus Between Energy and Technology

China's 12th FYP ushered in more than a change in energy policy priorities. It was designed to address the 'Four "Uns"', as articulated by former Premier Wen Jiabao, and change course for an economy that was deemed 'unstable, unbalanced, uncoordinated and unsustainable'. As such, it also saw the convergence of China's industrial upgrade plan and its energy priorities. In mid-2014, the country announced an 'energy revolution', which was later formalised in a publicly released policy paper setting out the main overall targets and strategies for China's energy sector through 2030 (NDRC 2016). The energy revolution spans energy consumption, by mandating demand-side management for industry and changing consumer habits; energy production—calling for enhancing efficiencies and reducing emissions from China's energy infrastructure and energy technology and also includes an effort to develop, commercialise and diffuse next-generation energy technologies through innovation and international cooperation.

In the context of China's industrial programme, climate change mitigation became an opportunity for underpinning China's economic transition and a potential means of advancing China's bid for global technology leadership (Geall 2017). And given the country's scale and strong ability to incentivise industrial outcomes, it has proven capable of rapidly driving change. The 12th FYP highlighted seven strategic emerging industries that would receive preferential support, including renewable energy technologies and electric cars. The subsequent plan, the 13th Five-Year Plan (13th FYP; 2016–2020), continues the emphasis on clean technologies, although it aims to give the market, rather than state subsidies, a determining role in selecting the most competitive green industries and technology leaders (Geall, 2017).

China has since become the global leaders in renewables. In 2012, China's installed capacity of wind and solar power was 61 GW and 3.4 GW, respectively, while the annual electricity generated by renewables was only 2.1% of China's total consumption. By 2017, China's wind and solar power capacity had increased to 168.5 GW and 130.06 GW, respectively, and renewables were generating 5.3% of China's electricity supply (Linster and Yang 2018). Installed solar capacity has outstripped the 110 GW targeted in the 13th FYP, with 186 GW installed in June 2019. Similarly, wind capacity is largely on track to meet its 13th FYP target of 210 GW of installed capacity, having reached 193 GW in June 2019.

On the back of increased manufacturing capabilities, the average price of global PV modules decreased by 79% from 2010 to 2017. At the same time, the subsidy programme was draining central government coffers, with the total amount of wind and PV subsidies in 2017 estimated at about 170 billion yuan (Lin and Yang 2018) and a source of global trade friction, as Chinese manufactured solar PV modules were the target of anti-dumping measures. But ultimately, Chinese companies' ability to reduce costs and support investments globally in the 'low carbon' economy have supported wider efforts to tackle climate change (Goron 2018).

Moreover, since 2015, China has invested over US$100 billion a year in domestic renewable energy projects, almost double the US's $64 billion in 2018 (BNEF 2019). Of over 8 million renewable energy jobs globally, 3.5 million were in China in 2015 and the Chinese government estimates that between 2016 and 2020, new investments in renewables will create 13 million jobs (Jaeger et al. 2017). China has therefore been driving global renewables consumption growth, both by installing capacity at home and exporting solar panels and wind turbines. As such, China's decarbonisation goals and commitment to the UN climate process are consistent with and supportive of its key economic and technological ambitions, namely, the domestic economic rebalancing away from energy-intensive heavy industries towards innovation and services. Moreover, increasing the share of renewables in the energy mix also supports China's energy security by reducing import dependence and limiting the effects of geopolitical conflict or price volatility on energy supply.

2 Climate Leader or Climate Villain?

China's commitment to its 'energy revolution' suggests that renewables will account for a growing share of the country's energy use while supercharged efforts to spur innovation are already turning Chinese companies into global leaders in the technologies underpinning the energy transition. To date, China's domestic efforts have altered the country's international position markedly, too. China was cast as the villain of global climate talks in Copenhagen in 2009, but has become an active participant, if not a de-factor leader, in climate diplomacy especially since President Trump's decision to withdraw from the Paris agreement created a leadership vacuum. The commitments to rebalance the Chinese economy, phase out coal gradually and develop energy technologies suggest that China will stick to its pledges. Whether or not China voluntarily assumes climate leadership globally is perhaps a moot point, but the technological and economic changes within China suggest that it will inevitably play a more prominent role globally. That said, how China is perceived globally will depend not only on its exports of clean energy but also on how it manages its own energy transition. There are already signs that China could become both a climate leader and villain, with respect to its domestic energy consumption and its overseas investments.

2.1   Electrification Before Decarbonisation

It is important to note that China's 'energy revolution' emphasises air quality, rather than carbon mitigation, with mandatory targets to reduce air pollutants such as SO_2 and NOx with less emphasis on greenhouse gas emissions more broadly. The plan also reiterates China's climate change commitments undertaken in the Paris framework, to peak CO_2 emissions around 2030 or earlier, and to reduce carbon emission per unit of GDP by 60–65% compared to 2005, without, however, setting an absolute cap for carbon emissions. Put simply, China's air quality and climate policies have been developing relatively autonomously from each other with air pollution the main source of concern for the Chinese government. Air pollution is perceived as an environmental problem, while climate change has been framed as a development issue and until March 2018, each policy was under the supervision of different parts of the state administration (Yamineva and Liu 2019).

In addition, China's energy transition is at the intersection of a number of policy priorities whose relative importance for decision-makers can fluctuate. In 2019, for example, given the decelerating economy and a weak industrial complex, air pollution woes are falling slightly in importance, in large part because the largest polluters are impacted by the economic moderation. So, costly efforts to mitigate air pollution, such as the coal-to-gas switch, are also slowed. At the same time, given the ongoing trade war with the US, concerns about supply security and import dependence are resurfacing. It is perhaps too early to tell, at the time of writing, whether China's economic slowdown and its frictions with the US are secular rather than structural trends and while a number of long-standing policy priorities, such as the economic rebalancing and the adoption of cleaner energy, will maintain their importance, the speed and scope with which they will be pursued will depend on these macroeconomic and geopolitical trends.

When considering China's future energy mix, there are a host of possible scenarios, which suggest dramatically different demand profiles for China. On one hand, China's National Renewable Energy Center (CNREC 2018) estimates that under current policy guidelines, China's fossil fuel consumption can peak in 2020 and decline gradually through 2035. By 2035, the share of coal in the energy mix would be just over 10% as its use in the power and industrial sectors would fall. At the same

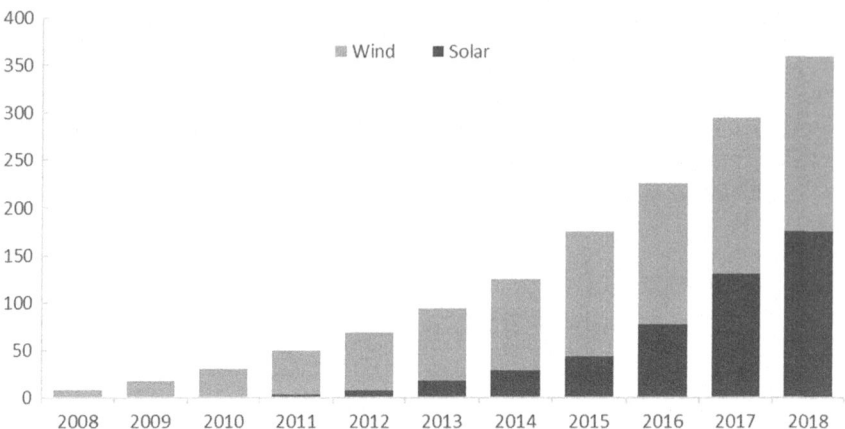

Fig. 4 Installed renewable capacity, GW. *Source* China Electricity Council

time, oil demand would also begin to decline due to higher levels of electrification in transport and industry. And with the rise of renewable energy, China would not need gas to serve as a bridge between coal and renewable energy. With improved economics, according to the CNREC, China could require an additional 80–160 GW of new solar PV installation and 70–140 GW of new wind capacity per year. Such an aggressive rollout likely assumes that China manages to overcome a number of challenges, related to the rigidity of the domestic electricity market (Chen 2017) and to develop the intelligent infrastructure required for its effective deployment (Fig. 4).

Indeed, CNREC considers an even more aggressive scenario whereby coal consumption falls even further, thanks to a faster adoption of renewables, but these scenarios should be contrasted with how China's fossil fuel industry views the country's energy future. CNPC, China's largest oil and gas company, in its 2050 outlook, expects primary energy demand to peak between 2035 and 2040 (CNPC 2016) and while the share of coal in the energy mix will continue to fall through to 2050, in 2035 coal will still account for roughly a third of primary energy demand. Indeed, absolute demand for coal is expected to remain at 2018 levels until 2025, given its importance in power generation. CNPC sees coal's share of the energy mix continue its decline but it will still remain the single largest supply source through to 2050. In power generation, according to CNPC, coal is set to peak only in 2030 and even as industrial coal use declines after 2030, coal in petrochemicals will offset some of that fall (Fig. 5). To be sure, falling costs of renewables and reforms in the Chinese power sector could allow for the more progressive scenarios in China to materialise, but in the near term, the country may well electrify much faster than it can decarbonise. China's international stance will look very different in each of the above scenarios, and its domestic demand for clean energy technologies and equipment will also determine the availability and price of exports.

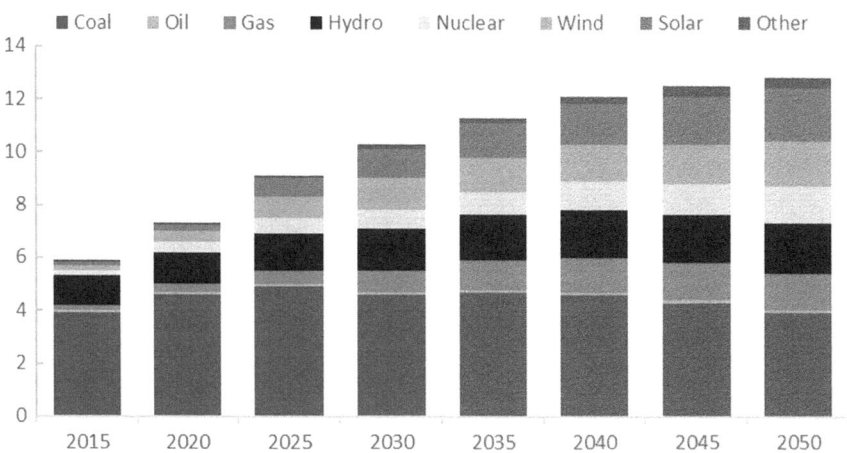

Fig. 5 China's power demand forecast, trillion kwh. *Source* CNPC

2.2 China Commercialises and Exports Clean Tech

By virtue of China's scale, the rapid ramp-up of renewables in the country has cata-pulted Chinese companies' status globally. Chinese solar manufacturers account for about 60% of global solar cell production, with a recent ranking suggesting also that these manufacturers are also among the highest quality manufacturers globally (Geall 2017). Similarly, China in 2018 was the largest wind market (for both onshore and offshore), leading Chinese manufacturers to capture almost half of the global market (Lacal-Arántegui 2019) yet when discounting installations in China, European man-ufacturers remain more dominant globally. Indeed, while Chinese companies have also helped cost reductions globally, research suggests that they experience difficul-ties when developing global products and competing in terms of innovation. Chinese firms mainly file patents in the domestic markets but seem to have limited innova-tion competitiveness globally (Cao et al. 2018; Pan et al. 2019). Whether China is a leader or a follower in renewable energy manufacturing remains a topic of debate (Tyfield et al. 2014), but it is still undoubtedly the biggest investor in cleantech and the largest market, capable of commercialising new energy products. Globally, China has invested an estimated $44 billion in clean energy projects in 2017 (IEEFA 2018), and the International Energy Agency forecasts China will build 40% of the world's wind energy capacity and 36% of global solar capacity from 2015 to 2021.

China's largest power utility, State Grid Corporation has followed in the renew-able manufacturers' footsteps to expand its global reach. Within China, State Grid's investment in long-distance high-voltage power transmission lines has been an impor-tant factor in alleviating some of China's curtailment issues. State Grid has also set out its vision of creating a global supergrid called the 'Global Energy Interconnec-tion' (GEI) that aims to link every continent with undersea transmission cables to power the world with green electricity (IRENA 2019). In the interim, however, State Grid focuses on smaller scale projects globally.

Beyond exports of renewable energy infrastructure, China's efforts to increase energy efficiency at home have brought it to the forefront of new technologies driving efficiency gains. China now accounts for roughly half the global market for energy service companies, including connected devices which allow real-time control of energy consumption. China is now the largest market for smart metres, having installed more than 500 million (IEEFA 2018), and ongoing progress in developing connected devices and smart energy management systems will gradually allow it to expand its global footprint, much like it is currently doing with lithium-ion batteries and electric vehicles.

China's lithium-ion battery production capacity is growing rapidly, supported by Beijing's goal of making China the leading country in the global EV supply chain. Chinese companies are expected to hold 121 gigawatt-hours (GWh) of battery production capacity by 2020, dwarfing Tesla's 35 GWh (IEEFA 2018). Already in 2016, an estimate of 200 energy storage manufacturers produced a combined capacity of 120 GWh of energy storage that is scheduled to be operational by 2018. That year, China's lithium-ion battery shipments increased by 80% year on year (Vorrath 2017).

More broadly, both Chinese companies and Beijing are focusing on energy storage solutions. In October 2017, the Chinese government issued its first national policy document guiding the energy storage industry, outlining two stages of development through 2027. The policy document views the first phase running through 2022, during which domestic storage technology production to reach advanced international standards, with preliminary accomplishments in creating a standards body. Subsequently, throughout the 14th FYP, China aims to develop a mature industry with world-class, internationally competitive technology. Subsequently in 2018, the government announced a major storage push to help address curtailment of domestic renewable energy generation, followed by an additional two-year action plan in 2019–2020 to support R&D in energy (NDRC 2019). Under the programme, the government will issue subsidies to encourage the construction of energy storage facilities and support trials of new storage technology testing pumped hydro storage, compressed air storage, magnetic energy storage and large-scale battery storage deployments. The Chinese government aims to maintain domestic manufacturing control over the whole supply chain for energy storage—from raw materials, battery technology and supercapacitors to micro-grid and smart-grid equipment.

While battery and storage technologies are critical elements of the energy transition, they are also a key component in electric mobility. Electric car manufacturers have been identified by the Chinese government as potential 'indigenous innovation' rivals for incumbent auto-manufacturers and with strong incentive schemes, China in 2018 was the largest electric vehicle market globally (IEA 2019) with a stock of 2.3 million units; it accounted for almost half of the global electric cars, having sold 1.1 million electric cars that year. China is also the largest market globally for electric two- and three-wheelers and for low-speed electric vehicles (LSEVs), estimated at 5 million units. Yet the EV fleet consumed an estimated 58 terawatt-hours (TWh) of electricity in 2018, with China accounting for 80% of that demand.

These latest data points highlight a number of unique facets of cleantech diffusion in China: regardless of whether Chinese companies are leaders or followers in innovation, strong government backing allows them to commercialise new products rapidly. Moreover, disruptive innovation has emerged in China from lower tech products, including electric two-wheelers and LSEVs, defying government policies (Zhou et al. 2018) and escaping the reach of high-tech innovation in large firms (Tyfield 2014). Chinese cleantech companies could therefore export a wide array of products to both developed and less affluent markets.

2.3   How Green Are the Belt and Road?

One case in point is China's Belt and Road initiative (BRI), the country's multibillion-dollar infrastructure investment plan. Overseas investments under the auspices of the BRI have the potential to transform global development towards lower carbon alternatives, as seen in the above examples, but it could also lock in high-carbon growth in developing countries. Ironically, while China's domestic oversupply of solar panels led to a fall in global prices (and related concerns about China's trade practices) through 2016, the country's record-breaking solar buildout in 2017 absorbed much of the country's module output, leading to higher prices for international customers (IEEFA 2018).

Recent research highlights that between 2000 and 2016, 66% of power sector lending from Chinese banks went into coal projects. China was involved in 240 coal-fired power projects in 25 of the 65 Belt and Road countries, with a total installed capacity of 251 GW. Chinese financial institutions have committed or offered funding for over one-quarter (102 gigawatts GW) of the 399GW of coal plants currently under development outside China (IEEFFA 2019), including investment in export coal mines, coal-fired power plants, and the associated rail and port infrastructure in countries including Vietnam, South Africa, Pakistan and Indonesia. Yet, there is a growing divergence between state-backed financing and private funding. From 2014 to 2017, more than two-thirds of China's energy-sector loans and investments (through banks, policy banks and state-owned companies) were in fossil fuels (Zhou et al. 2018), while nearly two-thirds (64%) of cross-border energy-sector investment by Chinese privately owned enterprises were in renewable energy.

Going forward, China's global energy investments will likely be determined by a number of factors, including the state of the domestic market, companies' desire to expand their global footprint, and the pull from host countries (Downs 2019), but barring a radical shift in global demand—i.e. that host countries push back against investments in coal-fired power plants as they have rejected hydropower investments—China's energy exports will both help and hinder the global energy transition.

A final factor determining China's exports will be its ongoing tussle with the US. As the US–China trade war escalates into a deeper geopolitical and commercial rivalry, China's clean energy investments, especially to the extent that they could begin to set global standards in energy management for example, could become a source of geopolitical friction.

2.4 The US and China: A Quest for Technological Dominance

The trade war between the US and China has highlighted a paradigm shift in US–China relations that is increasingly shaping China's energy policies and in extension, the global ramifications of its energy transition. The US Department of Defense's latest Indo-Pacific Strategy paper, for example, highlights the 'geopolitical rivalry between free and repressive world order visions' (US Department of Defense 2019). Officials in the US administration including Vice President Mike Pence and Secretary of State, Mike Pompeo have issued increasingly hawkish speeches regarding China's economic statecraft and accusing it of becoming increasingly aggressive and destabilising. This has vindicated those in China that have argued that the US is aiming to contain China's rise, and while the 'decoupling' narrative is not endorsed by Chinese officials, advisers and strategists (Wang 2019) are contemplating what a commercial Cold War could look like (Meidan 2019).

Up until the trade war, closer US–China relations led to an integration of goods, capital, technology and people, with a view that this economic integration would mitigate security competition. But the trade war and the US ban on Chinese telecom giant Huawei in May 2019 are threatening to decouple supply chains, especially those that use sensitive technology. Indeed, even if the Huawei ban is lifted, from China's perspective, the US's ability to cut off tech companies from their supply chains has been made abundantly clear.

A similar case in point is the US designation of Chinese companies for trading with Iran. US sanctions on two subsidiaries of China's largest shipping company, COSCO Shipping Energy, in October 2019 led Western companies to shun all COSCO assets, regardless of the ownership structure, for fear of being caught up in US secondary sanctions. Given these precedents, governments and businesses must now prepare for the possibility that a Chinese supplier or partner with which they work could be sanctioned by the US government, offering initial glimpse of de-globalisation of trade flows and even of technology (Rosset 2019). The takeaway in China has been that it must indigenise as much as it possibly can. So, even if the 'Iron curtain' on technology (Paulson 2018) will not materialise, it is no longer a mere fantasy. The race between the US and China is increasingly beyond a competition for better quality technological innovation, but also for standard setting.

2.5   Controlling Critical Resources

The rising mistrust of the US has also made China acutely more aware of its reliance on global resources and supply chains. Oil and gas markets are seen to be dominated by Western companies and US financial institutions, while sea lanes of navigation are effectively dominated by the US. Not only do commodities transact in US dollars (for the most part) but also US naval supremacy—which in many parts of the world as a provision of global goods (i.e. uninterrupted flows)—is an existential threat from Beijing's point of view, as it could be used by the US as a means of cutting off vital supplies from China. China's growing appetite for imported energy has confronted the country with the insecurity of its energy supplies. The Chinese government has deployed vast resources in a bid to hedge against these vulnerabilities with varying degrees of success.

The energy transition, however, brings with it demand for new types of resources. Rapid growth in EVs has boosted demand for metals used in power batteries, such as lithium, nickel, manganese, cobalt, tungsten, magnesium and rare earth. Production of these metals is heavily concentrated among a relatively small number of companies and in a handful of countries. Roughly, half of the global cobalt supplies, for example, are located in the conflict-prone Democratic Republic of Congo (DRC). China relies on the DRC for around 80% of its cobalt imports, but it has also effectively established a monopoly over output in the DRC as well as over intermediate and refined cobalt (Gulley et al. 2019).

Rare earths are another case in point. While most of the 17 rare earth minerals are not geologically rare, they are expensive and polluting to mine and produce, leading the US to limit production and allowing China to rise to predominance over rare earth production since the 1990s. Beijing has also shown its willingness to use rare earths exports as a geopolitical tool when in 2012, it issued export restrictions on rare earth. To be sure, the subsequent rise in prices and fears about China's dominance led to investments in additional resources globally. Moreover, efforts are being made to create cobalt-free batteries, and only a small minority of wind turbines (less than 2% in the US) are built with rare earth elements. Some minerals can also be recycled, re-used and stockpiled, thereby further reducing their perceived scarcity (Quiggin, 2017).

Nonetheless, by taking the lead on renewables, China has improved its geopolitical standing in several respects. By producing more of its own energy, China is reducing its reliance on fuel imports and the risks of energy disruption which could put a brake on its economic ambitions. Its technological expertise in renewables has established it as a leading exporter of clean energy technology and could help cement the country's technological dominance.

Moreover, China's investments along the BRI, namely, its investments in power grids could shape the geopolitics of energy in new and profound ways. While the US has shaped and safeguarded fossil fuel trading routes, China could shape power networks going forward. Infrastructure links and the Internet may become new battlegrounds for influence and control between competing powers (IRENA, 2019) with inter-state electricity cut-offs becoming a foreign policy tool, akin to pipeline politics or sanctions.

Notably, however, electricity trading tends to be more reciprocal than trade in oil and gas. So cross-border electricity trading will create opportunities for regional cooperation, and the creation of 'grid communities'.

3   A Brave New World

China's emergence as a global economic power and energy consumer in the early 2000s altered the geopolitical landscape of oil and gas. The scale of Chinese demand growth boosted fossil fuel production and trading, but also raised alarm bells in Beijing about the strategic vulnerabilities associated with its strong appetite for imports. Concerns about import dependence were gradually overshadowed by the rising cost of environmental degradation. Yet as China has sought to rebalance its economic structure and rise up the industrial value chain, government-supported efforts to spur innovation are already turning Chinese companies into global leaders in the technologies underpinning the energy transition. Beijing has also been able to capitalise on these gains to become a global leader on climate. Yet, China's track record is extremely mixed. It is the fastest growing renewables market globally and is also the biggest market for electric mobility in the world, but coal is now and will remain for the foreseeable future an intrinsic part of the Chinese energy mix.

In its overseas investments, China is fuel-agnostic and technology-agnostic, willing to finance and sell both coal-fired power plants and clean energy equipment and solutions. The scope and speed with which China chooses to pursue its own energy transition will remain a key variable in the global energy shift. Indeed, China is unique in its efforts to decarbonise before it has fully industrialised. As such, China could electrify its energy use before it decarbonises power. Indeed, despite China's commitment to tackling toxic air quality and climate change, it is still committed to domestic growth and development. To the extent that coal can offer stable, secure and cheap energy, it will remain a component of China's energy mix.

As the gulf between the US and China deepens, these contradictions will become increasingly visible. China may look to slow its shift away from coal while also accelerating its efforts to become a global leader in clean technologies. As part of its industrial policy, Beijing will continue to promote the indigenisation of clean technology as well as research into disruptive innovation. Chinese companies may become both technology followers and leaders, given the size and scope of the market. Some innovation will be catered to domestic needs and may find welcoming export markets in developing countries but increasingly, as the competition with the US intensifies, China is already controlling supply chains and shaping the infrastructure critical to the energy transition.

References

Asian Infrastructure Investment Bank (AIIB) (2017) President Jin Liqun, Opening address as prepared for delivery at the Meeting of the AIIB Board of Governors, 16 June 2017, https://www.aiib.org/en/news-events/news/2017/20170616_002.html

Bloomberg New Energy Finance (BNEF) (2019) Clean Energy Investment Exceeded $300 Billion Once Again in 2018, 16 January 2019, https://about.bnef.com/blog/clean-energy-investment-exceeded-300-billion-2018/ accessed 25 October 2019

Boqiang L (2018) China is a renewable energy champion. But it's time for a new approach, World Economic Forum, https://www.weforum.org/agenda/2018/05/china-is-a-renewable-energy-champion-but-its-time-for-a-new-approach/

BP Statistical Review of World Energy (2019) https://www.bp.com/en/global/corporate/energy-economics/statistical-review-of-world-energy.html. Last accessed 28 October 2019

Cao X, Rajarshi A, Tong J (2018) Technology Evolution of China's Export of Renewable Energy Products. *International Journal of Environmental Research and Public Health*, 15(8). https://www.ncbi.nlm.nih.gov/pmc/articles/PMC6121901/. Last accessed 15 October 2019

ChinaPower (2016) "Is air quality in China a social problem?", February 15, 2016. Updated October 16, 2018. Accessed October 28, 2019. https://chinapower.csis.org/air-quality/

Jinqiang Chen 2017, "The Challenges and Promises of Greening China's Economy", Belfer Center Discussion Paper, https://www.belfercenter.org/publication/challenges-and-promises-greening-chinas-economy, accessed August 2019

Erica Downs 2004, "The Chinese Energy Security Debate", *The China Quarterly*, no. 177

Erica Downs, 2019, "China-Pakistan Economic Corridor Power Projects: Insights into Environmental and Debt Sustainability", https://energypolicy.columbia.edu/research/report/china-pakistan-economic-corridor-power-projects-insights-environmental-and-debt-sustainability, accessed 23 October 2019

Ely Adrian, Geall Sam, Dai Yixin (2019) Low carbon China: emerging phenomena and implications for innovation governance - introduction to the special section of environmental innovation and societal transitions. Journal of Environmental Innovation & Societal Transitions 30:1–5

Institute for Energy Economics and Financial Analysis (IEEFA) 2018, "China 2017 Review: World's Second-Biggest Economy Continues to Drive Global Trends in Energy Investment", http://ieefa.org/wp-content/uploads/2018/01/China-Review-2017.pdf

Sam Geall, 2017, "Clear Waters and Green Mountains: Will Xi Jinping Take the Lead on Climate Change?", *Lowy Institute Analysis Paper*, https://www.lowyinstitute.org/publications/clear-waters-and-green-mountains-will-xi-jinping-take-lead-climate-change, accessed 10 September 2019

Goron Coraline (2018) Fighting against climate change and for fair trade: finding the EU's interest in the solar panels dispute with China. China-EU Law Journal 6(1–2):103–125

Andrew L.Gulley, Erin A.McCullough, Kim B.Shedd, 2019, "China's domestic and foreign influence in the global cobalt supply chain", Resources Policy, pp. 317–323

IEA (2019), "Global EV Outlook 2019", IEA, Paris, www.iea.org/publications/reports/globalevoutlook2019/

IRENA, 2019, "A New World: The Geopolitics of the Energy Transition", http://geopoliticsofrenewables.org/assets/geopolitics/Reports/wp-content/uploads/2019/01/Global_commission_renewable_energy_2019.pdf

Joel Jaeger, Paul Joffe and Ranping Song, 2017, "China is Leaving the U.S. Behind on Clean Energy Investment", January 06, 2017, https://www.wri.org/blog/2017/01/china-leaving-us-behind-clean-energy-investment

Linster and Yang, 2018, "China's Progress Towards Green Growth: an international perspective", OECD Green Growth Papers, No. 2018/05, OECD Publishing, Paris, http://www.oecd.org/env/country-reviews/PR-China-Green-Growth-Progress-Report-2018.pdf, last accessed September 2019

Meidan M (2014) The Implications of China's Energy-Import Boom. Survival 56(3):179–200

Meidan M, Andrews-Speed P, Xin Ma, 2009 Shaping China's Energy Policy: actors and processes, Journal of Contemporary China, 18:61, 591—616, *Shaping China's Energy Policy: actors and processes. Available from:* https://www.researchgate.net/publication/248999566_Shaping_China's_Energy_Policy_actors_and_processes *[accessed Oct 28 2019]*

Miyamoto A, Ishiguro C (2018) "The Outlook for Natural Gas and LNG in China in the War against Air Pollution", OIES Paper, NG 139, https://www.oxfordenergy.org/wpcms/wp-content/uploads/2018/12/The-Outlook-for-Natural-Gas-and-LNG-in-China-in-the-War-against-Air-Pollution-NG139.pdf?v=79cba1185463, accessed October 2019

Myllyvirta L (2019a) Why China's CO2 emissions grew 4 percent during first half of 2019, Carbon Brief, 5 September. https://www.carbonbrief.org/guest-post-why-chinas-co2-emissions-grew-4-during-first-half-of-2019, last accessed Oct 28, 2019

NDRC (2016) 能源生产和消费革命战略 (Energy Supply and Consumption Revolution Strategy (2016–2030). http://www.ndrc.gov.cn/gzdt/201704/W020170425548780357458.pdf. Last accessed 10 October 2019

NDRC (2019) 关于促进储能技术与产业发展的指导意见 2019–2020年行动计划2019-2020 Action Plan to Guide the Technological and Industrial Development of Energy Storage Sector, http://www.ndrc.gov.cn/zcfb/zcfbtz/201907/t20190701_940747.html

P M, Dillon YZ, Zhou K (2019) Comparing the innovation strategies of Chinese and European wind turbine firms through a patent lens. Environmental Innovation and Societal Transitions 30:6–18

Paulson (2018) Remarks by Henry M. Paulson, Jr., on the United States and China at a Crossroads', 6 November 2018, http://www.paulsoninstitute.org/news/2018/11/06/statement-by-henry-m-paulson-jr-on-the-united-states-and-china-at-a-crossroads/

Quiggin D (2017) Scrapping the combustion engine: the metals critical to success of EVs, 28 July 2017

Ramaswami A, Tong K, Fang A, Lal RM, Nagpure AS, Li Yang, Huajun Y, Jiang D, Russell AG, Shi L, Chertow M, Wang Y, Wang S (2017) Urban cross-sector actions for carbon mitigation with local health co-benefits in China. Nature Climate Change 7:736–742

Rosset C (2019) Huawei Ban Means the End of Global Tech. Foreign Policy, 17 May 2019. https://foreignpolicy.com/2019/05/17/huawei-ban-means-the-end-of-global-tech/, accessed 19 September 2019

Tyfield D, Zuev D, Ping L, Urry J (2014) Low Carbon Innovation in Chinese Urban Mobility: Prospects, Politics and Practices. STEPS Working Paper 71

US Department of Defense (2019) Indo-Pacific Strategy Report Preparedness, Partnerships, and Promoting a Networked Region. Accessed 1 June 2019 from https://media.defense.gov/2019/Jul/01/2002152311/-1/-1/1/DEPARTMENT-OF-DEFENSE-INDO-PACIFIC-STRATEGY-REPORT-2019.PDF, p. 4

Wang O (2019) Chinese economists warn Beijing to prepare for decoupling from US. South China Morning Post, 7 July 2019. https://www.scmp.com/news/china/article/3017550/chinese-economists-warn-beijing-prepare-decoupling-us

World Bank (2019). https://data.worldbank.org/country/china?view=chart

Yamineva Y, Liu Z (2019) Cleaning the air, protecting the climate: Policy, legal and institutional nexus to reduce black carbon emissions in China. Environ Sci Policy 95:1–10

Zhou L, Gilbert S, Wang Y, Cabré MM, Gallagher KP (2018) Moving the Green Belt and Road Initiative: From Words to Actions, World Resources Institute. https://www.wri.org/publication/moving-green-belt-and-road-initiative-from-words-to-actions

Implications of the Global Energy Transition on Russia

James Henderson and Tatiana Mitrova

1 What Is Energy Transition and How Does It Affect Different Countries?

The current energy transition can be viewed as the fourth in a series of similar fundamental structural transformations of the global energy sector. V. Smil defines the first energy transition—from biomass to coal—as the period between 1840 and 1890 during which the share of coal in the energy balance increased from 5 to 50% (Smil 2018). The second energy transition is associated with the fast penetration of oil—its share grew from 3% in 1915 to 45% by 1975—and the third transition involved the partial replacement of both coal and oil by natural gas, the share of which increased from 3% in 1930 to 23% in 2017. All these transitions were driven by the comparatively higher economic efficiency of the new energy sources. Currently, however, as we are talking about the beginning of the fourth energy transition (from fossil fuels to low-carbon energy sources), the situation is quite different. The share of Renewable Energy Sources (RES) (excluding hydro) in total primary energy consumption in 2017 was 3%, but it is now expanding very quickly. In this fourth energy transition, in contrast to the previous three, a qualitative new driver is becoming critically important, namely, combating global climate change, which has led to the establishment of compulsory energy sector decarbonization targets.

In a more specific sense, energy transition is a translation of the German term "Energiewende", which came into international use in the early 2010s after the accident at the Fukushima nuclear power plant (OSCE 2013; Trüby and Schiffer 2018). As one of the most ambitious decarbonization projects for the energy sector on a

J. Henderson (✉)
Oxford Institute of Energy Studies (OIES), Oxford, England
e-mail: james.henderson@oxfordenergy.org

T. Mitrova
Energy Center, Moscow School of Management SKOLKOVO, Moscow, Russia
e-mail: mitrovat@me.com

© The Author(s) 2020
M. Hafner and S. Tagliapietra (eds.), *The Geopolitics of the Global Energy Transition*,
Lecture Notes in Energy 73, https://doi.org/10.1007/978-3-030-39066-2_5

national scale (aiming for a reduction of greenhouse gas emissions by 40% by 2020 and by 80–95% by 2050 from 1990 levels), "Energiewende" became a benchmark for large-scale climate-driven transformations around the world.

Today, energy transition is driven by a complex set of different drivers: climate agenda, technological progress and the availability of brand new technological solutions which are able to dramatically increase the efficiency of the energy sector and to change the traditional way that it functions. It has the means to satisfy the desire of all countries to ensure the competitiveness of their national economies and to boost development with affordable energy. Last, but not least, it taps into the need to increase energy security, which, obviously, corresponds to the geopolitical agenda.

Basically, there are several ways in which energy transition affects different countries:

(1) Direct influence:

- Countries which sign up to international climate agreements are supposed to comply with their official targets and obligations, changing their energy mix; they have no choice other than to develop new low-carbon strategies with a strong focus on RES, energy efficiency and other ways to reduce emissions.
- Global innovation and technological development makes many new technologies cheaper and more attractive, so often countries—driven by local stakeholders— opt to promote these technologies voluntarily in order to decrease the cost of energy and to sustain their economic competitiveness.

(2) Indirect influence (refers mainly to countries which are lagging behind the energy transition):

- The changing global fuel mix with a growing share of RES limits the demand growth for fossil fuels, thus resulting in lower than expected export volumes for coal, oil and gas from resource-rich countries.
- New rules are under discussion in certain parts of the world concerning carbon tracking of internationally traded goods and the creation of border carbon adjustments (BCAs) as part of the carbon taxation mechanism (Morris 2018; Mehling et al. 2017). A high carbon footprint for all exported goods might become a long-term source of instability for economies relying on fossil fuels.
- Banks and financial institutions are assessing climate risks and becoming more reluctant to provide financing for fossil fuel projects (UNEP 2019). This trend is most visible in the coal industry (BANKTRACK 2019), and it will create more obstacles for the further development of conventional energy in resource-rich countries.

2   Russia's Role in the International Energy and Climate Change Landscape and Energy Geopolitics

Despite the fact that Russia produces only 3% of the world's GDP and accounts for only 2% of the world's population, it is the third largest producer and consumer of energy resources in the world after China and the US, providing 10% of world production and accounting for 5% of world energy consumption. Russia consistently ranks first for global gas exports, second for oil exports and third for coal exports. With energy production of about 1470 mtoe, Russia exports over half of the primary energy it produces, providing 16% of global cross-regional energy trade, which makes it the absolute world leader in energy exports (ERI RAS and SKOLKOVO 2019). Its strategic behaviour regarding energy transition is therefore important not only for the country itself, but also for the rest of the world.

From a climate perspective, the country ranks fourth globally in terms of carbon dioxide emissions. Russia continues to rely on fossil fuels, while its GDP energy intensity remains high amid relatively low energy prices and high capital costs. The share of RES (solar and wind power) in the energy mix is negligible, and officially is not projected to rise above 1% by 2035.

From an energy geopolitics viewpoint, Russia is traditionally regarded as a resource-rich country and is accused of abusing its power as an energy (especially natural gas) exporter. In the Energy Strategy of the Russian Federation up to 2030, it is explicitly stated that, "energy exports should help to promote the country's external policy". Russia's use of geopolitical power in the field of energy increased during the 2000s, mainly due to Russia's use of its energy exports in its political relationships with post-Soviet states, as well as the strategic and economic effects of new Russian export pipelines to Central and Eastern Europe. It was mainly applied to influence the political behaviour of those countries purchasing Russian energy. However, the metaphor of an "energy weapon" is misleading: Russia has not used tough means of influence—the so-called 'hard energy weapon'—in the context of Western Europe (Tynkkynen et al. 2017). Nevertheless, the image of a dangerous, unpredictable player in the global energy geopolitics game has defined Russia for the last two decades.

3   The Direct Influence of Energy Transition on Russia

As mentioned above, there are two direct ways in which energy transition can influence different countries: first, through the official targets and NDCs set by international climate change agreements which influence national strategies to reduce carbon, and, secondly, by commercial decisions taken by the main investors and other stakeholders active in the country in order to take advantage of innovations and green technologies.

4 Russian Climate Policy and the Paris Agreement

Decarbonization is the main global driver of energy transition. Individual regions, countries or their associations set goals for reducing the carbon footprint in the energy sector and introduce mechanisms to stimulate this process—carbon taxes, emissions trading systems, etc. According to the World Bank Group (World Bank and Ecofys 2018), fifty-one carbon pricing initiatives have been implemented or are scheduled for implementation in regional, national and subnational jurisdictions. Consequently, during the period 2008–2017, the carbon content of electricity decreased by 50–100 gCO2/kWh in the European Union, US, Canada, China, Australia, Kazakhstan and many other countries (Staffell et al. 2018).

Despite the global trend, for many years, the climate agenda and drive for decarbonization were not essential factors in the economic and energy strategy of the Russian Federation. Russia signed the Paris Agreement in 2016, with voluntary obligations to limit anthropogenic greenhouse gas emissions to 70–75% of 1990 emissions by 2030, provided that the role of forests was taken into account as much as possible. But even with this very low target (which does not require any significant effort given the country's economic stagnation), Russia has not ratified the Agreement although it did finally join the Agreement in September 2019 (without official ratification, just by the decree of Prime Minister Dmitry Medvedev).

It is still unclear exactly what Russia's climate goals will be under the Paris Agreement. A few important milestones are envisaged in 2020, including the development of a Low-Carbon Strategy for Russia up to 2030 and beyond, and the adoption of a carbon regulation framework. But it is worth noting that there are still many strong opponents to the Paris Agreement and carbon regulation in Russia, including some representatives of the authorities, fossil fuel businesses and the scientific community. Currently, discussion is mainly focused on delivering good reports and 'greenwashing', as opposed to real climate action.

The Paris Agreement is mentioned only once in the draft version of "Russian Energy Strategy Up to 2035", a key document defining the country's strategic priorities in this critically important industry, which was submitted to the government by the Energy Ministry in 2015, but which has still not been officially approved. It states that "in 2016, the Russian Federation signed the Paris Climate Agreement, which included, among other things, the development by 2020 of a strategy of socio-economic development with a low level of greenhouse gas emissions for the period until 2050. In order to minimize possible negative consequences for the Russian fuel and energy complex from the implementation of this agreement, an extremely balanced approach is needed to take into account some additional regulatory measures to counter climate change" (Ministry of Energy of the Russian Federation 2017).

This very cautious approach towards decarbonization is driven by several factors:

- Scepticism concerning the anthropogenic nature of climate change dominates among stakeholders—senior representatives of the Russian Academy of Sciences as well as many state officials publicly express their doubts over the very concept of anthropogenically created climate change. Many experts, academicians and policymakers see it as a concept manufactured by the West to undermine Russia.

- Secondly, following the economic downturn and restructuring in the 1990s, Russia de facto reduced greenhouse gas emissions sharply (by about 30%). Between 1998 and 2008 emissions increased in line with GDP growth, while in the period 2010–2016 Russia's GDP has grown by 73%, while the level of emissions has increased by only 12% due to further economic restructuring and faster growth of the financial and other non-energy-intensive sectors (KOMMERSANT 2016). As a result, Russia is currently well within its emission limits due to the high initial starting point in 1990 and economic stagnation since then.
- Lastly, Russia's energy sector has a lower carbon footprint than many other countries, including Poland, Germany, Australia, China, India, Kazakhstan, the Arab countries of the Persian Gulf, the US, Chile and South Africa. Around 35% of electricity is generated by carbon-free Nuclear Power Plants (NPPs) and large hydropower plants, and 48% comes from gas (IEA 2018), with gas gradually displacing petroleum products and coal from the Thermal Power Plant (TPP) fuel basket (the share of gas in TPP electricity generation has increased from 69 to 74% during 2006–2017).

This background explains why Russia has sidestepped the global decarbonization trend for so long. Its participation in international environmental cooperation has always been determined primarily by its external policy objectives. In Soviet times, participation in global environmental initiatives was a channel of collaboration with the West. In the 1990s, it was a means of integration into the international community and one of the major areas of cooperation with the US. In the 2000s, Russia used the environmental agenda to gain trade-offs from Western partners along with attracting foreign investment (Makarov 2016). At present, an understanding of the possibilities to reap benefits from the country's natural capital is slowly increasing among Russian political and business elites, so in the longer term Russia's involvement in international environmental cooperation may be catalysed by an increasing need for international re-integration.

5   Businesses Promoting Green Technologies in Russia

Businesses in Russia currently seem to be more preoccupied by the climate and energy transition agenda than the authorities. Initially, export-oriented companies started to realize the threat of changing perceptions and regulations for their traditional business, in particular, steel and aluminium making companies, as well as paper and chemicals businesses. Producers are now hurrying to implement ESG reporting to please their investors and to develop "green products" (green steel, green aluminium, etc.) using new energy technologies and different offset mechanisms in order to remain competitive in their core export markets. From their perspective, the decarbonization of the Russian energy sector and the expansion of RES is needed urgently.

There is also a rising cluster of companies interested in RES development as their main market, namely, equipment producers for solar and wind farms. Several oligarchs (such as Anatoly Chubais and Victor Vekselberg) have entered this business and have now become its strongest proponents. Their commercial targets are related to export expansion and so they are extremely interested in cooperation with other countries which could become export markets for their equipment. The primary targets in this respect are the former Soviet republics which prefer not to further increase their dependence on China. Current volumes of trade are negligible, but it is possible that at some point this could create tension between Russia and China.

6 National Technology Policy

While not paying particularly serious attention to climate policy, Russia is on the other hand very sensitive to its rate of technological development. The country's leadership clearly realizes that Russia runs the risk of falling behind in the development of new energy technologies that have become standard in most of the rest of the world. This is the reason behind its strict requirements on equipment localization for renewable energy and smart grids, and numerous import substitution programs. Despite this realization, energy transition technologies are definitely not the main focus of Russia's technology policy. In the key state document which defines priorities in this area, (the State Program for "Energy Development" approved in 2014 and amended in 2019), only the "promotion of innovative and digital development of the fuel and energy complex" is mentioned as a target, together with many new technologies in hydrocarbon production and processing. Nothing at all is mentioned concerning low-carbon technologies (Ministry of Energy of the Russian Federation 2019a).

The desire to overcome the technological gap and to reduce the potential need for imports if energy transition becomes mainstream has created some interest among Russian authorities concerning technologies for energy transition. Several huge grant programs and networks have been created for this purpose (RVK, Energynet, etc.), but surprisingly they are mainly focused on digital technologies, rather than low-carbon ones. Russian authorities regard the digital transformation of the energy sector as a technological challenge (bearing in mind the high level of current import dependence for all high-tech equipment and the potential threat of sanctions, which could create serious problems for national energy security) and this is the reason why digitalization has become the main driver of energy transition in Russia. In 2018, Vladimir Putin signed a decree establishing a special state program for the creation of a "Digital Economy" in which energy infrastructure is mentioned as one of the key components. The Energy Ministry has also developed its special project called "Digital Energy" (Ministry of Energy of the Russian Federation 2019b) which is focused primarily on the digitalization of regulation and the creation of a whole institutional framework for a wide-scale introduction of digital technologies in the energy sector.

As the main drivers of this movement are the fear of technological lag, import dependence on foreign equipment and, even more importantly, software (especially given the threat of new sanctions in the energy sector), no large-scale international cooperation can be expected in this area. Indeed, protectionism and the creation of various trade and economic barriers is highly likely.

7 Indirect Influence

The indirect consequences of the energy transition are more obvious and sensitive for Russia. Any changes in the demand for fossil fuels result in lower energy exports, while the potential introduction of BCA (Border Carbon Adjustments) might become a threat for all Russian exports, and new rules of behaviour by investors could further constrain the availability of funding for Russian energy projects.

8 Energy Transition Limits Demand for Fossil Fuels and Constrains Russian Energy Exports

Energy transition affects regional energy balances, specifically when RES implementation starts to limit growth or reduce overall demand for fossil fuels. For example, rapid reduction in coal use in the EU energy balance threatens Russian exports to Europe, which have already dropped considerably during the last decade. The share of electric vehicles in key markets (China, the US, EU) is forecast to grow rapidly, which is likely to reduce the demand for petroleum products. In 2018 in India, solar energy (PV) was 14% cheaper than coal generation, while China will achieve network parity in 2020 which will reduce demand for imported pipeline gas and LNG, all of which will impact regional energy balances.

For Russia, as for many other resource-rich and energy-exporting countries, energy transition creates new long-term challenges and calls into question the sustainability of the whole economy, which is highly dependent on hydrocarbon export revenues. In the period from the beginning of the 2000s, Russia managed to increase its energy exports dramatically: from 2000 through 2005, exports grew by an unprecedented 56% (ERI RAS and AC RF 2016), exceeding the total energy exports of the USSR, providing an incredible acceleration of the national economy and underpinning the country's position on the international arena as an "energy superpower". But as the global financial–economic crises came in 2008, the growth in energy exports halted. The post-crisis years of 2011–2014 witnessed very high oil prices but stagnant export volumes, and a lack of petro-dollar revenues resulted in GDP stagnation at an oil price of 110 $/bbl, which was a clear evidence of deep structural economic problems. More recently, oil and gas export revenues have declined from the heights of 2008–2012 under the impact of falling prices for hydrocarbons. Nonetheless, even in 2017, hydrocarbons provided 25% of GDP and 39% of the country's federal budget, 65% of foreign earnings from exports, and almost a quarter of overall investment in the national economy (Trading Economics 2018).

Obviously, the energy transition affects the prospects for Russian fossil fuel exports, particularly coal and oil, but natural gas exports might also be significantly affected by a further increase in emission reduction goals. Indeed, economic modelling has shown that climate-related actions outside Russia could lower Russia's GDP growth rate by about a half a percentage point (Makarov et al. 2017). ERI RAS-SKOLKOVO analysis (ERI RAS and SKOLKOVO 2019) has demonstrated that the role of the fuel and energy complex in the Russian economy will continue to decline from its peak in 2012–2013, affected by shifts in world energy markets. The technological transition of the world energy sector from the dominance of fossil fuels to low-carbon energy resources could lead to a 16% reduction in fuel exports and an 8% reduction in primary energy production in Russia over the next two decades. In general, by 2040 this could reduce the value added by the fuel and energy complex by a quarter and value added by supporting enterprises by another 2–3%, due to a decrease in capital investments within the sector. As a result, average GDP growth in the country is forecast to slow down between 2016 and 2040 from 1.7 to 0.6% per annum and the share of the energy sector in Russia's GDP will decline from 25% to just 14%. This signifies the end of the dominance of the fuel and energy complex in the national economy during the Energy Transition.

There is little if any hope within Russia that this downward movement will be mitigated by internal factors. GDP growth projections have been revised downwards to 1–2% per annum due to ongoing systemic economic crisis, international financial and technological sanctions and an unfavourable investment climate. Gone are the years of high GDP growth (7–8% per annum) in the first decade of this century. Russia is feeling the impact of a stagnating economy, flat domestic energy demand, the necessity of keeping domestic regulated prices unchanged and insufficient investment in the deployment of new technologies. This situation, which limits investment capacity, is further compounded by the high cost of capital in the domestic financial market and the negative impact of financial sanctions.

As a result, the global rise in RES targets and the transition towards a decarbonized energy economy are regarded in Russia as a significant threat for export revenues and thus for Russian economic, and therefore political, security (GARANT.ru 2017).

9   Carbon Tracking of Internationally Traded Goods and The Creation of Border Carbon Adjustments (BCA) Challenge Russia's Non-energy Exports

Russia also faces the risk that market barriers for its exports of energy-intensive goods, which constitute currently 30% of exports, could be introduced. These restrictions are currently under discussion in Europe and in other parts of the world and might soon become an important component of international trade. Under these circumstances, Russian energy-intensive, export-oriented industries—initially the metallurgical and chemical sectors—might face significant problems in protecting, never mind expanding, their niche in export markets.

10   Difficulties in Attracting International Financing for Fossil Fuel Projects

Another important implication, which makes Russia's perception of energy transition so negative, is a further increase in the difficulty of attracting international financing for domestic fossil fuel projects. Many global banks and investment funds have already removed coal projects from their portfolios, and some have started to refuse to finance oil projects. Financial sanctions currently in place are already a serious burden for the development of Russian energy projects, and so further restrictions due to climate considerations would make life even more difficult.

11   Russia's Potential for Energy Transition and Its Geopolitical Implications

As shown above, the climate agenda is not a major policy issue in Russia, while the competitiveness of the national economy, as well as its energy security, is already protected by cheap abundant hydrocarbons (primarily natural gas). As a result, for Russia the only important driver for energy transition is technological progress and a desire to prevent the emergence of a strong technological gap between it and other countries. Nevertheless, despite this limited motivation to promote energy transition in Russia, there are some areas where the potential benefits are huge and which could create substantial value for the Russian economy, attracting considerable investment if proper regulation was to be put in place. These key areas for Russia are the following:

– Energy efficiency,
– Renewables,
– Decentralization and distributed energy resources, and
– Hydrogen.

12 Energy Efficiency

Factors relating specifically to Russia—the cold climate, the vast distances, a huge endowment of natural resources, poor economic organization and marked technological backwardness—have resulted in its economy having a high level of energy intensity, 1.5 times higher than the world average and that of the US, and twice that of the leading European countries (ERI RAS and AC RF 2016). Across practically all industrial sectors, there is a substantial energy efficiency gap compared not only with the best available technologies but also with current performance in other countries. Even with comparatively low fuel and energy prices, the share of fuel and energy costs in overall production costs in Russia is higher than in developed and many developing countries (Bashmakov 2013). Before the 2009 economic crisis, Russia was one of the world leaders in terms of GDP energy intensity reduction rates, and the gap between Russia and developed countries was narrowing dramatically. A 40% reduction of GDP energy intensity within ten years was achieved between 1998 and 2008; however, since 2009 this reduction has slowed down and even reversed.

Obviously, for such an energy-intensive economy, issues such as energy efficiency and conservation are key for any Energy Transition plan: according to analysis from the IEA 30% of primary energy consumption and enormous amounts of hydrocarbons (180 bcm of gas, 600 kb/d of oil and oil products and more than 50 Mtce of coal per annum) could be saved in Russia if comparable OECD levels of efficiency were to be achieved (IEA 2011). A significant reduction in the growth of energy consumption could be provided by structural energy conservation (changing the industrial and production structure of the economy), with an increase in the share of non-energy-intensive industries and products. The next most important factor in constraining the growth of energy consumption could be improved technical application of conservation processes which could provide a total energy saving of 25–40%. However, it will be extremely difficult to close the gap with OECD countries—it is actually widening due to a lack of investment in processes which could quickly renew assets or improve energy efficiency. If we also include ongoing administrative barriers and, most importantly, a lack of availability of 'long money' and of credits for energy efficiency projects for small market participants, coupled with the persistence of relatively low natural gas prices in the long term, then Russia could well remain stuck in a state of high energy intensity. Strong policies are required to change this pattern, accompanied by a substantial increase in energy prices.

Unfortunately, at present, there appears to be little incentive for anything to change, unless Russia can develop energy efficiency technologies which could then be exported to the rest of the world. Internally, it would seem to be too politically risky either to raise prices or to force spending on energy efficiency at a time of economic stagnation.

13 Renewable Energy Sources

Russia's energy balance is strongly dominated by fossil fuels, with natural gas providing 52% of total primary energy demand, coal providing 18% and oil-based liquid fuels a further 18%. Carbon-free sources of energy in Russia are represented primarily by large-scale hydro and nuclear (which enjoys strong state support). The role of solar, wind, biomass and other sources of renewable energy is negligible—less than 1% of the total supply (ERI RAS & AC RF, 2016). The total share of RES (including hydro, solar, wind, biomass and geothermal) in Russia's total primary energy consumption was just 3.2% in 2015. By the end of 2015, total installed renewable power generation capacity was 53.5 gigawatts (GW), representing about 20% of Russia's total installed power generation capacity (253 GW) with hydropower providing nearly all of this capacity (51.5 GW), followed by bioenergy (1.35 GW). Installed capacity for solar and onshore wind by 2015 amounted to 460 MW and 111 MW, respectively (IRENA 2017).

According to the draft Energy Strategy of Russia for the period up to 2035 (Ministry of Energy of the Russian Federation 2017), the share of renewable energy in Russia's total primary energy consumption should increase from 3.2 to 4.9% by 2035. This includes Russia's approved plans to expand its total solar PV, onshore wind and geothermal capacity to 5.9 GW by the end of 2024. The foundation for the growth of RES deployment in Russia is Decree 449, passed in 2013, which created a legal framework to establish a renewable energy capacity system in the country. The decree is designed to encourage the development of renewable energy in Russia, particularly focusing on wind and solar photovoltaics, and to a lesser extent, small-scale hydropower. Under the new regulatory system, energy developers of projects with an output of at least 5 MW can bid for capacity supply contracts with Russia's Administrator of the Trading System via annual tenders. Winning suppliers are paid both for the capacity they add to the energy system and for the energy they supply, based on long-term, 15-year contracts with fixed tariffs. This regulation sets a predictable legal and regulatory environment that allows developers to commercialize capacity as a separate commodity to the power itself and ensures the economic attractiveness of these projects for investors. In return, RES developers are expected to ensure they can provide the promised capacity, within the right timescale and with sufficient localization of equipment (Power Technology 2018).

Since then annual renewable capacity additions have risen from 57 MW in 2015 to 376 MW in 2018 (320 MW solar, 56 MW wind). What is more significant, however, is the significant decline of capital expenditure in the renewables auctions which have taken place during the last two years: by 35% for wind and by 31% for solar, according to the Energy Ministry (Ministry of Energy of the Russian Federation 2019b). This process was not smooth, as some capacity auction rounds have struggled to attract bids. Just over 2GW of renewable capacity was awarded in tenders between 2013 and 2016, while the 2017 auction resulted in a total of 2.2GW of wind, solar and small hydro awarded in a single round. In 2018, 1.08GW of capacity was allocated between thirty-nine projects. Additionally, in 2017 five waste-to-energy projects were introduced to the capacity market scheme, with a total capacity of 335 MW. But in 2018, the tender for waste energy capacity failed, due to strict new requirements for bidders to provide performance guarantees.

As technological policy is the main driver of Russia's interest in renewables, the country is focused on building its own RES manufacturing capacity. Russia has set a relatively high level of local content which is required to qualify for the highest tariff rates, an essential component of many Russian RES projects' long-term feasibility. The percentage of Russian-made equipment required to avoid tariff penalties was relatively modest in the early days of the auction system but has now risen to 65% for wind farms and small hydro, and 70% for solar until 2020, with the long-term target level of localization set by the government at 80%. These high levels have been behind the failure of several tenders, especially in wind farm development, for which there has been little to no Russian-made equipment proposed. The requirements have encouraged foreign firms to partner with Russian power companies and manufacturers. Several international joint ventures have been established including Fortum and state-owned technology investor Rusnano's wind investment fund, and WRS Bashni, a partnership between Spanish developer Windar Renovables, Rusnano and Russian steel firm Severstal. Wind equipment was localized by Vestas Manufacturing Rus in the Nizhny Novgorod region, while SGRE (Siemens-Gamesa Renewables) and Lagerwey are also entering the Russian market (Power Technology 2018).

The problem is that the current support mechanism will expire in 2024. Russia's unambitious RES targets and ambitious localization targets will be nearly fulfilled by this time, and the influx of foreign renewables developers might stop if no new incentives for the RES market are created. But in order to create these incentives, the Russian government needs to first confirm the long-term role of renewables in its energy balance, which is rather difficult to do without a decarbonization agenda.

However, it seems that as a country with the world's largest natural gas reserves and the second-largest reserves of thermal coal, Russia does not see any real value in transitioning from fossil fuels to zero-carbon energy sources. Despite the country's massive potential in wind and solar resources and the virtually limitless land available for development, the availability of oil, gas and coal is suppressing clean energy development. Diversifying this energy mix towards carbon-free energy sources is a challenging task: low prices for hydrocarbons, the unfavourable geographical dispersal of potential renewable resources from the point of their utilization (they are mainly concentrated in non-populated areas with a long distance to the centres of consumption) and their comparatively high costs (caused mainly by the mandated requirements on localization, which often results in uncompetitive per-unit costs), are hindering the development of these energy sources in Russia.

According to IRENA (2017), Russia theoretically has the potential to increase its share of renewables from 4.9 to 11.3% of total primary energy consumption by 2030. However, without a reassessment of its energy strategy priorities and a wider transformation of its energy system, this will not be achieved. As a result, it seems that the only real incentive to develop any form of new technology will be the export market. If Russia could export renewable electricity or technology it might be worth pursuing, but for domestic use it is a real challenge, unless it becomes significantly cheaper. In addition, for the country as a whole, the real economic necessity is to maintain export revenues, which has resulted in enthusiasm for potential hydrogen export to Europe, as this is where the real incentive seems to lie.

14 Russia's Decentralization and Distributed Energy Resources Potential

Historically, the Russian energy system has always developed in an extremely centralized way. The Russian electricity sector, for example, relies on a huge centralized power system, while distributed energy resources, including microgrids based on renewables, are developing slowly and only in remote and isolated areas. Russia has one of the world's largest national centralized power systems with single dispatch control; as of 2017, the total length of its trunk networks was over 140 thousand km, its distribution networks were over 2 million km and the installed capacity of power plants was 246.9 GW. This energy system was created and historically developed on a hierarchical basis with centralized long-term planning bodies. For decades, the centralized model has been, and remains, the basis of its energy strategy. The role of distributed generation has historically been significant only in remote areas of the Far East, Siberia and the Arctic, which are too expensive to connect to a single network. However, the penetration of distributed energy resources (DER) into the centralized system has now begun, with potentially significant consequences for the incumbent actors.

Decentralization in the power sector began when the role of economies of scale in power generation globally ceased to be significant due to technological improvements. The catalyst for this change was the emergence in the 1980s of gas turbines and reciprocated gas engine technologies. It was the reciprocated gas engine global market that showed steady growth rate (CAGR 17%) until the late 2000s (Diesel and Gas Turbine Worldwide 2006). In the US, distributed generation has played a role in the electric power sector for several decades (Rhodium Group 2017). Historically, these DERs have consisted of dispatchable resources; however, the recent increase of non-dispatchable PV capacity marks a change in this trend. BNEF's forecast shows that the decentralization ratio will exceed 15% (as it did in Germany in 2017) in eight countries by 2040 (BloombergNEF 2017). Globally, annual distributed generation capacity additions have already exceeded centralized ones, and non-generation types of DER have even more potential than distributed generation. The estimated potential for Demand Response and Energy Efficiency in the US in 2014 (37 GW) was higher than for CHP (18 GW) or Solar (8 GW) (Rhodium Group 2017). In line with other countries, the integration of DER into the Russian electricity sector became noticeable in the 2000s, but in the past seventeen years it was limited to distributed generation only. The development of this process in Russia is driven not by global climate change or energy independence concerns, but by the economic considerations of the largest electricity consumers. Almost all the big Russian industrial companies (including oil and gas industry leaders like Gazprom, Rosneft, Lukoil, Novatek and Sakhalin Energy) are involved in distributed generation projects in order to get a more affordable power supply. Meanwhile, micro-generation using renewables for households in Russia is still largely confined to enthusiasts. There are just a few cases evident in the regions, all of them stimulated almost only by economic expediency reasons.

However, non-generation types of DER in Russia are only in the very early stage of development. Demand response technologies began to develop in the country in 2016–2017, but only a small proportion of power consumption is affected (for example, 54 MW in the second price zone of the wholesale power market, or 0.1% of total capacity in this zone). Demand response in retail electricity market is in its experimental stage, and energy efficiency policies have not yet achieved significant results. According to I. Bashmakov (Bashmakov 2018), GDP energy intensity in Russia in 2017 is just 10% lower than in 2007, a disappointing outcome given that the initial energy efficiency target set in 2008 was to reach a 40% decline in GDP energy intensity by 2020. Substantial federal budget subsidies were allocated but very limited change has occurred, and as a result the initial target has been significantly scaled down to 9.41% and federal funding has been discontinued (Ministry of Economic Development of the Russian Federation 2018).

Nevertheless, DER has significant potential in Russia. According to a study by SKOLKOVO Energy Centre (Khokhlov et al. 2018), it has the potential to cover over half of the needs for generating capacity (about 36 GW by 2035) if even a small part is actually utilized. The most promising type of DER in Russia is distributed co-generation (~17 GW). On-site self-generation units owned by electricity consumers can provide an additional 13 GW, demand response another 4 GW, energy efficiency technologies a further 1.5 GW and rooftop PV systems up to 0.6 GW. Indeed, if DER is fully utilized, the Skolkovo analysis shows that the entire generation gap in 2035 could be covered, although this would require significant acceleration from today's levels and a major push by the Russian government to introduce a favourable policy framework. However, although DER has been widely analysed in international literature, in Russia there has been no integrated assessment of its potential in response to future development needs of the national power system and as a result progress is likely to be slower than might be hoped.

In order to stimulate the maximum utilization of DER technologies, systemic changes are necessary in the architecture and policy of the Russian power sector, balancing the interests of new players with the existing model. An optimal combination of centralized generation and DER will need to be found, but in order to find such an outcome it will be necessary to develop principles and market mechanisms for the integration of centralized and decentralized parts and to ensure their reliable joint operation. The Russian authorities are some way from achieving this at present.

15 Nuclear

Russia is one of the world's largest producers of nuclear energy. Its nuclear industry has a role to play as an existing giant with decarbonization credentials. Russia is recognized for its nuclear expertise, and it is pursuing an ambitious plan to increase sales of Russian-built reactors overseas. Currently, it has thirty-nine reactors either under construction or planned overseas.

Moreover, Russia is attempting to create a new breakthrough in nuclear technology: nearly, all of the world's *reactors* operate with thermal (slow) *neutrons*, while *Russia* has developed *fast neutron reactors, through which it hopes to take the significant step of closing the fuel cycle.* Currently, Russia is a world leader in fast neutron reactor technology and is consolidating this through its *Proryv* ('Breakthrough') project. Starting in 2020–2025 it is envisaged that fast neutron power reactors will play an increasing role in Russia, with substantial recycling of fuel. Indeed, fast reactors are projected to account for some 14 GWe of capacity by 2030 and 34 GWe by 2050. If successful this new technology platform envisages a full recycling of fuel, balancing thermal and fast reactors, so that 100 GWe of total capacity requires only about 100 tonnes of input per year, from enrichment tails, natural uranium and thorium, with minor actinides being burned. About 100 t/yr of fission product waste would go to a geological repository (World Nuclear Association 2019).

16 Hydrogen

Russia remains isolated from international communities and partnerships which are developing hydrogen technologies and there is no national hydrogen program, and only in the very end of 2019 the first attempts to coordinate various research groups and interested parties appeared. At the same time, there are many resources in Russia capable of producing hydrogen, and there are a number of R&D activities in this area (most, however, far from commercialization) and also some prospective domestic demand niches for hydrogen. There is also design work and scientific development in the areas of production, storage and transportation of hydrogen, as well as its use in mobile transport. In addition, Russia has enormous potential to produce hydrogen and export it on a global scale. Therefore, hydrogen technologies are being spoken about in a positive way both at the largest Russian forums and within the largest Russian companies (Melnikov et al. 2019).

On the production side, there are proven technologies for producing "grey" hydrogen in Russia, similar to those used elsewhere in the world. They are deployed at oil and gas processing plants, metallurgical plants, etc. (methane conversion). All hydrogen produced is used on-site—for example, to improve the quality of hydrocarbon processing. Furthermore, Gazprom and Rosatom are working on technologies for producing hydrogen with a minimum carbon footprint by using adiabatic conversion of methane (Aksyutin et al. 2017) and high-temperature nuclear reactors

(Ponomarev–Stepnoi et al. 2018). These technologies are at a preliminary scientific research stage or (in the case of adiabatic methane conversion) being tested at an experimental laboratory unit. In addition, Russian developers are also conducting laboratory tests on hydrogen generation by aluminium–water reaction[1] and fuel processors for the conversion of natural gas and diesel into a hydrogen-rich fuel mixture[2] and the release of pure hydrogen from it.[3] The Kurchatov Institute and various research centres at the Russian Academy of Sciences are also engaged in scientific research in the field of electrolysis.

However, the present work on the transportation and storage of hydrogen is less developed because it currently tends to be consumed at the place of production. Gazprom, the owner and operator of Russia's gas transmission system, has conducted studies showing the possibility of adding up to 20% hydrogen to transported natural gas, but real experiments have yet to be conducted.

Nevertheless, the resources for hydrogen production in Russia are huge, mainly because the country has such vast hydrocarbon reserves and wind potential. In addition, existing gas transportation infrastructure (including new gas pipeline projects) and a growing natural gas (LNG) industry could provide a foundation in the long term for the development of "blue" hydrogen production and export given the low cost of raw materials and the possibility of hydrogen transportation both via pipelines and in a liquefied form.

According to Gazprom estimates, transporting hydrogen via export gas pipelines could entail a risk of violating long-term contractual obligations related to gas quality and would necessitate additional investments in the gas transmission system. Therefore, the company is considering an alternative, namely, the production of hydrogen from natural gas at the consumer site after methane has been transported through the trunk pipelines. Gazprom has valued the European market for hydrogen produced in this way at 153 billion euros by 2050, according to Bloomberg.

From a Russian perspective, it is also important to note that production of "blue" hydrogen from steam methane reforming can actually be a relatively green option because its generation industry has one of the lowest carbon footprints in the world. Gas-fired thermal power plants dominate in the generation structure (around 48%), while nuclear power plants (18%) and hydroelectric power plants (17%) exceed the share of coal-fired power plants (16%). As a result, the carbon content of electricity produced in Russia is less than in the US, China, Australia, India, Japan, Germany and other countries. In certain regions, particularly where hydro and nuclear dominate, this creates opportunities for the production of what is effectively "green" hydrogen via electrolysis with electricity supplied from the regional electricity grid, without the development of solar and wind power. This can give Russia a significant potential cost advantage. This has prompted interest from international players. For example, Kawasaki Heavy Industries plans to revisit a feasibility study on the export

---

[1] Joint Institute for High Temperatures of the Russian Academy of Sciences, JIHT RAS.

[2] "Central Research Institute of Ship Electrical Engineering and Technology" ("Central Research Institute SET"), Krylov State Research Centre.

[3] http://www.niiset.ru/index.php/vodedprod.

of hydrogen produced in the Magadan region to Japan. Although this project has not yet received a development go-ahead, interest in such initiatives is likely to increase as the infrastructure develops in the Far East and the cost of hydrogen electrolysis and logistics technologies goes down.

Even greater opportunities could open up for Russia if its renewable energy potential is ultimately realized. Currently, the share of green hydrogen produced at RES plants (electrolysis) is nearly zero. But although the share of wind power in Russia's energy balance is currently insignificant (less than 1%), the total potential from this sector is estimated at 17.1 thousand TWh, which is sixteen times higher than total generation in Russia in 2018. As a result, many studies of the global potential of "power-to-x" technology refer to Russia as one of the "hidden champions" as its huge potential is currently offset by a lack of interest from the state and the stakeholders.

17   Conclusions on Geopolitical Implications for Russia

It is clear from the analysis above that the global energy transition towards a lower carbon system presents some real threats for Russia. Perhaps the most obvious is financial, with lower hydrocarbon rents meaning lower budget revenues and slower economic growth, with implications for government spending and the wealth of the Russian population at large. This could have implications abroad, if reduced military spending limits Russia's hard power, and at home, if the political regime is undermined by its ability to satisfy the welfare demands of its population. Furthermore, these problems could be exacerbated by the fact that Russia may have a weaker position in international financial markets as restrictions on the availability of capital for carbon-intensive industries may well be increased. In addition, even Russia's non-energy exports may be impacted if carbon tax adjustments are made for imported goods in key markets. The combination of all these factors could weaken Russia's global negotiating position, which could be further undermined by the increased use of renewables in countries where Russia has previously exercised leverage through energy exports. For example, Russia's position in Southern and Eastern Europe is likely to be weakened as those countries become less reliant on imported energy and are able to diversify away from Russian oil, gas and coal. Equally, countries in NE Asia, where Russia is hoping to gain an increasing foothold, thanks to oil and gas exports, could also become less engaged with the Kremlin as their energy needs increasingly focus on alternative sources with lower carbon intensity.

However, despite the presence of these clear threats to Russia's geopolitical status, there are also reasons for optimism, thanks to the country's huge potential as a carbon sink and as a potential developer of new technology. Firstly, it is possible to envisage that Russia could become a leader in the sale of decarbonized oil and gas, in particular, because it has huge potential in forestry. If managed properly, then reforestation could be used by oil and gas producers as a means to offset the GHG emissions in the supply chain and also possibly the CO_2 emissions at the burner tip. Indeed, some oil and gas companies (Lukoil for example)[4] are already considering how this strategy might be used to offset their carbon impact in order to improve their global bargaining position. Secondly, Russia has huge wind power potential, especially in the Arctic, and if improvements in technology could allow DC lines to be connected to major consumers in European and China, then Russia could become a major exporter of green electricity (The Moscow Times 2019).

Thirdly, Russia is also attempting to develop a unique competence in a new generation of nuclear technology based on a closed nuclear fuel cycle. If it can become a world leader in this field, which could transform the outlook for nuclear energy, it could provide a massive carbon-free energy source for non-OECD countries where energy demand continues to grow rapidly. It would therefore allow Russia to build deep relationships with economies in Africa, the Middle East and South-East Asia, expanding its geopolitical influence significantly. Finally, and more traditionally, Russia could also exploit its gas resources to increase its presence in new market niches, such as the bunker market where LNG could become a much more desirable fuel following the introduction of stricter emissions rules by the International Maritime Organisation (IMO) in 2020.

In addition to these energy-related issues, Russia could also benefit from the opening of the Northern Sea Route as the Arctic ice continues to melt. Although the opening itself would actually be caused by global warming, one key benefit from increased utilization of a shorter route between Europe and Asia could be to reduce the carbon footprint of transport between the two regions, thus limiting further emissions impacts and potentially providing Russia with another source of geopolitical influence, given the importance of this emerging transport route. The implementation of appropriate national decisions on fuel regulation and environmental requirements would be required to maximize this potential as a "green" transport option, but it is certainly possible to see this new source of bargaining power for Russia emerging over the next decade.

18   Overall Conclusion

In the light of Russia's current position as a major hydrocarbon exporter and consumer, any notion of a rapid energy transformation is problematic. A large-scale shift from hydrocarbons to renewable energy sources provides energy consumers with more choices, meaning that Russia's control of energy flows becomes a less effective instrument of (geopolitical) power. Furthermore, since the Russian state budget is highly dependent on energy export revenues, a major change in this sector will have a negative impact in many other parts of the economy, including military funding. Lastly, although Russia has plenty of potential for renewable energy, the country does not have a heavy focus on the sector at present and is therefore unlikely to pioneer technological development in the wind and solar industries. Perhaps not

---

[4]See LUKOIL CSR policy at https://csr2017.lukoil.com/pdf/csr/en/hse.pdf.

surprisingly then, although Russia is involved in international climate policy, it does not work to promote it and instead and has to date used diplomacy to influence international energy and climate policy in a way that rather discourages change. One key reason for this inactivity is the fact that political power in Russia is ultimately linked with the control of strategic resources (most importantly, hydrocarbons) and the export revenues derived from these resources (Tynkkynen et al. 2017).

Russia's attitude towards Energy Transition is therefore quite controversial: while acknowledging some of the key trends, the country is basically refusing to accept the consequences of its main driver—decarbonization—and is focusing only on attempts to develop technological expertise in its usual centralized manner. Nevertheless, at a certain point the country will have to develop a long-term vision for both domestic energy market development and export strategy in order to adapt to the profound transformation of the global energy system.

However, the domestic market environment is not conducive to the development of transition-friendly energy sources. The institutional environment is too rigid, there is not enough capital available and there is a level of cynicism as to whether it is really needed. The key question is whether the issue of export revenues can be a catalyst, although there are a number of issues in this regard.

Firstly, the real threat from decarbonization concerns exports to Europe, where gas demand in particular could suffer with potentially significant consequences for Gazprom. This has resulted in an initial interest in hydrogen technology, specifically methane pyrolysis, which allows the production of hydrogen and black carbon and where Gazprom claims to have a competitive advantage. However, it remains to be seen whether and how quickly this technology moves beyond the laboratory stage.

In contrast, other markets are less at risk, and as a result it would be a perfectly justifiable strategy to try and re-focus hydrocarbon exports towards Asia, Africa, Latin America, and this has indeed become a long-term goal. However, Europe cannot be ignored because it is too important for short-term revenues, but Russia's real energy transition strategy for the next 20–30 years may just be to become the lowest cost hydrocarbon supplier to emerging economies.

It would also seem that nuclear development could be a part of a transition strategy. It is of course controversial, but arguably Rosatom could become a key export company, not only providing Russia with additional revenues but also strengthening its geopolitical influence, especially in non-OECD countries which still have fast growing energy demand.

Russia may also attempt to establish itself as an equipment and software producer for green energy, but it remains an open question as to whether it can realistically hope to compete with China and the US on a technology front. As a result, its best option may be to make incremental improvements at home while encouraging the coal-to-gas switching model in emerging economies by continuing to offer low-cost gas (and to an extent oil) to markets where the cost of energy supply and improving air quality is more important than any CO_2 emission targets.

Finally, one area of competitive advantage in a decarbonizing energy world may be the potential for reforestation across Russia's huge geography. Companies are gradually waking up to this potential, as even if they do not believe in anthropogenic emissions as an issue themselves, they can see that a business advantage could be gained from addressing the problem as perceived by other countries. As a result, use of reforestation as a carbon offset mechanism, either for direct gain or to add "green value" to Russia's hydrocarbon exports, may become a growing theme over the next decade.

References

Aksyutin OE et al (2017) The contribution of the gas industry to the formation of an energy model based on hydrogen. Vesti gas science—scientific and technical collection. Environmental protection, energy saving and labor protection in the oil and gas complex. Special edition - 2017, p. 12

BANKTRACK (2019) Bank moves out of coal: A guide to the latest new banking sector commitments on reducing coal financing", https://www.banktrack.org/campaign/bank_moves_out_of_coal?fbclid=IwAR2AwpQfYxEbIl7hXdQMfqiQhzrWPVaEJb1HzaZtAe1c3ueBwUIstVFhb1g. Accessed 17 November 2019

Bashmakov I (2018) What Happens to the Energy Intensity of Russia's GDP? Ecological Bulletin of Russia 7:8 (in Russian)

Bashmakov I. (2013), "Driving Industrial Energy Efficiency in Russia", Center of Energy Efficiency (CENEf), Moscow, March 2013. Available at: http://www.cenef.ru/file/Idustry-eng.pdf (2013). Accessed 17 November 2019

Bloomberg New Energy Finance (2017) "New Energy Outlook"

Diesel & Gas Turbine Worldwide (2006), "30th Power Generation Order Survey"

ERI RAS & AC RF (2016), "Global and Russian Energy Outlook Up To 2040", Available at: https://www.eriras.ru/files/forecast_2016.pdf Accessed 17 November 2019

ERI RAS & SKOLKOVO (2019), "Global and Russian Energy Outlook 2019", Energy Research Institute of the Russian Academy of Sciences & The Energy Centre, Moscow School of Management SKOLKOVO, Moscow 2019. Available at: https://energy.skolkovo.ru/downloads/documents/SEneC/Research/SKOLKOVO_EneC_Forecast_2019_EN.pdf. Accessed 17 November 2019

GARANT.ru (2017), Presidential Decree of May 13, 2017 No. 208 "On the Strategy of Economic Security of the Russian Federation for the Period until 2030", Available at: https://www.garant.ru/products/ipo/prime/doc/71572608/ (in Russian). Accessed 17 November 2019

International Energy Agency (IEA) (2018) Electricity Information. IEA/ OECD Publications, Paris

International Energy Agency (IEA) (2011) World Energy Outlook 2011. IEA/OECD Publications, Paris, p 2011

International Renewable Energy Agency (IRENA) (2017) "Renewable Energy Prospects for the Russian Federation (REmap working paper)", IRENA, Available at: https://www.irena.org/publications/2017/Apr/Renewable-Energy-Prospects-for-the-Russian-Federation-REmap-working-paper Accessed 17 November 2019

Khokhlov A., Melnikov Y., Veselov F., Kholkin D., Datsko K. (2018), "Distributed energy resources in Russia: Development Potential", SKOLKOVO Energy Centre, Moscow School of Management SKOLKOVO, October 2018. Available at: https://energy.skolkovo.ru/downloads/documents/SEneC/Research/SKOLKOVO_EneC_DER_2018.10.09_Eng.pdf Accessed 17 November 2019

KOMMERSANT (2016), "Favorable Climate" Kommersant Business Guide, 26 May 2016. Available at: https://www.kommersant.ru/doc/2988887 (in Russian). Accessed 17 November 2019

Makarov I.A. (2016), "Russia's Participation in International Environmental Cooperation", Journal Strategic Analysis, Volume 40, Issue 6. Available at: https://doi.org/10.1080/09700161.2016.1224062. Accessed 17 November 2019

Makarov I. A., Chen H. Y., Paltsev S. (2017), "Finding Itself in the Post-Paris World: Russia in the New Global Energy Landscape", MIT Center for Energy and Environmental Policy Research, No. WP-2017-022, Joint Program Report Series Report 324, December 2017

Mehling M., van Asselt H., Das K., Droege S., Verkuijl C. (2017), "Designing Border Carbon Adjustments for Enhanced Climate Action", Climate Strategies, December 2017. Available at: https://climatestrategies.org/wp-content/uploads/2017/12/CS_report-Dec-2017-4.pdf

Melnikov Y., Mitrova T., Chugunov D. (2019), "The hydrogen economy: a path towards low carbon development", SKOLKOVO Energy Centre, Moscow School of Management SKOLKOVO, June 2019. Available at: https://energy.skolkovo.ru/downloads/documents/SEneC/Research/SKOLKOVO_EneC_Hydrogen-economy_Eng.pdf. Accessed 17 November 2019

Ministry of Economic Development of the Russian Federation (2018), "State Report on the State of Energy Savings and Energy Efficiency in the Russian Federation in 2017" Moscow, (in Russian)

Ministry of Energy of the Russian Federation (2019a) State Strategy "Energy Development", (in Russian). https://minenergo.gov.ru/node/323 (2019). Accessed 17 November 2019

Ministry of Energy of the Russian Federation (2019b), "Presentation on the results of the Fuel and Energy Complex functioning in 2018 and its targets for 2019", Moscow, (in Russian)

Ministry of Energy of the Russian Federation (2017), "Draft Energy Strategy Up to 2035", Available at: https://minenergo.gov.ru/node/1920 (in Russian). Accessed 17 November 2019

Morris A. C. (2018), "Making Border Carbon Adjustments Work in Law and Practice", Tax Policy Center, Urban & Brookings Institution, 26 July 2018. Available at: https://www.brookings.edu/wp-content/uploads/2018/07/TPC_20180726_Morris-Making-Border-Carbon-Adjustments-Work.pdf

OSCE (2013) Energy Concept for an Environmentally Sound, Reliable and Affordable Energy Supply, Germany's Federal Ministry of Economics and Technology, Federal Ministry for the Environment, Nature Conservation and Nuclear Safety, 24 April 2013. https://www.osce.org/eea/101047 Accessed 17 November 2019

Ponomarev—Stepnoi NN et al (2018) Nuclear energy technological complex with high temperature gas cooled reactors for large scale environmentally friendly hydrogen production from water and natural gas. Gas Industry, 11, 2018

Power Technology (2018) Is Russia finally ready to embrace renewable energy? https://www.power-technology.com/features/russia-renewable-energy/ Accessed 17 November 2019

Rhodium Group (2017) What Is It Worth? The State of the Art in Valuing Distributed Energy Resources. https://rhg.com/research/what-is-it-worth-the-state-of-the-art-in-valuing-distributed-energy-resources/ Accessed 17 November 2019

Smil V (2018) Energy and Civilization: A History. MIT Press

Staffell I, Jansen M, Chase A, Cotton E, Lewis C (2018) Energy Revolution: Global Outlook. Drax, 5 December 2018.https://www.drax.com/energy-policy/energy-revolution-global-outlook/

The Moscow Times (2019) Murmansk launches construction of Russia's largest wind power park, 24 September 2019. https://www.themoscowtimes.com/2019/09/24/murmansk-launches-construction-of-russias-largest-wind-power-park-a67406

Trading Economics (2018) Russia GDP Growth Rate. https://tradingeconomics.com/russia/gdp-growth. Accessed 17 November 2019

Trüby J, Schiffer HW (2018) A review of the German energy transition: taking stock, looking ahead, and drawing conclusions for the Middle East and North Africa. Energy Transitions, Volume 2, Issue 1–2, pp. 1–14, December 2018. https://doi.org/10.1007/s41825-018-0010-2 Accessed 17 November 2019

Tynkkynen V, Pynnöniemi K, Höysniemi S (2017) Global energy transitions and Russia's energy influence in Finland, Policy Brief. 19/2017. https://tietokayttoon.fi/documents/1927382/2116852/19_Global+energy+transitions+and+Russias+energy+influence+in+Finland.pdf/67f16b9c-daa7-445d-9ad6-76a6c1b30a99/19_Global+energy+transitions+and+Russias+energy+influence+in+Finland.pdf?version=1.0

UNEP (2019) Global Launch of the UN Principles for Responsible Banking, Press-release, 24 September 2019. https://www.unenvironment.org/news-and-stories/speech/global-launch-un-principles-responsible-banking

World Bank & Ecofys (2018), "State and Trends of Carbon Pricing 2018", World Bank, Washington, DC, May 2018. Available at: https://openknowledge.worldbank.org/bitstream/handle/10986/29687/9781464812927.pdf?sequence=5&isAllowed=y

World Nuclear Association (2019) Nuclear Power in Russia. https://www.world-nuclear.org/information-library/country-profiles/countries-o-s/russia-nuclear-power.aspx. Accessed 17 November 2019

A Fine Balance: The Geopolitics of the Global Energy Transition in MENA

Robin Mills

1 Introduction

The Middle East and North Africa (MENA) region[1] presents a curious challenge for the global energy transition. It remains the global centre of oil exports and is also a very important gas-exporting zone. Fast-growing populations, energy-intensive industrialisation, a series of oil-led booms, and provision of subsidised fuels, have led it to be amongst the highest energy-consuming regions in the world, whether per capita or per unit of GDP. Its infrastructure, economies, political systems and international relations have been profoundly shaped by hydrocarbon wealth, even in those regional countries with a smaller or no hydrocarbon endowment.

At the same time, MENA, with hot and mostly semi-arid or arid climates, and large concentrations of coastal urban development, is particularly vulnerable to climate change. Weak and/or repressive states are experiencing continuing political upheaval and conflict. Although climate change has not yet been a key driver or shaper of these conflicts, it could become an increasing future stressor, particularly in combination with economic decline driven by a loss of hydrocarbon rents. MENA, though, has excellent potential for low-carbon energy through renewables (mostly solar), and through a geological and industrial endowment suited for carbon capture, storage and use (CCUS).

---

[1]Definitions vary; here the Middle East includes the six Gulf Cooperation Council (GCC) countries (Bahrain, Kuwait, Oman, Qatar, Saudi Arabia, United Arab Emirates), Yemen, Iran, Iraq, Syria, Lebanon, Jordan and Israel/Palestinian Territories, while North Africa includes Egypt, Libya, Tunisia, Algeria and Morocco with Western Sahara.

R. Mills (✉)
Qamar Energy, Dubai, UAE
e-mail: robin@qamarenergy.com

© The Author(s) 2020   115
M. Hafner and S. Tagliapietra (eds.), *The Geopolitics of the Global Energy Transition*,
Lecture Notes in Energy 73, https://doi.org/10.1007/978-3-030-39066-2_6

The MENA countries are bound together by shared languages, religions and history. Yet, there are also profound differences between them, including their immediate neighbourhood, geography, colonial legacy, hydrocarbon endowment, non-hydrocarbon resources, level of economic development, size of sovereign wealth holdings, ethno-sectarian make-up, political system, ideology, international alliances and others. These idiosyncratic factors, as well as different choices and personalities in their political and business leadership, have shaped so far significantly different approaches to the global energy transition.

The current 'energy transition' globally can broadly be understood as a response to the imperative of climate change. Previous transitions, as from wood to coal, and from coal to oil and gas, have been driven by the availability, lower cost and improved convenience and utility of the new energy source. The current transition includes the attempt to promote low- or non-carbon energies (renewables, including solar, wind, biomass, hydropower, geothermal and others; nuclear; carbon capture and storage); non-carbon energy carriers (mostly electricity, but also hydrogen), for transport (electric vehicles), energy storage (batteries, thermal storage), industry and other uses; and improved energy efficiency and productivity (Bazilian and Howells 2019). The attempt to move away from fossil fuels has also been driven by local environmental and social impacts, by moves by consuming countries to increase energy security and reduce dependence on possibly unfriendly or unstable hydrocarbon exporters, as well as by (largely unfounded) concerns about resource scarcity. The energy transition also includes the quest for universal access to modern energy. It is not simply the introduction of new technologies, but also involves the changes in markets, institutions and regulations that allow or are induced by technological changes.

The energy transition brings with it shifts in corporate and national power. The geopolitics of the transition have begun to be considered, for instance, by the International Renewable Energy Agency (IRENA) (IRENA 2019), and other scholars.[2] In general, this work so far sees a strongly negative impact of the transition on current major fossil fuel exporters, including many MENA states as well as countries such as Russia, Venezuela and Nigeria.

In MENA, this transition is also bound up with a geoeconomic transition, which centres on the shift of oil and gas markets away from traditional customers in North America, Europe and developed Asia, and towards developing Asian countries, notably China and India. This is accompanied by trends in political power, notably the possible diminution of the US role in the region (particularly in the Persian/Arabian Gulf), the greater self-assertiveness of regional powers including Saudi Arabia, the UAE, Turkey and Iran, and the rising involvement of Russia and in future likely China and India. The MENA region faces these developments against the backdrop of global trends, such as a turn against free trade, the rise of populist economic and social policies, and a fracturing into geopolitical blocs, all further complicating the transition (Bazilian et al. 2019).

---

[2]For example, https://www.belfercenter.org/publication/geopolitics-renewable-energy, http://www.dieterhelm.co.uk/energy/energy/burn-out-the-endgame-for-fossil-fuels-2/.

MENA countries' responses to the energy transition revolve around four axes. The first is the restructuring of their economies to cope with lower prices for now, and permanently reduced hydrocarbon rents in the long term. The second is the attempt to safeguard the future of their hydrocarbon industries. The third is a gradual but accelerating move to retool their domestic energy systems for a lower carbon future. And the fourth is to deal with the geoeconomic transition.

2 The Nature of the Regional Energy Economy

The MENA region contains the majority of the world's oil reserves: 48.3% of world oil reserves in the Middle East, with a further 3.7% in North Africa (BP 2019). Its position in gas is less dominant but still very important, with 38.4% of world gas reserves in the Middle East and 4% in North Africa. More significantly, its resources are predominantly in large, conventional high-quality reservoirs with well-developed infrastructure and close to export routes, resulting in much lower production costs than the big but costly resources of US shale/tight hydrocarbons, Canadian oil sands and Venezuelan extra-heavy oil.

As a result, MENA provided 38% of gross world oil exports in 2018, much larger than any other region (Russia, with 12.8%, was next). It supplied 32.8% of world LNG exports (from a total 431 billion cubic metres (BCM), but a smaller share of gas exports by pipeline (4% from the Middle East, mostly within the region, and 5.4% from North Africa, mostly to Europe, from global international pipeline trade of 805.4 BCM). Even though North Africa's share of world exports is relatively small, it is important in the European market by supplying 13% of the continent's natural gas consumption and 10% of its oil.

The region has also become a much more significant consuming centre in its own right. It accounts for 17.3% of world gas consumption, 10.9% of oil consumption, 7.7% of carbon dioxide emissions from fuel combustion and 6.1% of electricity generation (BP 2019), although comprising only 5.7% of world population[3] and 3.8% of GDP.[4] This has been driven by a number of factors: the paucity of other traditional energy sources (hydropower and coal); the hot, arid climate with a high requirement for air-conditioning and desalination; the oil-driven economic boom of 2003–14; policies of energy-intensive industrialisation (oil refining, petrochemicals, aluminium, steel, cement, ceramics); and low, subsidised prices for energy which have encouraged inefficiency and waste.

The region remains almost entirely dependent on hydrocarbons for electricity (Fig. 1). However, this is likely to change substantially in the coming decades, as discussed in Sect. 6. The limited use of hydroelectricity today is virtually all in Iran, Iraq, Egypt and Morocco; that of nuclear in Iran; and coal in Israel and Morocco.

This hydrocarbon bounty is distributed unevenly across the region (Fig. 2). Some states (mostly the GCC and Libya) have small populations and large resources;

---

[3] World Bank, 2018 or closest available year.
[4] Nominal, 2018 or closest available year.

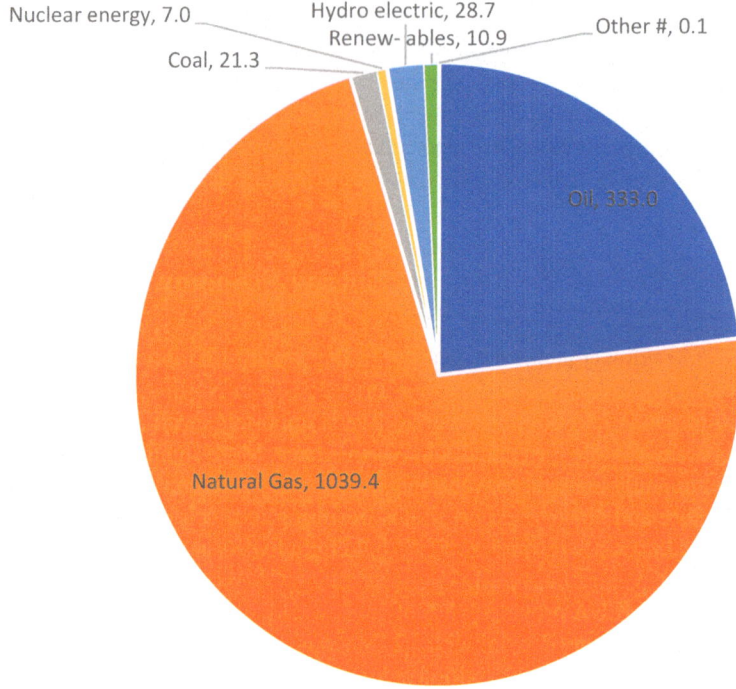

Fig. 1 Power generation by source, Middle East + Egypt (TWh/year 2018). *Source* BP Statistical Review of World Energy 2019

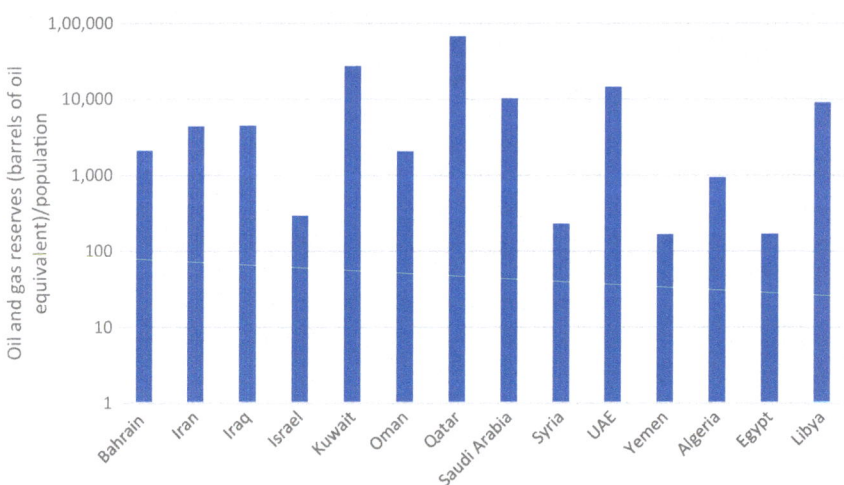

Fig. 2 Hydrocarbon endowment per capita (log scale). *Source* BP Statistical Review of World Energy 2019; Wood Mackenzie. Includes only total proved reserves of oil and gas; oil reserves include gas condensate and natural gas liquids (NGLs) as well as crude oil; Other countries not included due to negligible proved reserves

this is even more so when considering only the citizen populations, given shares of expatriates around 90% in Qatar (Snoj 2019) and the UAE (CIA 2019). Some, such as Iraq, Iran, Algeria, Yemen, Syria and Egypt, have significant resources but also relatively large populations. And others, notably Lebanon, Jordan, Morocco and Tunisia, have very little or no hydrocarbon production. Israel is in an anomalous position by virtue of its recent discovery of major gas fields, and its relative political and economic isolation from its neighbours. Yet even the oil and gas importers of the region are linked to their hydrocarbon-exporting neighbours by flows of labour, remittances and investment (Mohaddes and Raissi 2019).

The region is not highly integrated in trade or energy terms. This is due to the rather similar nature of exports and comparative advantage between most of the region states; political and security barriers; and a significant degree of protectionism.

Despite long negotiations, particularly with the UAE and Oman, Iran has not been able to start gas exports to its Gulf neighbours, due to unrealistic commercial expectations, political disputes, sanctions and high domestic use. The Arab Gas Pipeline (AGP) from Egypt to Jordan, Syria and Lebanon, and a related pipeline to Israel, the Arish–Ashkelon pipeline,[5] functioned from 2008 to 2012 but were disrupted by sabotage in Sinai following the 2011 Egyptian Revolution, and gas shortages within Egypt itself. The most successful regional pipeline, Dolphin, runs from Qatar to the UAE and Oman, and started deliveries in 2007 to Oman and 2008 to the UAE (EIA 2011). Its operations have not been interrupted by the boycott of Qatar imposed by the UAE, Saudi Arabia and Bahrain in June 2017, but it has not been expanded either. The GCC Interconnection Authority, for electricity, has also continued to function, but it trades electricity only on an emergency or quantity basis, not on market terms, and utilisation has only been around 5–6% of capacity (KAPSARC 2018).

In the absence of reliable access to pipeline gas despite the proximity of huge resources, Kuwait (2009), Dubai in the UAE (2010), Abu Dhabi in the UAE (2016), Egypt (2015) (TAQA Arabia 2019), Jordan (2015) (The Jordan Times 2016), Israel (2013) (MEES 2013), Bahrain (2019) and possibly in future Sharjah in the UAE (2020), Lebanon (2021), Saudi Arabia (no firm date), Morocco (perhaps as late as 2028) (African Energy 2019) and possibly Tunisia (Songhurst 2019), have turned to LNG imports. Although far from economically optimal, this has reflected a pragmatic approach.

Yet, more recently, there have been some more welcoming suggestions for energy cooperation, with contracts on gas exports between Israel, Egypt and Jordan; agreements on electricity interconnections of Iraq with Jordan and the GCC; and discussions of cross-border gas and carbon dioxide pipelines between Saudi Arabia, Bahrain, the UAE and Oman.

---

[5]https://www.sourcewatch.org/index.php/Arish%E2%80%93Ashkelon_Pipeline.

3   Regional Conflict and Weak States

MENA has been a region of particular geopolitical volatility and contestation since the Second, perhaps even the First World War. Scholars continue to dispute the causes, but MENA has seen an unusually high number of inter-state and, more recently, intra-state conflicts, while other areas that were troubled in the 1960s and 1970s, such as Latin America and south-east Asia, are now relatively stable and peaceful.

Outright MENA wars include the various Arab–Israeli wars; the Lebanese civil war (1975–1990); the Iran–Iraq war (1980–88); Saddam Hussein's invasion of Kuwait and US-led expulsion (1990–91); the Algerian civil war (1991–2002); civil war in Syria (2011-present) with various foreign interventions; the Libyan revolution and subsequent civil war (2011-present); civil war and regional intervention in Yemen (2015-present); and the US invasion of Iraq, subsequent civil war and conflict with the 'Islamic State' (2003-present). Accompanying these has been a level of insurgencies, territorial disputes, revolutions and coups (notably in 2011's 'Arab Spring'), severe international sanctions (at various times, on Iran, Iraq and Libya) and state repression. Major protests, sometimes to the stage of enforced changes of leadership, recurred in 2019 in Algeria, Sudan, Lebanon and Iraq.

Oil and gas resources have been the direct target of some of these conflicts, most clearly with the invasion of Kuwait. Oil rents have helped underpin well-armed autocratic governments with a tendency towards military approaches to disputes. The importance of the region for the global economy, and the prevalence of small, wealthy states who have relied on external patrons for security, has led to repeated intervention by the US, Russia and European countries. Intra-state 'resource regionalism' (Mills and Alhashemi 2018) has encouraged conflicts and separatism over resource-rich areas, such as the Kurdistan region of Iraq. Regional states have also developed 'proxies' and non-state allies in their weaker neighbours. Most notably Iran has patronised Hezbollah in Lebanon and the Houthi movement in Yemen, but Saudi Arabia, the UAE, Qatar and Turkey have played varying roles in these countries as well as in Libya, Egypt, Syria and Iraq.

Resource wealth has in many cases underpinned the development of modern states. It has also tended to entrench the power of authoritarian elites, whether republican, military or monarchical. Under internal stress, these states have proved to be brittle, with state failure particularly extreme in Yemen, Libya, Syria and Iraq.

The theory of the rentier state was developed by the Iranian economist Hussein Mahdavy (Mahdavy 1970), the Egyptian economist and former prime minister Hazem Beblawi, and the Italian economist Giacomo Luciani (Beblawi and Luciani 1987). They argued that states with large natural resource endowments or other 'unearned' income (such as foreign aid) were not reliant on taxing their people, but instead served to allocate or distribute the revenues, and that this would create a large unnecessary bureaucracy while reducing the incentives for hard work and government accountability.

Rentierism may be one part of the wider phenomenon of the 'resource curse',[6] which has been used to describe the various negative effects of a heavy dependence on natural resource revenues: a lack of competitiveness in the non-resource economy and in manufacturing exports; slower overall GDP growth rates; an overvalued exchange rate; high macroeconomic volatility; weak and corrupt institutions; a tendency to authoritarianism and foreign and civil–military conflict, and a deficit in democracy, human rights and gender equality.

Some work (Brunnschweiler and Bulte 2006) has challenged the very existence of the resource curse; dismissed it as confusing the direction of causality (poorer countries have a higher share of natural resources in GDP, not vice versa); argued that it applies mostly to oil (not gas, minerals or agricultural resources); suggested that it applies mostly to the post-1973 period of elevated oil prices (Ross and Andersen 2012)[7]; or argued that it affects states with poor-quality initial institutions, and that better institutions bring a 'resource blessing'. MENA countries also have a complex political and institutional history,[8] including the legacy of Ottoman and Western imperialism. Nevertheless, many of the posited negative effects of rentierism and the resource curse, along with some of the benefits, are highly visible in the region.

MENA conflict has repeatedly threatened world oil and gas supplies, going back to the 1951–1953 nationalisation of Iranian oil and the subsequent US/British-backed coup; the 1956 Suez crisis; 1973–1974 oil embargo; and the 1979 Iranian Revolution. More recently, the Libyan Revolution and the sanctions placed on Iran by the Obama administration pushed oil prices to a range of $100–110 per barrel. There have been repeated concerns over oil transit through the Strait of Hormuz in particular, and also through the northern and southern passages of the Red Sea, the Suez Canal and the Bab El Mandeb.

In contrast, tensions in the Gulf during 2018–2019 were severe, triggered by the Trump administration's decision to abandon the Joint Comprehensive Plan of Action (JCPOA) nuclear deal with Iran, followed by sanctions intended to cut off its oil exports entirely. Yet, the September 2019 missile and drone attack on Saudi Arabia's Abqaiq oil processing plant, and other strikes blamed on Iran or its allies, as well as attacks on Iran's own shipping, had by October 2019 had virtually no lasting impact on oil prices, despite briefly taking out half Saudi production capacity.

This more relaxed attitude by markets can be explained by a negative outlook on the world economy and hence oil demand; by the availability of oil in strategic storage; by the confidence in US shale producers to ramp up production quickly in the event of a price spike; and, perhaps, to a sense that oil is losing its geopolitical importance in the longer term, as non-oil alternatives enter the mainstream.

Historically, and even up to 2008, OPEC cartel power, oil supply disruptions and fears of 'peak oil supply' were a strong justification for countermeasures by the developed countries: the International Energy Agency, founded in 1974 (IEA

---

[6]Reviewed in https://www.jstor.org/stable/25054077?seq=1#page_scan_tab_contents.

[7]With regards to transitions to democracy, https://papers.ssrn.com/sol3/papers.cfm?abstract_id=2104708.

[8]See, for instance, Lapidus (2018), *A History of Islamic Societies*, Cambridge University Press.

2019), and its coordination of strategic stocks; programmes of alternative energy (nuclear, coal, biofuels and the beginnings of modern wind and solar); high fuel taxes; efficiency standards and conservation; and incentives for new oil development (Alaska and the North Sea in the 1970s and 1980s) and unconventional fossil fuels. With lower world oil prices post-2014, and a much more relaxed attitude to future supply, environmental drivers have supplanted security concerns as the main impetus for support for electric vehicles and other non-oil technologies.

4   Economic Restructuring and Resilience

For all the talk of renewables in the energy transition, the most intense effect on MENA oil and gas has so far come from competition from another hydrocarbon producer—the US. The rise of shale and tight oil and gas output has led the US from being the world's biggest oil importer as recently as 2013, to being the world's largest producer, ahead of Saudi Arabia and Russia, and at times a net oil exporter during 2019. Simultaneously, in seeking to export its surplus of cheap gas, the US also challenges Qatar and Australia as the largest LNG exporter over the next few years, with Russia coming up in fourth place. Finally, abundant low-cost feedstock has also caused a renaissance in US petrochemical production, a further challenge to Gulf industry.

In late 2014, oil prices fell from $100 per barrel to as low as $26 in January 2016, and have only partially recovered since to around $60 per barrel as of October 2019, despite the virtual elimination of Iranian and Venezuelan exports. After a surge in LNG prices in 2018 (Singapore LNG Index reached $11.66 per million British thermal units (MMBtu) in August 2018[9]), primarily due to China's coal-to-gas switching, spot prices fell back in mid-2019 to around $4 per MMBtu, though long-term contract prices remained higher.

This has put profound pressure on fiscal and current account balances for the MENA hydrocarbon exporters (Fig. 3). Investment budgets and subsidies have been cut, while non-oil revenues have been raised by introducing various fees and taxes, such as value-added tax (VAT) in some of the GCC countries. Budget gaps have been plugged by debt issuance, both at the sovereign level and by state firms, often the national oil company, with Saudi Aramco, Abu Dhabi National Oil Company (ADNOC) and Petroleum Development Oman (PDO) all floating sizeable bonds.

However, there has been relatively little progress in reducing the large public-sector wage bill, or on privatisation. Indeed, protests in countries such as Oman in the wake of the Arab revolutions in 2011, and in Iraq in 2019, have been met with promises of even more fiscally unsustainable state handouts. The much-heralded proposed initial public offering (IPO) of Saudi Aramco might realised $29.4 billion for a 1.7% stake, and the sale of 10% of ADNOC Distribution (the company's fuel retail arm) on the local market raised $851 million in 2017 (Gulf Business 2017),

---

[9]Energy Market Company Pte Ltd (EMCSG).

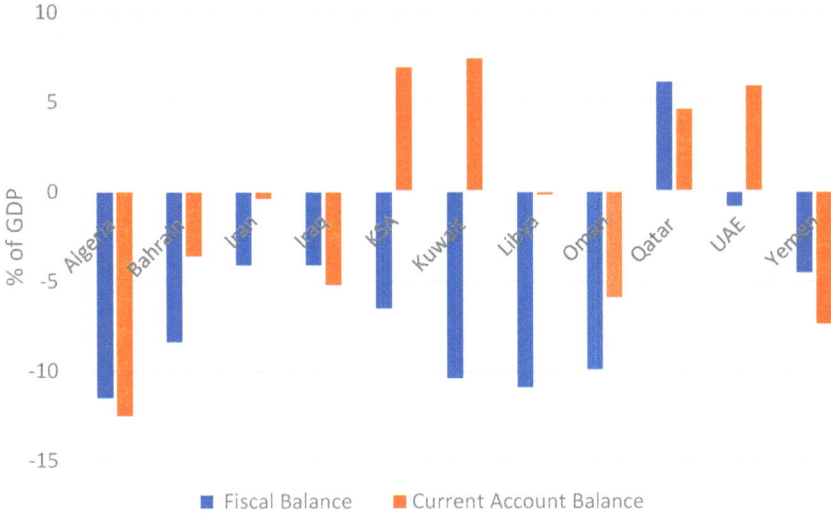

Fig. 3 Budgetary and current account balances, MENA oil exporters, 2019 estimate. *Source* Arabia Monitor; IMF

but these amounts are not particularly material in the big picture, and the companies involved are already well-run. Their importance is more for injecting a commercial mindset and stimulating progress in privatising other less effective entities.

Attempts at diversification from oil date back to the 1970s; in the case of Iran, to the 1960s. Diversification has passed through a number of eras (Hvidt 2013). In the 1970s, the fear was that resources would deplete in the relatively near future. The high oil prices of the time gave an abundant source of funds for investing in new industries, which of course were also tempting to bureaucrats and rent-seeking businesspeople. In the 1980s and 1990s, low oil prices created awareness of the need to find other sources of income.

In the 2000s, a renewed influx of petroleum revenues allowed investment in infrastructure, real estate and energy-intensive industries such as aluminium, steel, cement and petrochemicals, usually supported by cheap, subsidised energy. Saudi Arabia, in particular, developed more advanced petrochemical activities via firms such as Saudi Arabia Basic Industries Corporation (SABIC). Iran, under the pressure of sanctions, also built up industry to serve domestic needs and create export products that were less vulnerable than crude oil to interruption. Although not exactly a diversification away from energy, these industries do offer more technological sophistication and a reduced correlation with world oil prices.

Surplus revenues were used to build up large sovereign wealth funds, particularly in the UAE, Kuwait and Qatar. These have a mix of objectives: avoiding upwards pressure on the pegged currencies; saving to reduce macroeconomic and budget volatility driven by oil price changes; saving for a future of diminished oil revenues; and investing strategically in the domestic and international economy. At the same time, some states, notably the UAE, Oman and Bahrain, diversified into tourism, aviation, logistics, financial services and other non-oil activities, even if these did not contribute much to budgetary revenues[10] and remained largely dependent on the wider regional economy.

In the latest era of diversification, widespread regional political unrest (since 2011) and lower oil prices (since 2014) have again concentrated attention on the development of the non-oil economy. Another imperative has been added: the prospect of peak oil demand. Although most Gulf policymakers still do not expect this to be soon, a slowing of demand and falling prices within the next 20 years or so still represents an enormous challenge. In 2015, Abu Dhabi's crown prince Mohammed bin Zayed stated in a speech that, "In 50 years, when we might have the last barrel of oil, the question is: when it is shipped abroad, will we be sad? If we are investing today in the right sectors, I can tell you we will celebrate at that moment" (TheNational.ae 2015). In April 2016, Saudi Crown Prince Mohammed bin Salman (MBS) reflected on a much shorter timeline: "I think by 2020, if oil stops we can survive. We need it, we need it, but I think in 2020 we can live without oil" (QUARTZ 2016).

Most of the MENA countries have an 'economic vision' for some future date, 2030 or another. Saudi Arabia's 'Vision 2030',[11] launched under MBS, is one of the most recent and influential. Yet its attention to the energy sector is surprisingly limited, with more focus on technology, tourism and social transformation. The major energy goals outlined in the Vision 2030 are to maintain oil production capacity at 12.5 million bbl/day by 2020, increase raw gas production capacity from 12 billion cubic feet per day (Bcf/d) to 17.8 Bcf/d in 2020 (with a further target of 23 Bcf/d by 2027), increase domestic refining capacity from 2.9 million bbl/day to 3.3 million bbl/day by 2020, and phase out energy subsidies by 2025 (were previously planned to be phased out by 2020). This reflects a tension between the oil and non-oil sectors. Should hydrocarbon rents simply be a cash cow, funding investment into diversification? Or should the existing strengths of the oil sector be the foundation for a more value-added and environmentally sustainable energy industry, at home and abroad? In September 2019, this question seemed to be answered, at least for now, when experienced technocrat Khalid Al Falih was replaced as energy minister by MBS's half-brother, Prince Abdelaziz bin Salman, and as Aramco chairman by Yasir Al Rumayyan, MBS confidant and head of the key Public Investment Fund (PIF), with holdings in Tesla, Uber and other future-oriented companies.

The challenges of rapidly retooling entire economies remain enormous. In a recent paper, Steffen Hertog concludes that, "Even under ideal conditions, it will be impossible to become 'post-rentier' by 2030 and hard to imagine even by 2050. The maths are quite similar for other high-rent countries, including those of the GCC" (POMEPS 2019). While the Gulf countries have at least articulated a vision of diversification and taken some significant steps towards it, other MENA hydrocarbon-exporting

---

[10]Dubai being an exception.

[11]https://vision2030.gov.sa/en.

countries have struggled. Iran has already a fairly diversified economy, with sanctions pushing it towards further self-sufficiency and lower dependence on oil revenues, but at the cost of a severe recession. Iraq's cumbersome bureaucratic and sectarian system is straining to deliver results in the oil, gas and electricity sectors, but progress beyond these is very limited. The constant state of crisis management militates against longer term planning, despite the schemes laid out in studies by the World Bank (World Bank 2012) and International Energy Agency (IEA 2012). Algeria too struggles to sustain its petroleum exports, let alone diversify, with an excessively bureaucratic system and hostility to foreign investment. And the internationalised civil wars in Syria, Libya and Yemen of course make long-term economic progress impossible.

A further question is whether the expected fall in hydrocarbon revenues, at least relative to the rest of the economy, will reverse the 'resource curse' and rentierism described above.[12] More likely, this will not be automatic, and true diversification will rest on building up strong and capable institutions. Ultimately, less dependence on resource rents may diminish military competition and within-state conflict. But in the medium term, economic stress and a struggle over the remaining rents may lead to more violent confrontations, as with Saddam Hussein's decision in 1990 to invade Kuwait to try to solve his regime's financial problems (Karsh and Rautsi 2008).

5 Future-Proofing the Hydrocarbon Industry

5.1 Current and Future Challenges

The future of the MENA hydrocarbon industry reflects a delicate balance between meeting current challenges and ensuring its future resilience. The nature of this task, and the methods used, vary according to the exigency of each country's situation.

5.1.1 Climate

The primary challenge of the large GCC oil producers—Saudi Arabia, the UAE and Kuwait—is how to maximise the value of their resources in a climate-constrained world, where they face significant non-OPEC competition. The other big oil resource holders—Iraq, Iran and Libya—ought to be thinking similarly, but are battling political and security problems that prevent their planning effectively for the long term, as are Syria and Yemen. The lesser producers, such as Algeria, Egypt, Oman, Bahrain

---

[12]Addressed in, for instance, https://bruegel.org/2017/08/towards-eu-mena-shared-prosperity/.

and Tunisia, have in a way the more straightforward challenge of maximising output (or at least slowing declines), rationalising domestic consumption and controlling costs in the medium term, including new exploration, enhanced recovery and unconventional resources.

The major gas resource holders—Qatar and Iran—have a somewhat rosier situation, given the fuel's much lower carbon content and relatively clean environmental performance. Yet, in the long term, they too face the challenge of assuring demand: ensuring gas remains cost-competitive against renewables, while reducing or eliminating its carbon footprint.

Israel and, possibly, Lebanon, have to manage the transition to being gas producers and exporters, in a market constrained by limited local demand and tricky borders. Should debt-ridden Lebanon find significant amounts of gas, its corrupt and dysfunctional political system will be strained to use it effectively.

Finally, the non-producers, Morocco and Jordan, deal with issues more analogous to other energy importers, with their outlook improved by the growing availability of reasonably priced gas, and competitive renewables.

The region's energy-intensive industrial sector is also challenged by climate measures taken in major markets, particularly Europe. This could include border carbon taxes on imports from countries without an equivalent carbon price (Lowe 2019), or even outright bans on imports of oil, gas or other products with a carbon footprint in production that exceeds some specified limit.

Some regional countries have taken a relatively proactive approach to the climate challenge. Morocco has been a leader in deploying renewable energy and pioneering concentrated solar power. Abu Dhabi launched the Masdar clean energy initiative in 2006, which featured an intended 'zero-carbon' (now carbon-neutral) city, and domestic and international renewables investments. It competed hard to win the headquarters of the International Renewable Energy Agency (IRENA) in 2009 (TheNational.ae 2009a). The UAE made climate change an important part of its international diplomacy and backed the inclusion of carbon capture and storage (CCS), suited to its oil industry, in the Clean Development Mechanism, a move criticised by IRENA's first director-general Hélène Pelosse (TheNational.ae 2009b), in a reminder of some of the complexities in hosting international organisations. Dubai, through its 'Sustainable City',[13] electric vehicle and hydrogen pilots, and most practically by its world-record solar power prices, has also made the environment a key part of its brand and future orientation, despite the continuing high-carbon footprint and other negative environmental features of the Gulf model of urbanisation.

In contrast, Saudi Arabia has long had a rather negative and obstructionist approach to climate change negotiations (Depledge 2008), and at COP24 in Poland in December 2018, in alliance with Kuwait, Russia and the US, refused officially to "welcome" the Intergovernmental Panel on Climate Change's report on limiting warming to below 1.5 °C (Bradshaw et al. 2019). Saudi Arabia's position has been that it should be compensated for losses arising from restrictions on fossil fuels (TheNational.ae 2009c), and that climate policies should not 'unfairly' target oil.

---

[13] https://www.thesustainablecity.ae/.

Yet, as noted below, Saudi Aramco has implemented a number of policies to reduce its emissions and prove its business against climate policies.

Every MENA country except Libya and Yemen (for understandable reasons), Iraq and Iran has submitted a document on Nationally Determined Contributions (NDCs) towards meeting its Paris goals (Yemen, Iran and Iraq have submitted Intended NDCs). These generally include improvements in energy efficiency both in generation and use; gains in renewable and sometimes nuclear energy; fuel substitution; boosting public transport; reducing gas flaring and leakage; changing land use; and an overall reduction in greenhouse gas emissions below business as usual (BAU). For example, Iraq will cut 90 million tonnes from 2020 to 2035, 14% below BAU (UNFCCC 2015a); Iran will cut 4% below BAU unilaterally and up to 12% with international aid (UNFCCC 2015b) and Morocco will cut 42% below BAU (UNFCCC 2015c). Carbon capture and storage is mentioned by the UAE, along with carbon sequestration in marine and coastal environments (UNFCCC 2015d). Saudi Arabia also mentions CCS (UNFCCC 2015e).

Within the MENA region, Climate Action Tracker assesses only three countries against their Paris goals: Saudi Arabia (rated Critically Insufficient), the UAE (rated Highly Insufficient) and Morocco (rated 1.5 °C Paris Agreement Compatible).[14] In fact, Morocco is one of only two countries in the world rated so highly. Most MENA countries have major technical and economic potential for greenhouse gas cuts at relatively low (or even negative) cost, but sustained political will and commitment is lacking, though the region is far from unique in this regard.

MENA countries are themselves severely threatened by the effects of climate change, particularly given that most are already water-stressed and experience high temperatures. Urban development and important agricultural areas in coastal areas are threatened by sea-level rise and saline water infiltration. Weak regional states, and countries neighbouring MENA such as Afghanistan, Pakistan and the Sahel, may come under further stress as results of drought, desertification, groundwater depletion, upstream dams (notably on the Nile, Tigris and Euphrates), intensified summer heatwaves, interruptions and price spikes in world food trade, migration and other climatic impacts. These stresses may exacerbate the economic impact of falling resource rents (discussed below).

5.1.2 Peak Oil Demand

Recent years have seen growing attention to the concept of 'peak oil demand', that in the relatively near term, global oil demand will begin to decline, because of the expansion of electric vehicles, improvements in efficiency, environmental pressure on plastics and greenhouse gas emissions limits. Estimates for when this might occur range from the mid-2020s to the 2040s or even beyond (Fig. 4).

Yet, even the 2040s are not that far away, when it comes to the question of retooling an entire economy and MENA's expensive and long-lived energy assets. MENA

---

[14]https://climateactiontracker.org/countries/.

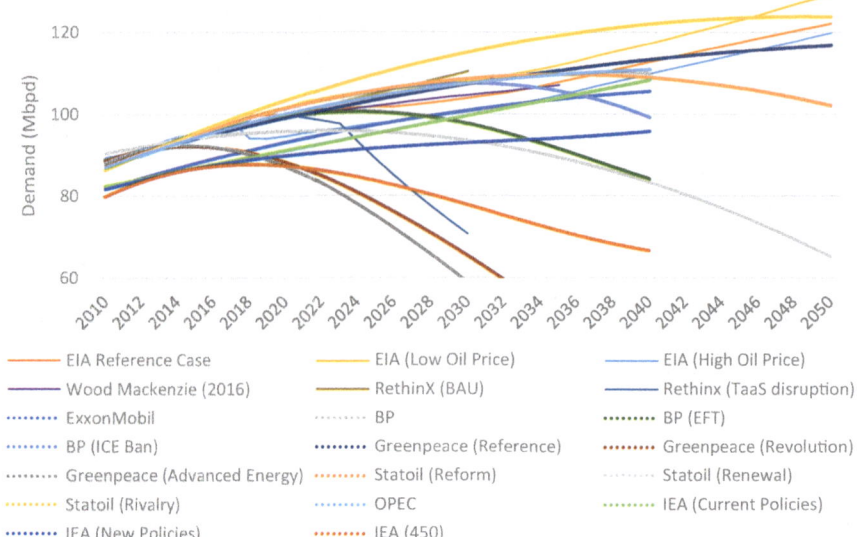

Fig. 4 Estimates for peak oil demand. *Source* International Energy Agency World Energy Outlook 2016; http://corporate.exxonmobil.com/en/energy/energy-outlook/download-outlook-for-energy-reports; BP Energy Outlook 2018; energy [r]evolution, Greenpeace https://www.greenpeace.org/archive-international/Global/international/publications/climate/2015/Energy-Revolution-2015-Full.pdf; Statoil; RethinX; Energy Information Administration; OPEC. RethinX does not give exact figures so they have been approximated from its graphs. Not all forecasts give figures for every year, so a smooth polynomial has been used to interpolate

national oil companies (NOCs) are aware of the peak oil demand question, but have generally not subscribed to the idea of an early peak, nor fully internalised its implications. The immediate pressure has been to survive the post-2014 period of 'low' prices, without the consideration that this might be a permanent state of affairs, or indeed that prices may be in a secular downtrend.

During 2015, Saudi Arabia and other producers substantially ramped up output to fight a 'price war' and gain market share, hoping to deter competitors. But non-OPEC production, particularly US shale, proved more resilient than expected. In 2016, they changed tack.

The 'Declaration of Cooperation' (OPEC 2016) in December 2016 between OPEC countries (in MENA: Saudi Arabia, Iraq, the UAE, Kuwait, Iran, Libya and Algeria; Qatar withdrew from OPEC in December 2018 (CNBC 2018)) and a group of non-OPEC producers, led by Russia and including Oman, Kazakhstan, Mexico and others, called for oil production cuts of about 3.5% by each OPEC country and lower cuts by the non-OPEC adherents to the 'OPEC+' or 'Vienna Group' pact. In the event, Saudi Arabia has substantially over-complied, Russia has under-complied, but the agreement has helped to raise prices and reduce excess inventories for now.

Even if US shale output slows into the 2020s, the question remains about the sustainability of this arrangement. Some other non-OPEC producers, such as Brazil and new entrant Guyana, are expanding, while promising shale resources in Argentina, Australia, Russia and elsewhere are attracting interest. Given decline rates from existing fields, a substantial amount of new oil will be required even in a world of peak demand. Prices could be volatile on the way down, and there will still probably be price spikes triggered by phases of underinvestment or security problems.

The low production costs of most MENA countries mean they will still be able to generate rents, though at overall reduced levels. Long-term production restraint risks 'stranding' large quantities of viable resources, while keeping prices elevated makes both competing production and non-oil technologies more viable. But Fattouh and Dale argue that OPEC countries' short-term budgetary needs mean they will be unable to pursue a high-volume/low-price strategy to maximise output and squeeze out competitors in the run-up to 'peak demand' (Dale and Fattouh 2018). Indeed, OPEC's own 2019 World Oil Outlook (OPEC 2019) expects its production of oil and other petroleum liquids to fall from 35 million barrels per day (bbl/day) in 2019, to 32.8 million bbl/day by 2024. The economic impact on oil exporters is determined less by the exact date of the peak, which will anyway be visible only in hindsight, and more by the pace of demand growth prior to the peak, the absolute volume at peak and the subsequent rate of decline.

5.2 Value Generation, Internalisation and Demand Defence

In response to these pressures, the main MENA oil producers are following essentially two linked strategies. The first is to generate more value from their existing production. The second is to create new markets, or at least defend existing ones.

Investment in the oil upstream has become a lesser priority. In some countries, this is because of domestic political or security problems. In Saudi Arabia, it reflects a lack of space in the market for major production gains, particularly while the OPEC + quotas remain. ADNOC has plans for a substantial rise in capacity; Kuwait's ambitious expansion plans have made little progress over the last couple of decades; while Iraq's production capacity continues to grow despite bottlenecks. Iran is capable of maintaining production around current levels, diminished by sanctions, but significant growth would require international investment in a post-sanctions situation. Libya, too, would require several years of improved security and investment to regain its pre-Revolution level of 1.6 million bbl/day, or the aspirational 2 million bbl/day. Oil output in Algeria, Bahrain, Oman, Egypt and Qatar is in slow long-term decline unless there are major new discoveries, or investment in EOR or unconventional resources.

The Middle Eastern NOCs have invested relatively little in international upstream, because of their vast and low-cost domestic resources and, in some cases, lack of adequate technical and managerial skills. Kuwait Foreign Petroleum Exploration Company has been the most notable exception, building up a sizeable if scattered global portfolio. Mubadala, a sovereign wealth-type vehicle of the Abu Dhabi government, has made a smaller but more concentrated number of upstream investments in Qatar, Oman, south-east Asia, Russia and Egypt. Recently, Qatar Petroleum has enjoyed overseas exploration success, in partnership with IOC supermajors, in South Africa, Cyprus and Guyana. But these projects remain small compared to domestic production.

So value generation has primarily come outside upstream oil. Value generation is not a new strategy—it dates back at least to the 1960 s in Iran, the 1970 s in Iraq and the early 1980 s in Saudi Arabia—but has been pursued more vigorously and consistently in recent years. Saudi Arabia, mainly under Aramco and Saudi Arabia Basic Indus- tries Corporation (SABIC), has been the leader, with ADNOC catching up under the leadership of Sultan Al Jaber (appointed CEO in 2016), along with sovereign wealth vehicle Mubadala (which took over the International Petroleum Investment Corpo- ration (IPIC), another state fund, in 2017[15]). Qatar Petroleum, under CEO Saad Al Kaabi, also named energy minister in November 2018, has adopted a more aggressive domestic and international growth strategy, including taking over the stakes of IOCs in expiring Qatari fields. Kuwait Petroleum Corporation (KPC), Oman Oil Company and Sonatrach (Algeria) have been active, but to a lesser degree.

Part of enhanced value generation includes 'in-country value creation', trying to improve the capacity of local firms or joint ventures to supply the domestic oil industry with equipment. Eventually, this might produce internationally competitive oil services and engineering firms, as Norway has done. However, it runs the risk of protectionism and raising local production costs, and the skills developed have to be transferable to new sectors. Localisation—boosting the skills development and hiring of citizens instead of expatriates—has also been pursued. It can provide high-skilled, high-paid work but, given the low and falling labour intensity of the oil industry, will not make a dent in unemployment except in the smaller GCC countries.

The other part of value creation includes a move 'downstream'—to refining, petrochemicals, tankers, storage, trading and fuel retail. This is taking place both at home and abroad. Aramco launched trading in 2012 (Saudi Aramco 2017), while ADNOC Trading was set up in 2018 (Arab News 2018), and the company has been promoting its Murban crude grade as a regional price benchmark (TheNational.ae 2019). With the launch of China's International Energy Exchange's crude oil contract in March 2018, there is also some pressure on the Gulf producers to ensure that control of their commodity's pricing does not move to their main customer (WSJ 2019).

Petrochemicals are seen as one of the most promising areas for future oil demand, since demand typically grows faster than GDP in emerging economies, and alter- natives such as biomaterials remain more costly and limited in supply. Producing petrochemicals releases some greenhouse gases, but the products themselves do not until and unless they decompose or are burnt. The Gulf has developed a large basic chemical industry, based mostly on previously flared gas, and using methane and ethane as feedstocks to yield fertilisers, methanol, polyethylene and polypropy- lene. Now, new petrochemical strategies focus on mega-scale integrated refining and petrochemical complexes, culminating in Aramco/SABIC's proposed direct crude- to-chemicals conversion plant at Yanbu' on the Red Sea coast. These offer improved

---

[15]https://www.mubadala.com/en/mubadala-investment-company-old.

margins and market adaptability but are much more complex to construct and operate. Speciality chemicals move up the value chain and can support local industries such as automobile components but do not have the previous feedstock cost advantage.

The second part of the strategy, demand defence and creation, overlaps with downstream development. Internationally, the particular focus has been Asia—the traditional markets of Japan and South Korea, now with China, India and others such as Pakistan, Vietnam and Indonesia. So far, there has been little attention to Africa, though this could change with growth in countries such as Ethiopia. Saudi Aramco has joint-venture refineries in China, Japan and South Korea, and it and ADNOC are exploring a giant greenfield refinery in western India with a consortium of Indian state firms.

Creation of gas demand has received less attention but is perhaps more powerful, given the competition between gas and coal in Asia, and the need for substantial infrastructure for importation and use. Qatar Petroleum has extended its LNG portfolio into the US. Qatar Petroleum is a majority owner (70%) of the Golden Pass LNG terminal in Texas, with Exxon Mobil (17.6%) and ConocoPhillips (12.4%) and is planning to invest $20 billion in US conventional and unconventional oil and gas assets (Reuters 2018a). Algeria, by contrast, once a gas superpower, is struggling: its core European market is seeing low prices amid Russian and American competition, while exports are dropping because of underinvestment in new fields and rising domestic consumption.

Aramco and ADNOC, whose gas businesses today essentially serve only domestic demand (ADNOC has a relatively small LNG export unit), have targeted gas as a key growth area, as has Mubadala which was already a significant gas player via its 51% holding in the Dolphin pipeline. But this is more challenging for them than Qatar, as they have to enter a crowded international space rather than relying on a domestic surplus. Aramco has taken a stake in Sempra Energy's US Gulf Coast export facility (25% of Port Arthur LNG) (Saudi Aramco 2019) and has had discussions to enter Russia's Arctic-2 LNG project.

The northern tier of the Middle East may see the most development in gas, with a complex mix of commercial and political factors. Iran began exporting gas to Iraq in June 2017 (EIA 2019), although Baghdad has come under US pressure to find alternatives. The Kurdistan region of Iraq has large gas resources which could serve the crowded Turkish market, or the rest of Iraq. Iraq is gradually making headway on reducing gas flaring and boosting power generation; it could eventually link up to the GCC and become the gas and power conduit between Iran and the Gulf to the east, Turkey to the north and the eastern Mediterranean to the west.

But oil and even gas still have to contend with the climate imperative. Aramco is the most advanced of the region's NOCs in thinking about demand defence in terms of climate compatibility. It has invested substantial research and development funding to improve engine efficiency, including developing some radical new designs, which could keep oil-driven transport competitive with electric vehicles for longer, while reducing emissions. Its gasoline compression ignition technology has been accepted by Japanese carmaker Mazda Motor, and it is working with Achates Power on opposed-piston engines (PR News Wire 2018). These are said to be 30-50%

more efficient than conventional diesel and gasoline engines, have lower soot and NOx emissions than diesel and can run on low-octane gasoline (Green Car Congress 2018). It is also researching octane-on-demand to lower consumption and emissions of high-octane fuel used mostly during acceleration (SCMP 2019). Aramco has also worked on mobile carbon capture and storage for vehicles, though the economic viability of this seems doubtful. It has sought to develop non-metallic products from petrochemicals, such as oil-field pipes.

It touts the low-carbon footprint of its production (a result of low flaring, reducing methane leakage, improving energy efficiency of operations, and inherent advantages such as prolific reservoirs and limited water cut). This could become important as low-carbon fuel benchmarks are adopted by consuming countries and push out high-carbon crudes such as Canada's oil sands.

Aramco, ADNOC and QP have all launched large carbon capture and storage facilities, which inject carbon dioxide from gas processing or industrial facilities into underground reservoirs for safe disposal or enhanced oil recovery (EOR). Aramco is a member, and ADNOC considering becoming a member (Reuters 2017), of the Oil and Gas Climate Initiative (OGCI), along with leading international oil companies such as Shell and Occidental, with CCUS as a key focus.

In the longer term, two other technologies offer promise for sustainable hydrocarbon use. Direct air capture (DAC) extracts carbon dioxide from ambient air. It can then be used for improving plant growth in greenhouses, creating products such as low-emission cement, plastics or synthetic fuels, or injecting underground for EOR. As all realistic emissions paths compatible with limiting global warming to 1.5 °C or 2 °C involve substantial amounts of 'negative emissions', DAC could have a growing role, and MENA countries offer ideal geological, economic and social conditions.

Hydrogen has also attracted growing recent interest as a clean fuel. Its prospects for light vehicles appear doubtful, because of the major improvements of battery cars. However, hydrogen could be viable in long-distance trucking, ships and air travel. It is likely to be even more important for seasonal energy storage, balancing variable renewables, and for decarbonising heavy industry by providing high-temperature heat, and as a reducing agent such as in steelmaking.

'Blue hydrogen', made from fossil fuels with carbon capture and storage, or 'green hydrogen', produced by electrolysing water with low-carbon electricity, both have promise in MENA. Japan has expressed particular interest in developing a 'hydrogen economy' that could include imports from MENA and other suitable producers such as Australia. Dubai has set up a pilot hydrogen electrolysis facility at its Expo site.

6 Retooling the Domestic Energy System

Most MENA countries are engaged, at varying paces, in restructuring their domestic energy—primarily electricity—sectors. This has primarily been driven by economic and security-of-supply, not environmental, imperatives.

Many regional countries struggled with limited gas availability during the 2000s, as the economic boom outpaced production or insecurity cut off imports. In Egypt, Iraq, Iran and Sharjah, this manifested itself in power cuts and interruptions of gas to industry. Exports from Algeria and Oman were limited by feedstock. Jordan, Saudi Arabia, Kuwait and the UAE burnt large quantities of expensive liquid fuels to meet power demand.

Sector reform has manifested itself around three pillars: reductions in subsidies, increased private involvement and greater use of non-hydrocarbon energy.

Exposure to world LNG market prices, whether as an importer or exporter, has helped set an upper benchmark for local gas prices and encourage reduction in subsidies. Dubai, in 2008 (Boersma and Griffiths 2016), and Iran, in 2010 (IMF 2011), were two of the first to begin significant reforms of electricity, gas, water and petrol/diesel prices, and most regional countries have since followed suit. This has restrained demand growth, though it is difficult to isolate the exact impact because of the concurrent economic slowdown from 2014 onwards.

The breadth and social acceptance of the reforms differs from one country to another in the MENA region; nevertheless, energy prices are generally still below international or fully cost-reflective levels and in some cases reform has been post-poned or cancelled due to lack of parliamentary support (seen in Iran, Bahrain and Kuwait). In some cases in the GCC, particularly Saudi Arabia, and in Iran, rises in energy prices for the consumer were followed by a cash transfer programme or other type of in-kind assistance by the government, to help protect lower income citizens and to minimise any risk of civil discontent.

Independent power producers (IPPs) have been introduced from the early 2000 s onwards in most regional countries, breaking the model of the vertically integrated, state-owned monopoly utility. However, privatisation of distribution has remained very limited, and true electricity markets do not exist, with the 'single buyer' model persisting and a state monopoly remaining in charge of transmission. Oman has made some steps towards a more liberalised power market especially once it has opened private sector participation in electricity and water production. The GCC country also plans to privatise its transmission and distribution companies by selling up to 70% of its stakes in Muscat Electricity Distribution, and 49% of its stake in the Oman Electricity Transmission Company by 2020. Plans will likely be delayed due to their complexity (Reuters 2019a). Improvements in dispatch, and the replacement of outdated plants, have gradually improved average fleet thermal efficiencies. But identifying the true cost of electricity, and the true value of different sources of generation, remains difficult.

6.1 Alternative Energy Sources

The emergence of alternative electricity generation is the most striking and, in the long term, consequential of these changes. Over the next one to two decades, the use of oil in the MENA power system is likely to be largely phased out, while growth in gas use in power slows sharply.

These alternatives to hydrocarbons include coal, nuclear and renewables.

6.1.1   Coal

The turn to coal clearly illustrates the prioritisation of security-of-supply and economic motivations over environmental ones, even if some of the coal plants are promoted as 'carbon capture-ready'. It, and nuclear, indicates the preference for large, centralised facilities under the firm control of the national utility. There is no coal mining in MENA (outside a small amount in Iran[16]), so the use of domestic resources, or the preservation of domestic employment, is not a factor as it is in parts of Europe, the US, India, China or South Africa. But coal is cheap and readily available from a range of suppliers (US, Australia, South Africa and Indonesia) who do not bring the same security-of-supply concerns as some oil and gas exporters.

This explains the interest in coal from Dubai, Egypt and Morocco, and its historical use in Israel. Cement plants in Egypt and the UAE have turned to coal to replace expensive or unavailable gas or fuel oil. Dubai is constructing a large (2.4 GW) ultra-supercritical coal-fired power plant at Hassyan in the south of the emirate (Arabian Industry 2018). Egypt's plans for a sizeable clean coal power plant (2640 MW) with Chinese and Emirati backing were apparently called off by the Egyptian Ministry of Electricity in 2019 (IEA Clean Coal Centre 2019). Conversely, coal use in Israel is being replaced for environmental reasons by its new offshore gas, and renewables in Morocco will erode the share of its coal plants.

6.1.2   Nuclear

Nuclear power is more complicated, and the timelines and motivations of each country vary. Nuclear power globally has been struggling to maintain its share of the energy mix, due to the shutdown of old plants, nuclear reductions or phaseouts in countries such as Germany over safety concerns following the 2011 Fukushima accident in Japan, and the high cost and long timelines for new builds. The MENA region, though, had no operating nuclear power plants (there were some small research reactors) until recently but has now emerged as a relative bright spot for the industry.

Iran has one operational plant, 1 GW at Bushehr, on the Gulf, which came into service in 2011 (The Japan Times 2019a) after originally starting construction in 1975. Nuclear power has been seen as a source of national pride and technological advancement (Vaez and Sadjapdour 2013). In the 1970 s, there was a fear that Iran's oil resources would be exhausted relatively quickly; in the 2000 s, the concern was over the lagging pace of gas field development compared to fast-rising consumption. Now,

---

[16]Turkey, not included here in MENA, has substantial lignite mining and has made it a core part of electricity diversification.

an increase in gas output makes this less of an immediate worry, but the programme has a momentum of its own. In November 2019, Iran began construction on another 1 GW reactor at Bushehr (DW 2019), with Ali Akbar Salehi, head of the Atomic Energy Organisation of Iran, commenting that each reactor would save 11 million barrels of oil annually (a reasonable figure assuming that nuclear power solely displaces oil).

The link to the country's uranium enrichment programme is not straightforward, since Russia is supplying the fuel for Bushehr, though Iran does have ambitions to construct additional, domestically designed reactors. The US-led sanctions under the Obama administration were specifically predicated on limiting the country's enrichment, given that this could eventually produce weapons-grade uranium, not on the civil nuclear power programme per se. The full scale of Iran's nuclear plans might reach 10 GW by 2035,[17] out of a total installed capacity of some 137 GW, generating some 20% of the country's power. However, likely nuclear capacity will be much less than this, and the high cost and long construction times contrast with the relative ease and cheapness for Iran of turning to modern solar and wind.

In December 2006, the GCC states announced they had commissioned a study on civil nuclear power. Of these, Kuwait explicitly ruled out nuclear in July 2011 following Fukushima in March 2011, but Abu Dhabi had already begun a nuclear power programme in 2009 with the award of the construction bid to a consortium led by Korea Electric Power Company (KEPCO) and involving Samsung, Hyundai, Doosan and Westinghouse. France's Areva, with Suez and Total, competed intensely with the Korean consortium for the contract, given the importance of their plans to develop export-oriented nuclear power industries. Originally, the first of four reactors, totalling 5.6 gigawatts (GW), was intended to start in 2017, but construction and training delays have pushed full commercial operation back probably to 2021.

Its use of nuclear power was predicated on the country's rapidly growing electricity consumption, the struggles of gas production to keep up with demand, the lack of progress on expanding the Dolphin gas import contract with Qatar and the relatively high cost of alternatives, including renewables. Subsequently, demand growth has slowed, domestic gas output is set to rise, and solar power in the UAE has achieved dramatic gains in size and cost-competitiveness. Therefore, the UAE Energy Strategy 2050 counts nuclear as part of its 'clean' energy (i.e. zero-carbon) target but does not foresee that any more reactors will be constructed by 2050. Greenhouse gas emissions were not an important part of the initial motivation for the programme but have been claimed retrospectively.

Saudi Arabia has had a number of abortive plans for nuclear power, with a programme announced in August 2009. In April 2010, the King Abdullah Centre for Atomic and Renewable Energy (KA.CARE) was established, and during 2011-2016 it produced various targets, for between 16 and 17 GW of reactors, and signed study agreements with GE Hitachi Nuclear Energy, Toshiba/Westinghouse, Exelon, the Korean Atomic Energy Research Institute, China Nuclear Engineering Corporation and Rosatom. Plans included full-scale reactors as well as small modular reactors

---

[17] Author's estimates.

integrated with desalination. Saudi Arabia has also sought to mine its domestic uranium resources.

As with renewable energy, KA.CARE's intentions were stymied by turf wars with the energy ministry, and its own lack of ability to finance plants. However, from 2017, nuclear plans appear to have revived under the Ministry's banner, with a framework agreement with the International Atomic Energy Authority, selection of two reactor sites and the start of construction on a 30 kW research reactor at Riyadh.

Jordan has long investigated nuclear power because of its lack of domestic oil and gas resources, and its uranium resources (in phosphate deposits which are difficult to extract (World Nuclear Association 2019)). It has set up a Committee for Nuclear Strategy in 2007, and it set out a programme for nuclear power to provide 30% of electricity by 2030. The shut-off of Egyptian gas supplies due to sabotage and falling production after 2011 gave further impetus to its search for energy security. But its small electricity market would find it hard to accommodate a large reactor, and the relatively poor kingdom would struggle to finance it. With a short coastline, cooling water is not readily available (the inland Al Amra site chosen in 2010 was to be cooled with water from the Khirbet Samra wastewater treatment plant (World Nuclear Association 2019)). With the new availability of regional gas, the development of its indigenous oil shale, and the expansion of Jordan's successful solar and wind programmes, nuclear power is unlikely to proceed.

Egypt is a more promising case given the large and fast-growing domestic electricity market. Again, nuclear power plans have been floated for many years, with a 150 MW nuclear plant being proposed as early as 1964, but have been repeatedly stalled. In 2004 and 2008, Egypt signed new nuclear cooperation agreements with Russia's Rosatom for a 1 GW nuclear reactor, but lack of action forced a renewal of the cooperation agreement in 2013. In 2015, a new agreement was signed with Rosatom to finance and build four 1.2 GW reactors at El-Dabaa on the Mediterranean coast (World Nuclear Association 2020).

The high cost and technical sophistication of nuclear power programmes raises the question of the motivations for pursuing them, beyond simple economic and environmental goals.

The UAE signed a '123' agreement with the US, giving it access to American nuclear suppliers, which committed to a 'gold standard' on regulation and transparency, including a commitment not to enrich uranium or reprocess nuclear fuel. This was an important part of assuaging concerns that a civil nuclear power programme could be a cover for a weapons programme, given the US objective to prevent further proliferation, and in particular any threat to Israel's regional monopoly on nuclear weapons. The UAE's agreement, though, does give it the right to match the conditions offered by the US to any other regional state in subsequent agreements. Although there is no sign that the UAE's power programme is intended to be part of 'nuclear hedging', it could develop skills that might eventually be used for that purpose. Perhaps, more significantly, it creates a large and high-profile asset, and long-term agreements with a variety of influential countries, which would encourage them to come to the UAE's aid in the event of a threat.

Saudi Arabia has not signed a '123' agreement, partly because of its desire to retain the option for domestic enrichment of its uranium resources. It has, however, been lobbying for US support of its programme. Its leadership has given ambiguous signals on proliferation (though it is party to the Treaty on the Non-Proliferation of Nuclear Weapons (NPT), and has a Comprehensive Safeguards Agreement with the International Atomic Energy Agency (IAEA)), with crown prince Mohammed bin Salman noting that Saudi Arabia would seek to acquire nuclear weapons if arch-rival Iran did so (Reuters 2018b).

6.1.3   Renewables

In contrast to the decentralised renewables models pioneered in Europe, and being promoted for African countries, MENA countries have preferred to introduce large-scale centralised renewable projects. These have typically been awarded by tender by the state-owned utility, ministry or energy regulator, with an offtake guarantee. Land is usually free and grid connections often provided. Commercial banks have proved willing to finance these projects up to 84-90% of project value at low rates, enabling world-record levelized cost of energy (LCOEs) to be bid. Dubai, Abu Dhabi and Saudi Arabia have set records in solar photovoltaic (PV), Dubai in concentrated solar thermal power (CSP) and Saudi Arabia in onshore wind. Oman, meanwhile, has inaugurated a giant solar thermal project to provide steam for enhanced recovery of heavy oil.

The less creditworthy countries, notably Jordan, Egypt and Morocco, have not achieved such low bid prices, but nevertheless they have made impressive progress, at costs well below that of burning oil or imported LNG. Morocco, in particular, has been a regional leader in CSP through its renewable agency MASEN.

Acwa Power, a private Saudi company held 45% by the Public Investment Fund (PIF) (Al-Monitor 2019; Reuters 2019b), and Masdar, the clean energy subsidiary of Mubadala, have emerged as regional leaders. Chinese, Japanese and European companies (such as Jinko Solar, Marubeni Corporation, French energy giant EDF) have also been well-represented among winning consortia. For instance, Abu Dhabi's 1.17 GW Noor Abu Dhabi Solar Plant is a joint venture between the Abu Dhabi Government and a consortium of Marubeni Corporation and Jinko Solar Holding.

Progress is patchy across the region, though some of the laggard countries are now catching up. Saudi Arabia struggled from turf wars between the King Abdullah Centre for Atomic and Renewable Energy (KA.CARE), and the Ministry of Energy; and more recently between the ministry's Renewable Energy Project Delivery Office (REPDO) and the PIF (Power Technology 2019; Bloomberg Environment 2019). Iran has some domestic capability but struggles to attract international investment under the burden of sanctions and its own opaque domestic environment. It has, though, carried out a large programme of hydroelectric dam construction, problematic under drought conditions and part of a wider struggle to meet local water needs for agriculture and ecology. Iraq struggles to offer acceptable payment guarantees and to overcome its internal bureaucracy, but a recent offer of 755 MW solar PV[18] is encouraging. Qatar, with its abundant low-cost gas, has not prioritised renewables but has tendered for a 700 MW solar PV project, which could be expanded to 800 MW, to be operational by the fourth quarter of 2021 (PV Magazine 2019a).

Perhaps more surprising is the slow progress made in Oman, Bahrain, Kuwait, Lebanon and Algeria to date. These countries have struggled with combinations of domestic gas shortfalls, reliance of the power sector on high-cost oil or imported LNG and reductions in gas exports because of lack of feedstock. Idiosyncratic factors are at play, though with some common elements.

Oman, for instance, struggled with a lack of institutional capability, though it is now picking up on a series of renewable projects and recently commissioned its first wind farm. Lebanon's power sector has been hampered by politicisation, corruption and heavy losses from selling subsidised power, though there are tentative signs of recent progress. Poor public services, including lack of electricity, was a major factor behind the October–November 2019 protests (Reuters 2019c). Kuwait's energy projects in general have been badly delayed by continual gridlock between the government and parliament, with MPs wielding allegations of corruption.

Some countries, notably Saudi Arabia and Morocco (Oxford Business Group 2018), have sought to drive local economic development through renewables by specifying local content requirements. The UAE, by contrast, has focussed on achieving the lowest possible costs.

Alongside the large utility-scale projects, 'rooftop' and distributed renewables have made progress in some cases where the regulatory framework has been supportive. In particular, net metering programmes in Israel (PV Magazine 2018a), Jordan (PV Magazine 2018b), Abu Dhabi, 'Shams Dubai' (with 106 MW installed as of September 2019) (PV Magazine 2019b) and the Sahim scheme in Oman[19] have provided relatively attractive economics for larger scale (industrial and large commercial) installations. Despite subsidy reform, residential power prices in the region are still usually too low to encourage householders to install rooftop panels. However, solar water heaters have been mandated in Dubai and are popular in Jordan and Israel (Green Tech Media 2018).

Renewable power also raises the need and potential for electricity trading within MENA countries, such as the GCC, which has a pre-existing interconnection grid. Large-scale wind, solar PV and CSP projects would support expansion of the grid's current capacity (2.4 GW) and potential extensions to neighbouring countries. For instance, Iraq's Ministry of Electricity recently signed an electricity purchase agreement of up to 2 GW with the GCCIA, the leading advocate for regional power trading. Iraq shall receive power supplied by GCC countries (Kuwait, Saudi Arabia, and the UAE) from transmission lines from Kuwait. Saudi Arabia was in talks with Iraq last year for providing electricity from a 3 GW solar plant at a steep discount compared to what it imports from Iran ($21/MWh against $84/MWh from Iran), while Egypt

---

[18] https://moelc.gov.iq/index.php?name=News&file=article&sid=4558.

[19] https://www.aer.om/en/sahim.

has shown interest in a linking its national grid with Saudi Arabia's to meet peak demand with imports (up to 3 GW) (APICORP 2018).

Countries that have successfully developed large amounts of variable renewables may seek to export surpluses at certain times to their neighbours. They may also rely on dispatchable capacity in other countries to reduce their need for balancing variable renewables. And time differences across the region can be exploited; for instance, solar power in Jordan or Egypt can cover for demand in eastern Saudi Arabia during the early evening there. However, large-scale electricity exports from the region to Europe are unlikely in the medium term. From the Middle East, they would have to cross unstable areas in Syria or Lebanon. From North Africa, the distance is shorter and easier but the investment climate in Algeria and Libya has been unfriendly, and all the North African countries have been prioritising meeting their own demand. Europe would also, for reasons of local employment and security of supply, not wish to depend too heavily on its Mediterranean neighbours. Electricity exports from the GCC to South Asia via undersea cables could be a possibility. But the EU preference is likely to be more for imports of decarbonised industrial materials, such as steel, aluminium, fertilisers, cement and hydrogen, made in MENA with renewables, and fossil fuels with CCS (van Wijk and Wouters 2019).

Renewables are unlikely to perpetuate the 'rentier' model or contribute to the 'resource curse'. An attractive investment model is required to make MENA renewables competitive against neighbours, given the intense competition to lower financing costs. Although MENA has some of the world's best solar conditions, they are not superior enough to generate large rents, when including long-distance transmission.

7 Shifting Strategies in the Geoeconomic Transition

The US has since the 1970 s been by far the dominant outside power in the Middle East (if not so much in North Africa), replacing or expelling the Soviet Union as the sponsor of several client regimes. The Carter Doctrine, promulgated in 1980, declared that the US would use any means, including military force, to prevent outside domination of the Persian Gulf,[20] a warning squarely directed at the USSR following its intervention in Afghanistan. Though the Carter Doctrine was not directed at regional states, the US has also sought to prevent the rise of a regional monopolist which might dominate Middle Eastern oil supplies, leading the coalition to expel Saddam Hussein's Iraqi forces from Kuwait in 1991.

Though direct US imports of Middle Eastern oil have been relatively modest (Venezuela, Mexico and Canada being more important suppliers), this role has been played in recognition of the centrality of secure and reasonably priced energy to US allies in Western Europe and East Asia, notably Japan and South Korea. Another

---

[20]https://www.presidency.ucsb.edu/documents/the-state-the-union-address-delivered-before-joint-session-the-congress.

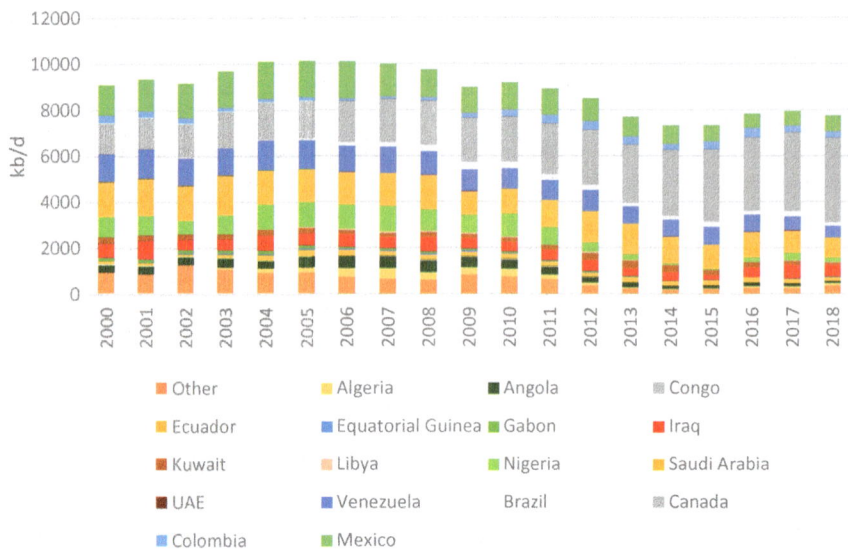

Fig. 5 US oil imports by country

objective, usually implicit, has become the most salient in recent years: the ability to deny oil in the event of conflict to US rivals, specifically China.

However, over the past two decades, even though US military forces in the Middle East remain strong, that role has gradually eroded. As the US has moved towards being a net exporter of oil, its imports from MENA have plunged (Fig. 5). The long wars in Iraq and Afghanistan, and the chaotic aftermath of the intervention in Libya, have drained domestic appetite for continuing involvement. The recent blundering withdrawal from parts of north-eastern Syria is symptomatic of this desire, and of the lack of a clear articulation of the continuing and compelling strategic rationale for remaining. Meanwhile, the US's main regional partners have been alarmed by its unpredictability, and have begun to take some steps to diversify their dependence on its security and diplomatic cover.

The key rising players have been China and Russia, China more in the economic sphere, Russia in the military and diplomatic. **Russia** has extended cooperation with various Middle Eastern countries, often across contradictory geopolitical alignments. Though it has given Iran diplomatic support in the sanctions standoff with Washington, it has not been able—or has not tried particularly hard—to shield Iran, an oil market competitor, from the consequences.

The OPEC + agreement has deepened the relationship with Saudi Arabia, making Moscow an oil market player that cannot be ignored. For the first time, Russia has delivered on production curbs—though it has not committed to cut as much as the OPEC countries, nor has it fully implemented those cuts. Russia's influence was also been important in convincing Iran to join the accord in December 2016 (Reuters 2016), though Iran now chafes at Saudi-Russia domination of OPEC policy. As a relatively high-cost producer, with more limited reserves than the main Gulf countries, and whose production was rising only slowly, Russia benefits more from the limitations. Its relative gains will be even more significant should oil demand growth slow or reverse in the near future.

Russian companies' investments in the region remain relatively limited, but Lukoil, Rosneft and Gazprom Neft have important assets in Iraq, Gazprom has long been involved in Libya, and Lukoil recently entered Abu Dhabi's offshore sour gas project. Rosneft entered the Kurdistan region of Iraq in September 2017, just ahead of the ill-fated referendum on independence which led to the loss of Kurdish control of the Kirkuk region and its important oil production. Rosneft signed for various oil-field developments, bought a stake in the main oil export pipeline through Turkey and agreed to construct a large gas pipeline to Turkey. However, its strategy in the area, whether as a competitor or a complement to compatriot Gazprom, is unclear.

Rosneft also bought a stake from ENI in the giant offshore Zohr gas field, while LNG player Novatek is involved in exploration in Lebanon. Russian companies, such as Kremlin-linked Stroytransgaz, hope to benefit from resources in Syria as a reward for supporting the Assad regime (Financial Times 2019). Though Gazprom Neft, Lukoil and Zarubezhneft have studied fields in Iran, none has been willing to invest under the penalty of American sanctions.

Nuclear power, via Rosatom, is another important area of cooperation. As discussed, Russia competed the long-moribund Bushehr reactor in Iran, has agreed to construct the Dabaa plant in northern Egypt (ROSATOM 2019), has been involved in Jordan's plans (now unlikely to progress) (Reuters 2018c) and discussed participation in Saudi Arabia's nuclear power programme (S&P Global Platts 2019). It is also constructing the Akkuyu plant in Turkey. Rosatom's integration across the nuclear value chain, its worldwide experience and its lack of political limitations give it advantages over its American, French, Chinese and South Korean competitors.

Overall in the region, Russian firms bring some reasonable technical skills and are desired for the diplomatic diversification. But they are not rich in capital, nor do they come with the vital domestic energy markets of the main Asian players. This will continue to limit them to a secondary role in the region's petroleum industry.

Some energy investments have gone the other way. These are not very large as yet, but Moscow does have an objective to replace the Western capital deterred by sanctions imposed over its annexation of Crimea in 2014, and not to become too dependent on Chinese funds. In 2018, Mubadala purchased a 44% stake in a unit of Gazprom Neft's operating fields in the Omsk and Tomsk districts of West Siberia (MUBADALA 2018). Saudi Aramco has explored the purchase of a 30% stake in Arctic-2 LNG from Novatek (Reuters 2019d). But the limited technical synergies and the barriers of political risk and lack of transparency that have held back foreign investment in general will probably limit the scope for Middle Eastern investment in Russian petroleum.

China has been a more important economic player. After playing second fiddle to Russia since 2017, Saudi Arabia has regained its slot as the country's largest crude oil supplier; Iraq is third, Oman sixth, while Iran was seventh in 2018 but has fallen to low levels (though China is now essentially its only paying customer). The MENA role in gas is less dominant, with China relying on pipelines from Central Asia, Russia and Burma, but Qatar was its second-largest supplier of LNG with 13.1% of the market in the first nine months of 2019; Oman and Egypt supplied small amounts.

Its large state oil firms have gained in prominence over the past decade, particularly in their leading role in Iraq, acquiring assets in Egypt (Egypt Today 2017), and more recently by acquiring stakes in Abu Dhabi's renewed concessions.

In return, China has become the largest exporter of manufactured goods to most of the Middle East. It has been the leading representative of Iran's few remaining trading partners and investors, and has defended the JCPOA. But it has not been willing to oppose US sanctions too vigorously, given the practical difficulties, and the fact this is just one issue in a wider confrontation with America over trade and other contentious topics.

China's Belt and Road Initiative (BRI) relates to the Middle East, particularly the maritime route through the Indian Ocean to the Arabian Sea, Red Sea and Suez Canal (Gurol and Shahmohammadi 2019). The Caspian route through Iran, the Caucasus and Turkey to Europe is also important, while the Gulf has played relatively less of a role (Mills et al. 2017). The BRI's primarily economic goals still bring with it diplomatic, strategic and perhaps military obligations and opportunities.

China's diplomacy has been relatively low-key and it has not established a significant regional military presence, wishing to avoid any direct appearance of challenging the US (Lons et al. 2019). However, it has a base in Djibouti, close to the Bab El Mandeb exit from the Red Sea (The Diplomat 2018), and Chinese military analysts have argued for extending this network of bases to the UAE and Pakistan (Duchâtel 2019). The Chinese-developed Gwadar port in Pakistan is close to the egress from the Gulf (and to the Indian-backed port at Chabahar in the strategic southern Mokran region of Iran). Beijing has sought to remain neutral in the region's conflicts, maintaining good relations with Saudi Arabia, the UAE and Israel (Efron et al. 2019) as well as Iran. At some point, though, as its regional interests deepen, it may find itself in a situation where this neutrality cannot be maintained.

India too has traditionally been more of an economic player, but the recent flurry of deals signed with ADNOC and Saudi Aramco have moved it up the league table of key Middle East partners. Less geopolitically intimidating than the Chinese but less risk-averse than the Japanese and Koreans, it has also seemed a more promising long-term market as Chinese growth has slowed. Early in 2019, ADNOC signed an oil storage agreement with the Indian Strategic Petroleum Reserves (ISPRL) to store 5.86 million barrels of Abu Dhabi oil, which was followed by ONGC Videsh, Bharat Petroleum and Indian Oil Corporation (IOC) being awarded stakes in Abu Dhabi's production licenses. Aramco has planned a $60 B refinery and petrochemicals complex in India.

India has not yet played the diplomatic and security role in the Middle East its proximity, contribution of migrant labour, and elements of shared culture, history and religion might suggest. In September 2019, it deployed warships to protect its tankers in the Gulf area following a spate of attacks on tankers blamed on Iran (The Economic Times 2019). The US has at some times attempted to cultivate India as a counterweight to China (Janardhan 2019), which could eventually include a heightened military role in the Gulf, as it did in the time of the British Raj (Gupta 2019).

Japan and **South Korea** are long-time economic partners of the Middle East, and particularly the Gulf, as importers of hydrocarbons and providers of manufactured goods and engineering services. They have increased their upstream investments in the region recently, particularly in Abu Dhabi and Iraq. Japan considered in September sending military forces to the Gulf to safeguard shipping, but this was opposed because of fears it would violate restrictions on the overseas deployment of the Self-Defence Forces (SDF) (The Japan Times 2019b). Ultimately, both will probably be too concerned about their immediate neighbourhood and the threat from China to commit large forces to the Middle East.

Europe's economic and diplomatic relationship with North Africa is more significant than with most of the Middle East, with France and Italy, and to a lesser extent Spain, especially involved as investors and customers. France and Italy have at times been at cross purposes in the civil war in Libya. The UK and France, in particular, have deepened their military and economic engagement with some members of the GCC, which has been welcomed as a counterbalance to worries over declining US attention. On the other hand, the 'E3' (UK, France and Germany) have been the leaders of the attempt to preserve the JCPOA with Iran in the face of the US violation of it, despite the GCC's dislike of the agreement.

The prospect of a transition to lower oil demand raises the question of whether the MENA region's geopolitical importance will decline. It will still, of course, have a central geographic position. Its gas exports, mostly from Qatar but possibly in future also from Iran, are expected to remain important for a longer period than for oil. Declining exports from Algeria and political instability in Libya have reduced their role as suppliers of gas to Europe, thus increasing the relative share of Russian gas, long an issue of concern for EU policymakers, though somewhat ameliorated by the rise of renewables and the increasing availability of LNG from diverse sources. However, in the short and medium term, the economic damage and instability from declining hydrocarbon rents could lead to a greater prominence for Middle Eastern political events. A secular decline in prices could be periodically interrupted by supply disruptions and temporary price spikes, which would themselves hasten the move towards alternatives. However, if the MENA region can avoid severe political upsets, it could even gain in geoeconomic importance for a while, as high-cost oil and gas producers are squeezed out of the market, and MENA exceeds the 42% share of world oil production it held in the early 1970 s, before the first oil shock.

8 Conclusions

The global energy transition centres on the shift from carbon-based energy sources and carriers to low/zero-carbon or decarbonised forms. Initially driven by environmental goals, this transition is now strongly encouraged by the growing cost-competitiveness and technical performance of many new energy technologies, particularly solar, wind, batteries and electric vehicles. However, hydrocarbons have also improved in cost, abundance and environmental performance in recent years, notably in the expansion of US shale oil and gas production, and the growth of the global LNG business.

From the MENA point of view, this energy transition coincides and partly overlaps with a geopolitical and geoeconomic transition. This involves severe political instability in many regional states; a diminished and less predictable role for the US and a rising position for Russia, China and some other states, which does not yet compensate for the American regress; and an ongoing shift of key markets away from the US and Europe towards emerging Asia.

The major MENA hydrocarbon exporters have, on the whole, coped with the fall in oil and gas prices since 2014, though with some economic difficulty and painful restructuring. So far, OPEC, with the new cooperation with Russia, has not just survived, but gained in coherence, given the stronger position of Saudi Arabia and its allies. But if oil demand growth continues to be weak or reverses, and/or non-OPEC competition remains strong, the loss of market share (or at least, failure to gain market share) will increase tensions between the OPEC + group's members.

Several leading MENA countries have found reasonable success in strengthening their hydrocarbon sectors to cope with new challenges, making their national oil companies more commercially minded and competitive, developing value-added trading, petrochemical and refining industries, and in particular in reorienting their markets towards Asia. Qatar has played its role in reshaping the world LNG business, and in adapting to the changes spurred by the appearance of US LNG exports. Some elements of a future climate-compatible industry are emerging—including a greater focus on petrochemicals and non-metallic materials, CCS, direct air capture and hydrogen—but remain quite nascent. The GCC countries in particular need to be bolder in investing in commercial-scale decarbonised projects.

Although their carbon footprints and levels of energy intensity remain very high by world standards, some regional states have also made increasingly swift and impressive progress towards integrating new energy into their domestic supply. Morocco, Jordan and the UAE stand out particularly. The dramatic new cost-competitiveness of solar PV, and improvements in wind and solar CSP, should see them rapidly gain a significant position in the energy sector of most MENA countries. Policy and government will and capability is catching up but still lags behind.

Some of the GCC states have made progress towards diversification and articulating 'post-oil' economic visions. So far, these do not involve a radical rethink of political and social systems. Iran, under the pressure of sanctions, has been forced towards diversification, though its economy has other severe structural flaws. Other major hydrocarbon producers, such as Algeria, Libya and Iraq, have been unable even to begin on serious economic reform. The challenge of moving to an economy not heavily dependent on oil and gas revenues within one to two decades remains enormous. Though states such as Dubai, Malaysia and Norway offer some lessons, there are almost no regional or even global success stories that MENA countries could emulate, meaning they will largely have to find their own path.

Success in these three endeavours—retooling domestic hydrocarbons, boosting the new energy economy and diversifying economically—will be essential for MENA states to cope with the wider geoeconomic transition. So far somewhat passive consumers of the transition, they need to take a more active role in developing and deploying the key technologies and business models, and translating these into an effective, proactive and positive role in climate diplomacy. The institutions built up during the resource-rich era have strengths, but also serious rigidities that will hinder reform towards the post-oil era. Some have made significant progress in strengthening and widening their diplomatic relationships, remaining close partners of the US while dealing successfully with Russia, China, European and other Asian countries. Others have already been torn apart when their domestic political fractures have been exploited by outside powers. The future for resource-rich MENA economies, and their energy-poor neighbours and co-dependents, in an era of energy transformation is not as uniformly gloomy as the early scholarship has painted it. But it will certainly not be easy, and not all states will achieve the transition successfully, or even survive.

References

African Energy (2019) Morocco scales back LNG import plans as gas strategy evolves, Newsletter Issue 389, 28 March 2019. https://www.africa-energy.com/article/morocco-scales-back-lng-import-plans-gas-strategy-evolves

Al-Monitor (2019) Where are the other IPOs in the Mideast?, 13 November 2019. https://www.al-monitor.com/pulse/originals/2019/11/ipo-aramco-middle-east-saudi-arabia-business.html

APICORP (2018) Electricity trading in MENA—huge potential but far behind, Vol. 03 No. 04, January 2018. http://www.apicorp-arabia.com/Research/EnergyReseach/2018/APICORP_Energy_Research_V03_N04_2018.pdf

Arabian Industry (2018) UAE makes steps towards second coal-fired power plant, 5 March 2018. https://www.arabianindustry.com/utilities/news/2018/mar/5/uae-makes-steps-towards-second-coal-fired-power-plant-5893979/

Arab News (2018) ADNOC's new trading unit to play 'critical role' in expansion plans, 23 April 2018. https://www.arabnews.com/node/1290106/business-economy

Bazilian M, Howells M (2019) Preface to the special issue on 'The geopolitics of the energy transition', Energy Strategy Reviews, Vol. 26, November 2019. https://www.sciencedirect.com/science/article/pii/S2211467X19301063

Bazilian M, Bradshaw M, Goldthau A, Westphal K (2019) Model and manage the changing geopolitics of energy. Nature Comment, 1 May 2019. https://www.nature.com/articles/d41586-019-01312-5

Beblawi H, Luciani G (1987) The Rentier State. Croom Helm, London

Bloomberg Environment (2019) "INSIGHT: Sovereign Wealth Fund Key to Saudi's Renewable Energy", 14 June 2019. https://news.bloombergenvironment.com/environment-and-energy/insight-sovereign-wealth-fund-key-to-saudis-renewable-energy

Boersma T, Griffiths S (2016) "Reforming Energy Subsidies. Initial Lessons from the United Arab Emirates", Brookings & Masdar Institute, January 2016. https://www.brookings.edu/wp-content/uploads/2016/07/brookings_masdar_reforming_energy_subsidies_uae.pdf

BP (2019) "Statistical Review of World Energy 2019"

Bradshaw M, Van de Graaf T, Connoly R (2019), Preparing for the new oil order? Saudi Arabia and Russia, Energy Strategy Reviews, Vol. 26, November 2019. https://www.sciencedirect.com/science/article/pii/S2211467X19300677

Brunnschweiler CN, Bulte EH (2006) The Resource Curse Revisited and Revised: A Tale of Paradoxes and Red Herrings, CER-ETH Working Paper 06/61, December 2006. https://papers.ssrn.com/sol3/papers.cfm?abstract_id=959149

CIA (2019) United Arab Emirates, CIA's the World Factbook. https://www.cia.gov/Library/publications/the-world-factbook/geos/ae.html

CNBC (2018) "Qatar quitting OPEC means the oil cartel is now just a 'two-member organization', oil analyst says", 3 December 2018. https://www.cnbc.com/2018/12/03/qatar-quitting-opec-leaves-oil-cartel-a-two-member-organization.html

Dale S, Fattouh B (2018) "Peak Oil Demand and Long-Run Oil Prices", The Oxford Institute for Energy Studies, Energy Insight: 25, January 2018. https://www.oxfordenergy.org/wpcms/wp-content/uploads/2018/01/Peak-Oil-Demand-and-Long-Run-Oil-Prices-Insight-25.pdf?v=ea8a1a99f6c9

Depledge J (2008) Striving for No: Saudi Arabia in the Climate Change Regime, Global Environmental Politics 8(4), pp. 9–35, November 2008. https://www.researchgate.net/publication/23546969_Striving_for_No_Saudi_Arabia_in_the_Climate_Change_Regime

Duchâtel M. (2019) "China Trends #2 – Naval Bases: From Djibouti to a Global Network?", Institut Montaigne, Blog, 26 June 2019. https://www.institutmontaigne.org/en/blog/china-trends-2-naval-bases-djibouti-global-network

DW (2019) "Iran starts building new nuclear reactor at Bushehr", 10 November 2019. https://www.dw.com/en/iran-starts-building-new-nuclear-reactor-at-bushehr/a-51192986

Efron S, Shatz HJ, Chan A, Haskel E, Morris LJ, Scobell A (2019) "The Evolving Israel-China Relationship", RAND Corporation, Research Report. https://www.rand.org/pubs/research_reports/RR2641.html

Egypt Today (2017) China's Sinopec looks to expand petroleum works in Egypt", 28 August 2017. https://www.egypttoday.com/Article/3/19830/China%E2%80%99s-Sinopec-looks-to-expand-petroleum-works-in-Egypt

EIA (2019) Background Reference: Iran, 7 January 2019. https://www.eia.gov/international/content/analysis/countries_long/Iran/background.htm

EIA (2011) United Arab Emirates, Country Analysis Briefs. https://www.eia.gov/international/content/analysis/countries_long/United_Arab_Emirates/archive/pdf/uae_2011.pdf

Financial Times (2019) "Moscow collects its spoils of war in Assad's Syria", 1 September 2019. https://www.ft.com/content/30ddfdd0-b83e-11e9-96bd-8e884d3ea203

Green Car Congress (2018) "Achates Power and Aramco Services Company partner on opposed-piston engine projects; Opposed-Piston Gasoline Compression Ignition (OPGCI)", 15 January 2018. https://www.greencarcongress.com/2018/01/20180115-achates.html

Green Tech Media (2018) Israel to develop 6GW of solar capacity over the next decade", 10 December 2018. https://www.greentechmedia.com/articles/read/israel-to-develop-6-gw-of-solar-capacity-over-next-decade

Gulf Business (2017) Abu Dhabi's ADNOC Distribution fuel firm IPO raises $851 m, 10 December 2017. https://gulfbusiness.com/abu-dhabis-adnoc-distribution-fuel-firm-ipo-raises-851m/

Gupta A. (2019), "India. India's evolving ties with the Middle East", Asia Society Policy Institute, 8 August 2019. https://asiasociety.org/asias-new-pivot/india

Gurol J. & Shahmohammadi P. (2019), "Projecting Power Westwards. China's Maritime Strategy in the Arabian Sea and its Potential Ramifications for the Region", CARPO Study 7, 11 November 2019. https://carpo-bonn.org/wp-content/uploads/2019/11/carpo_study_07.pdf

Hvidt M., (2013), Economic diversification in GCC countries: Past record and future trends, LSE Kuwait Programme on Development, Governance and Globalisation in the Gulf States, Number 27. http://eprints.lse.ac.uk/55252/1/Hvidt%20final%20paper%2020.11.17_v0.2.pdf

IEA (2019), History. From oil security to steering the world toward secure and sustainable energy transitions. https://www.iea.org/about/history

IEA (2012) Iraq Energy Outlook – WEO-2012 Special Report. https://webstore.iea.org/weo-2012-special-report-iraq-energy-outlook

IEA Clean Coal Centre (2019), "Egypt: EEHC delays coal-fired plant due to production surplus, News, 26 July 2019. https://www.iea-coal.org/egypt-eehc-delays-coal-fired-plant-due-to-production-surplus/

IMF (2011), "Iran – The Chronicles of the Subsidy Reform", IMF Working Paper WP/11/167, July 2011. https://www.imf.org/external/pubs/ft/wp/2011/wp11167.pdf

IRENA (2019), A New World. The Geopolitics of the Energy Transformation. https://www.irena.org/publications/2019/Jan/A-New-World-The-Geopolitics-of-the-Energy-Transformation

Janardhan N. (2019), "Saudi-Indian Ties: Looking Beyond the Bilateral", The Arab Gulf States Institute in Washington, Blog Post, 5 November 2019. https://agsiw.org/saudi-india-ties-looking-beyond-the-bilateral/

KAPSARC (2018), Electricity Market Integration in the GCC and MENA: Imperatives and Challenges, July 2018. https://www.kapsarc.org/research/publications/electricity-market-integration-in-the-gcc-and-mena-imperatives-and-challenges/

Karsh E. & Rautsi I., (2008) Why Saddam Hussein invaded Kuwait, Survival, 33:1, 18–30, DOI: https://doi.org/10.1080/00396339108442571

Lons C., Fulton J., Sun D., Al-Tamimi N. (2019), "China's great game in the Middle East", European Council on foreign Relations, Policy Brief, 21 October 2019. https://www.ecfr.eu/publications/summary/china_great_game_middle_east

Lowe S., (2019) Border Carbon Adjustment: How to get it right, Centre for European Reform (CER), 7 October 2019. https://www.cer.eu/in-the-press/border-carbon-adjustment-how-get-it-right

Mahdavy H., (1970), "Patterns and Problems of Economic Development in Rentier States: the Case of Iran." in: M.A. Cook, ed., *Studies in Economic History of the Middle East* (London: Oxford University Press), p. 466

MEES (2013), Israel Starts LNG Imports, Issue 56/5, 1 February 2013. https://www.mees.com/index.php/2013/2/1/transportation/israel-starts-lng-imports/78ad4480-87fc-11e7-a2f9-e5ff901bc260

Mills R. & Alhashemi F., (2018) Resource regionalism in the Middle East and North Africa: Rich lands, neglected people, Brookings Doha Center Analysis Paper, Number 20, April 2018. https://www.brookings.edu/wp-content/uploads/2018/03/resource-regionalism-in-the-mena_english_web.pdf

Mills R., Ishfaq S., Ibrahim R. & Reese A. (2017), "China's Road to the Gulf. Opportunities for the GCC in the Belt and Road Inititative", Emerge85, Delma Institute, Qamar Energy, October 2017. https://emerge85.io/wp-content/uploads/2017/10/Chinas-Road-to-the-Gulf.pdf

Mohaddes K. & Raissi M. (2019) The US oil supply revolution and the global economy, Empirical Economics 57, pp. 1515–1546. https://link.springer.com/article/10.1007%2Fs00181-018-1505-9

MUDADALA (2018) "Gazprom NEft, Mubadala Petroleum and the Russian Direct Investment Fund (RDIF) announce the completion of a transaction to jointly develop oil fields in Western Siberia" Press Releases, 5 September 2018. https://www.mubadala.com/en/news/gazprom-neft-mubadala-petroleum-and-russian-direct-investment-fund-rdif-announce-completion

OPEC (2019), World Oil Outlook 2019. https://www.opec.org/opec_web/en/publications/340.htm

OPEC (2016), "Declaration of Cooperation OPEC and non-OPEC". https://www.opec.org/opec_web/static_files_project/media/downloads/publications/Declaration%20of%20Cooperation.pdf

O'Sullivan M.L., Overland I., Sandalow D. (2017), "The Geopolitics of Renewable Energy" Paper, Belfer Center for Science and International Affairs, Harvard Kennedy School, 28 June 2017. https://www.belfercenter.org/publication/geopolitics-renewable-energy

Oxford Business Group (2018), "Morocco plans to add 10 GW of power from renewable energy sources by 2030". https://oxfordbusinessgroup.com/analysis/viable-alternative-plans-add-10-gw-power-renewable-sources-2030

POMEPS (2019) The Politics of Rentier States in the Gulf, POMEPS Studies 33, pp. 29–33, January 2019. https://pomeps.org/wp-content/uploads/2019/02/POMEPS_Studies_33.pdf

Power Technology (2019), "Riyadh sets out its stall as a global renewables hub", 27 February 2019. https://www.power-technology.com/comment/riyadh-sets-out-its-stall-as-a-global-renewables-hub/

PR News Wire (2018) "Achates Power and Aramco Services Company Announce Joint Development Agreement for Opposed-Piston Engine projects, 14 January 2018. https://www.prnewswire.com/news-releases/achates-power-and-aramco-services-company-announce-joint-development-agreement-for-opposed-piston-engine-projects-300582375.html

PV Magazine (2019a), "Qatari utility mulls jump to 800 MW tender as consortia submit bids", 6 August 2019. https://www.pv-magazine.com/2019/08/06/qatari-utility-mulls-jump-to-800-mw-tender-as-consortia-submit-bids/

PV Magazine (2019b), "Dubai has 106 MW under net metering", 16 September 2019. https://www.pv-magazine.com/2019/09/16/dubai-has-106-mw-under-net-metering/

PV Magazine (2018a), "Israel issues 1.6 GW scheme for rooftop solar", 29 March 2018. https://www.pv-magazine.com/2018/03/29/israel-issues-1-6-gw-scheme-for-rooftop-solar/

PV Magazine (2018b), "Jordan: Effective funding schemes for small-scale PV", 27 November 2018. https://www.pv-magazine.com/2018/11/27/jordan-effective-funding-schemes-for-small-scale-pv/

QUARTZ (2016) A powerful Saudi prince says his country will be ready to survive without oil by 2020, 25 April 2016. https://qz.com/669467/a-powerful-saudi-prince-says-his-country-will-be-ready-to-survive-without-oil-by-2020/

Reuters (2019a), "Oman says 25 investors interested in two power assets", 11 March 2019. https://www.reuters.com/article/oman-privatisation-electricity/oman-says-25-investors-interested-in-two-power-assets-idUSL8N20Y1RH

Reuters (2019b), "UPDATE 1-Saudi's PIF to raise stake in ACWA Power, to expand overseas", 13 February 2019. https://www.reuters.com/article/saudi-pif-investment/update-1-saudis-pif-to-raise-stake-in-acwa-power-to-expand-overseas-idUSL5N2080XF

Reuters (2019c), "Electricity, mobiles and cash: a snapshot of Lebanese grievances", 8 November 2019. https://www.reuters.com/article/us-lebanon-protests-grievances/electricity-mobiles-and-cash-a-snapshot-of-lebanese-grievances-idUSKBN1XI1VN

Reuters (2019d), "Falih: Saudi Aramco extends offer to buy stake in Arctic LNG 2 – TASS", 10 June 2019. https://www.reuters.com/article/lng-novatek-saudi-aramco/falih-saudi-aramco-extends-offer-to-buy-stake-in-arctic-lng-2-tass-idUSR4N23D08Z

Reuters (2018a), "Qatar Petroleum to invest $20 billion in U.S. in major expansion", 16 December 2018. https://www.reuters.com/article/us-qatar-gas-qp/qatar-petroleum-to-invest-20-billion-in-us-in-major-expansion-idUSKBN1OF07X

Reuters (2018b), "Saudi crown prince says will develop nuclear bomb if Iran does: CBS TV", 15 March 2018. https://www.reuters.com/article/us-saudi-iran-nuclear/saudi-crown-prince-says-will-develop-nuclear-bomb-if-iran-does-cbs-tv-idUSKCN1GR1MN

Reuters (2018c), "Rosatom looks to nuclear newcomers to cement dominance", 11 July 2018. https://www.reuters.com/article/us-russia-nuclear-rosatom/rosatom-looks-to-nuclear-newcomers-to-cement-dominance-idUSKBN1K11OS

Reuters (2017), "Big Oil's $1 billion climate fund picks investments, eyes new members", 27 October 2017. https://www.reuters.com/article/us-climatechange-ogci/big-oils-1-billion-climate-fund-picks-investments-eyes-new-members-idUSKBN1CW10Q

Reuters (2016), "Exclusive: How Putin, Khamenei and Saudi prince got OPEC deal done", 1 December 2016. https://www.reuters.com/article/us-opec-meeting/exclusive-how-putin-khamenei-and-saudi-prince-got-opec-deal-done-idUSKBN13Q4WG

ROSATOM (2019), "Egypt's Dabaa nuclear plant granted site permit", 9 April 2019. https://www.rosatom.ru/en/press-centre/industry-in-media/egypt-s-dabaa-nuclear-plant-granted-site-permit/

Ross M.L., Andersen J.J., (2012) The Big Oil Change: a closer look at the Haber-Menaldo analysis, APSA 2012 Annual Meeting Paper. https://papers.ssrn.com/sol3/papers.cfm?abstract_id=2104708

Saudi Aramco (2019), "Sempra LNG and Aramco Services Company sign heads of agreement for Port Arthur LNG", News, 22 May 2019. https://www.saudiaramco.com/en/news-media/news/2019/sempra-lng-asc-port-arthur#

Saudi Aramco (2017), "Aramco Trading inaugurates first international office in Singapore", News, 7 December 2017. https://www.saudiaramco.com/en/news-media/news/2017/aramco-trading-inaugurates-first-international-office-in-singapo

SCMP (2019), "Saudi Aramco invests in cleaner internal combustion engines as it doubles down on oil ahead of jumbo IPO", 14 October 2019. https://www.scmp.com/business/companies/article/3032737/saudi-aramco-invests-making-cleaner-internal-combustion-engines

S&P Global Platts (2019), "Russia, Saudi Arabia in talks on new oil, gas, petchems, nuclear cooperation", 15 July 2019. https://www.spglobal.com/platts/en/market-insights/latest-news/oil/071519-russia-saudi-arabia-in-talks-on-new-oil-gas-petchems-nuclear-cooperation

Snoj J. (2019), Population of Qatar by nationality – 2019 report, Priya Dsouza Communications, 15 August 2019. http://priyadsouza.com/population-of-qatar-by-nationality-in-2017/

Songhurst B., (2019) Floating LNG Update – Liquefaction and Import Terminals, The Oxford Institute for Energy Studies, OIES PAPER: NG149, September 2019. https://www.oxfordenergy.org/wpcms/wp-content/uploads/2019/09/Floating-LNG-Update-Liquesfaction-and-Import-Terminals-NG149.pdf

TAQA Arabia (2019), Egypt's Natural Gas Market Overview, 6th Sessions of the UNECE Group of Experts on Gas, Geneva, March 2019. https://www.unece.org/fileadmin/DAM/energy/se/pp/geg/geg6_March.2019/Item_5_Pakinam_Kafafi_Egypt.pdf

The Diplomat (2018), "China's Djibouti Base: A One Year Update", 4 December 2018. https://thediplomat.com/2018/12/chinas-djibouti-base-a-one-year-update/

The Economic Times (2019) "To ensure security of Indian vessels, Navy deploys warships in Gulf of Oman, Persian Gulf", 20 June 2019. https://economictimes.indiatimes.com/news/defence/to-ensure-security-of-indian-vessels-navy-deploys-warships-in-gulf-of-oman-persian-gulf/articleshow/69879830.cms?from=mdr

The Japan Times (2019a), "Iran and Russia launch new phase of nuclear power reactor construction", 11 November 2019. https://www.japantimes.co.jp/news/2019/11/11/business/iran-russia-launch-new-phase-nuclear-power-reactor-construction/#.XjKTUjJKgdV

The Japan Times (2019b), "With hazy plan for SDF dispatch to Mideast, is Japan pursuing contradictory goals?", 29 October 2019. https://www.japantimes.co.jp/news/2019/10/29/national/politics-diplomacy/japan-sdf-mideast-contradictory-goals/#.XjKpeTJKgdV

The Jordan Times (2016), LNG imports reach 66 bcf – ministry, 2 August 2016. http://www.jordantimes.com/news/local/lng-imports-reach-66-billion-cubic-feet-%E2%80%94-ministry

TheNational.ae (2019), "Adnoc's Murban crude to be listed on futures exchange in 'historic' step-change for Abu Dhabi's oil pricing", 4 November 2019. https://www.thenational.ae/business/energy/adnoc-s-murban-crude-to-be-listed-on-futures-exchange-in-historic-step-change-for-abu-dhabi-s-oil-pricing-1.933015

TheNational.ae (2015), Sheikh Mohammed bin Zayed's inspirational vision for a post-oil UAE, 10 February 2015. https://www.thenational.ae/opinion/sheikh-mohammed-bin-zayed-s-inspirational-vision-for-a-post-oil-uae-1.8710

TheNational.ae (2009a), UAE wins bid to house Irena headquarters, 29 June 2009. https://www.thenational.ae/uae/uae-wins-bid-to-house-irena-headquarters-1.512489

TheNational.ae (2009b), Irena chief resists UAE project, 24 September 2009. https://www.
thenational.ae/uae/environment/irena-chief-resists-uae-project-1.493032
TheNational.ae (2009c) Oil producers join Saudi in carbon reduction compensation
call, 17 October 2009. https://www.thenational.ae/business/oil-producers-join-saudi-in-carbon-
reduction-compensation-call-1.542465
UFCCC (2015a), "Intended Nationally Determined Contribution – Iraq", 10 Novem-
ber 2015. https://www4.unfccc.int/sites/submissions/INDC/Published%20Documents/Iraq/1/
INDC-Iraq.pdf
UFCCC (2015b), "Intended Nationally Determined Contribution – Iran", 19 Novem-
ber 2015. https://www4.unfccc.int/sites/submissions/INDC/Published%20Documents/Iran/1/
INDC%20Iran%20Final%20Text.pdf
UFCCC (2015c), "Intended Nationally Determined Contribution – Morocco", Novem-
ber 2015. https://www4.unfccc.int/sites/ndcstaging/PublishedDocuments/Morocco%20First/
Morocco%20First%20NDC-English.pdf
UFCCC (2015d), "Intended Nationally Determined Contribution – United Arab Emirates",
22 October 2015. https://www4.unfccc.int/sites/ndcstaging/PublishedDocuments/United%
20Arab%20Emirates%20First/UAE%20INDC%20-%2022%20October.pdf
UFCCC (2015e), "Intended Nationally Determined Contribution – Kingdom of Saudi Arabia", November 2015. https://www4.unfccc.int/sites/ndcstaging/PublishedDocuments/Saudi%
20Arabia%20First/KSA-INDCs%20English.pdf
Vaez A. & Sadjapdour K. (2013) "Iran's Nuclear Odyssey. Costs and Risks", Carnegie Endowment
for International Peace. https://fas.org/wp-content/uploads/2013/04/iran_nuclear_odyssey.pdf
van Wijk A.J.M. & Wouters F. (2019), "Hydrogen – The Bridge between Africa and Europe",
September 2019. http://profadvanwijk.com/wp-content/uploads/2019/09/Hydrogen-the-bridge-
between-Africa-and-Europe-5-9-2019.pdf
World Bank (2012) Integrated National Energy Strategy (INES): final report, Working Paper, 25
September 2012. http://documents.worldbank.org/curated/en/406941467995791680/Integrated-
National-Energy-Strategy-INES-final-report
World Nuclear Association (2020), "Nuclear Power in Egypt", Country Profile, January 2020.
https://www.world-nuclear.org/information-library/country-profiles/countries-a-f/egypt.aspx
World Nuclear Association (2019), "Nuclear Power in Jordan", Country Profile, June 2019. https://
www.world-nuclear.org/information-library/country-profiles/countries-g-n/jordan.aspx
WSJ (2019), "China's Oil Futures Give New York and London a Run for Their Money", 27
March 2019. https://www.wsj.com/articles/chinas-oil-futures-give-new-york-and-london-a-run-
for-their-money-11553679002

Addressing Africa's Energy Dilemma

Lapo Pistelli

1 Introduction

The transition to cleaner forms of energy production is happening in sub-Saharan Africa (SSA) with distinct characteristics compared to other world regions. African countries are gifted with a huge—and still largely untapped—energy potential but energy access is among the lowest in the world, mainly due to structural constraints (such as the poor efficiency of the power sector and the non-capillary diffusion of grids) as well as lack of significant investment in energy infrastructure, human resources and technology.

Considering that energy is the main enabler of economic development and an essential component of several Sustainable Development Goals (SDGs), if universal energy access in sub-Saharan Africa continues to be a distant objective, a large part of the 2030 Agenda for Sustainable Development will remain out of reach, depriving the region of the possibility to build a prosperous future. Thus, guaranteeing energy access is *the* priority of sub-Saharan countries, while the transition from fossil fuels to cleaner forms of energy is *a* complementary priority that manifests itself very differently across countries in terms of timing, modality and impact.

This chapter discusses how the ongoing low-carbon energy transformation could reshape geopolitics within Africa and between the continent and the rest of the world. The chapter first attempts to define what 'transition' means in African contexts and if the concept applies at all to African dynamics. It then delves into the drivers and modalities of Africa's alleged shift to finally explore geopolitical dynamics, questioning whether Africa is still the locus for the global supply of natural resources, introducing patterns of engagement between Africa and international/regional actors, and finally presenting the socio-economic implications of the shift.

L. Pistelli (✉)
Eni, Rome, Italy
e-mail: lapo.pistelli@eni.com

© The Author(s) 2020

M. Hafner and S. Tagliapietra (eds.), *The Geopolitics of the Global Energy Transition*,
Lecture Notes in Energy 73, https://doi.org/10.1007/978-3-030-39066-2_7

The analysis pivots between discussions of contemporary developments and future trends. The former accounts inevitably differ from those that elaborate on the 'likely' implications of renewables on geopolitics in the coming decades. While the study of current settings is based on historical and factual accounts, investigations of 'likely' futures are based on assumptions and scenarios.

We conclude that while the *venues and sources* of geopolitical interest might change in the new geopolitical order that the transition to renewable energy implies, the *content and modalities* of interaction may see a continuity with the past, namely, dependence on external financing and technology. With yet at least one novelty: increased relevance of regional interdependencies.

These latter may signal a counter-tendency at a time when increased tariffs and sanctions are everyday practice at the global level. The recent signing of the landmark African Continental Free Trade Area (AfCFTA) agreement in July 2019 emphasises Africa's willingness to reduce reliance over external partners and products while increasing regional integration. It is, however, a trend that has yet to materialise while it already presents pitfalls. According to the United Nations Conference on Trade and Development (UNCTAD), increased intra-African trade will depend, among others, on how 'rules of origin'[1] will be negotiated and implemented. Unless such rules are well thought through, future regional value chains on the continent will continue receiving little input from Africa, and a lot from abroad.

Importantly, at the time of revision of this book chapter, the world is experiencing an unprecedented pandemic, unforeseeable at the time of writing. The global economy is headed for recession and the impact on the African continent is very likely to be severe with potential social, economic and political disruptions. However, although investments and new electricity connections will slow down as a result of the crisis, our analysis and conclusions have so far remained solid: the pandemic will not reduce the urgency for universal energy access whose achievement underpins the success of the Sustainable Development Agenda; global lockdowns have highlighted the critical value of power infrastructure and this applies also to Africa, especially when it comes to the development of regional energy markets; the role of clean energy sources will continue to be strategic thanks to the crescent competitiveness of renewables and the complementary role of natural gas; furthermore, if there is one thing that the coronavirus pandemic has once again confirmed is the reliance of Africa on external actors, and yet its increasing willingness and capacity to coordinate at the continental and sub-continental levels.

---

[1] By defining the nationality of a product, rules of origin dictate the conditions for the application of tariff concessions, delimiting the range of products eligible for preferential treatment (UNCTAD 2019, pp. 3-4).

2 Understanding the Energy Transition in Sub-Saharan Africa

When we talk about the energy transition in sub-Saharan Africa, we cannot avoid problematising the concept of 'transition', which generally implies moving from a socio-economic and technological arrangement to another. The assumption behind the concept is that an arrangement does indeed exist. In sub-Saharan Africa, however, 600 million people (IEA 2018), over a total 1 billion (World Bank 2018a),[2] lack energy access. The 'transition' to cleaner forms of energy is thus only a complementary concern, second to that of providing access to energy. In other words, if the majority of the population does not have access to energy, it does not have anything to transit away from.

Nonetheless, as governments conceive national development plans, the idea of leapfrogging the conventional path to energy development and shift directly to renewables is starting to be contemplated. The 'National Determined Contributions' (NDCs)[3] to the Paris Agreement and the energy plans of many sub-Saharan countries indicate an initial move towards the energy transition. Differences are, however, great across the continent. Countries like Kenya and South Africa have commissioned a significant number of renewable energy projects in the last years (African Energy Live Data 2019),[4] though the inefficiencies of the power sector and the inadequacy of the energy infrastructure (common structural constraints in the sub-Saharan context) currently make the substantial penetration of additional capacity a tough challenge. However, the vast majority of sub-Saharan countries are still very far from expanding the share of renewables in the power mix.

Biomass and waste[5] cover about 60% of sub-Saharan Africa's *energy* mix, followed by oil and coal (16% each), gas (4%), nuclear and hydro (1% each), while renewables combined represent a mere 1% (Enerdata 2019). With a 51% share in the *electricity* mix,[6] coal is the dominant fuel in electricity production.[7] Gas holds 11% of the share, oil 7%, hydro 24% and solar, wind and geothermal combined 3% of electricity generation (Enerdata 2019). On average then, renewable energy in sub-Saharan Africa represents a mere 1% reality—yet with revolutionary potential.

---

[2]The IEA 2018 Outlook uses the estimate of 600 million people without access referring to the year 2017, in which the overall population in Sub-Saharan Africa was 1.05 bn.

[3]National Determined Contributions (NDCs) identify the post-2020 climate targets, including mitigation and adaptation, which countries committed to ratifying the Paris Agreement.

[4]According to African Energy Live Data, as of August 2019, installed electricity capacity from wind and solar totals 411 MW in Kenya and 1565 MW in South Africa.

[5]Primarily used in the residential and commercials sector where more than 70% of total biomass consumption is concentrated.

[6]Energy mix and electricity mix data are referred to 2017.

[7]Excluding from the analysis South Africa, Botswana and Zimbabwe, three countries mostly dependent on coal the power generation mix becomes hydro 51%, natural gas 26%, oil 18%, coal 1% and renewables 4%.

2.1   Energy Access Is the Priority of the Region

Africa's energy landscape is extremely diverse across the continent. It presents different energy structures (based on reliance on various natural resources), different levels of infrastructure development and consequential varied vested interests in specific sources, such as oil in Nigeria and Angola, coal in southern Africa. Nonetheless, a distinctive feature shared by all sub-Saharan countries is the world's lowest rate of energy access.

Although the overall electrification rate in sub-Saharan Africa has 'almost doubled since 2000, rising by 20% points to 43%' (IEA 2018, p. 78), today six people out of ten remain without electricity and have no promising prospects for improving their conditions any time soon. The sub-continent is still home to 20 countries with the world's lowest electrification rates and almost nine people out of ten (890 million over a billion) have no clean cooking access while the number of people relying on biomass, coal and kerosene for their main household cooking needs has increased by 270 million since 2000 (IEA 2018). Moreover, progress on energy access, uneven as it has been across the continent, concentrated in just four countries—Kenya, Ethiopia, Tanzania and Nigeria—which make up more than half of those gaining access since 2011.

This is particularly problematic not just for the achievement of Sustainable Development Goal #7, but also for the achievement of several other development goals, given that energy is the main enabler of economic development. It is estimated that if universal energy access in sub-Saharan Africa continues to be a distant objective, most of the 2030 Agenda for Sustainable Development will remain out of reach (IEA, IRENA, United Nations Statistics Division, World Bank, World Health Organization [WHO] 2018), subsequently depriving the region of the possibility to build a prosperous future.

To revert the trend and ensure universal energy access, the International Energy Agency foresees that growing electricity needs in sub-Saharan Africa will have to be met by using a combination of domestic gas and renewable energy sources, reducing as much as possible the use of oil and coal in power generation. Also, the use of biomass will have to be substituted with improved biomass cookstoves and Liquefied Petroleum Gas (LPG) as clean cooking options.

Considering that clean cooking represents only a minor part of the incremental investment required to ensure universal energy access in Africa, this article will mainly focus on the power sector, looking at the substitution of coal and oil with natural gas and renewables. These latter two resources are widely available in Africa, and the economy of the region would very much benefit from their increased use for power generation.

2.2 Structural Constraints to Gas and Renewable Energy Penetration in the Power Mix

Increasing the use of natural gas and renewables for power generation is not, unfortunately, easily achievable, though desirable, in sub-Saharan Africa. Highly efficient and economically sustainable power systems, together with costly infrastructure investments, are required to ensure the substantial penetration of those resources in the electricity mix. The absence of these two conditions across the continent constitutes the main reason why sub-Saharan power markets are underdeveloped, and gas-to-power and renewable energy projects are still relatively limited in the region. In many cases, this is also a problem for the oil and gas sector, where the development of gas projects are, in many cases, conditioned to the availability of liquefaction facilities for the sale of gas in the international gas markets.[8]

There is, however, one more underlying reason as to why the power sector faces challenges across the continent: the lack of industrial capacity, which leads to relatively limited energy demand, and hence the uselessness of increased power generation.

2.2.1 Efficiency and Economic Sustainability

Concerning efficiency and economic sustainability, a study of the World Bank (Kojima and Trimble 2016) analysed the relationship between costs and revenues of electric utilities in 39 sub-Saharan countries highlighting that only Uganda and Seychelles fully recover their operational and capital costs with the cash collected from customers. Such large deficits[9]—averaging 1.5% of gross domestic product (GDP)—'prevent power sectors from delivering reliable electricity to existing customers, let alone expanding supply to new consumers at an optimal pace' (Kojima and Trimble 2016, p. vii). The commercial losses faced by utilities raise the issue of 'creditworthiness': the deficit puts national utilities under financial stress—also affecting public finances since power utilities are usually state-owned entities—and exacerbates the off-take risk.[10] This sub-optimal situation hinders private investments in power generation even where the power grid would allow additional generation capacity to reach the final customer.

---

[8]According to Enerdata, total gas production in sub-Saharan countries was 67.5 bcm in 2017 and local markets consumed only 32.8 bcm. Thus, more than half (51.4%) of the production was exported. In the last two decades, gas production in sub-Saharan Africa has constantly been above gas consumption and the level of net export in the coming years is expected to be growing due to new Liquefied Natural Gas (LNG) developments (e.g. Mozambique).

[9]The revenue–expenditure gap of a public utility company is referred to as a 'quasi-fiscal deficit' because utility companies are usually state-owned and commercial losses often represent a government-sponsored collective subsidy.

[10]Under a power purchase agreement between a power generation project and the national utility, the non-payment risk incurred by an independent power producer.

2.2.2 Infrastructure Investment

However, power generation is only one part of the electricity sector value chain, and one of the most critical elements of sub-Saharan Africa's energy sector is indeed related to the development of the power grid. The inadequacy of the transmission and distribution network prevents the energy supply from reaching final demand representing one of the main causes of the low levels of energy access in the region.

While in the oil and gas industry the possibility to export and sell the resource on international markets constitutes a sufficient incentive for the private sector to invest, investments in electricity generation can only be paid back by the local (or regional) market. Hence, a well-functioning power grid is essential to attract private investments.

The 'natural monopoly' characteristics of transmission and distribution make the intervention of the public sector necessary, but the budget constraints suffered by the vast majority of sub-Saharan countries limit the scale of their intervention. For gas-to-power projects, the challenge is even tougher than renewables since the former requires infrastructure investments also in the midstream sector for the provision of gas-to-power power plants.[11]

2.2.3 Lack of Industrial Capacity

Though access to energy is certainly a necessary condition for economic development, economic development is, in turn, necessary to increase energy demand. A country that has limited industrial capacity is likely to need a relatively limited quantity of electricity.

Ghana, for example, has a rather strong economy if compared to the sub-Saharan region, and has made considerable gains in the expansion of energy access in recent years—providing 79% of the population with electricity in 2017 (World Bank 2018b). After suffering power shortages in 2014 and 2015, the government reactivated fast-tracking private power plants. However, today the country faces the opposite problem: excess electricity.[12] In other words, demand is lower than supply, posing severe financial risks to Ghana's economy. One might well ask why the power supply does not automatically generate its own demand among the 79% of the population already connected to electricity—especially considering the low GDP[13] per capita of the country (World Bank 2019a)—but energy supply will never ensure industrial development if adequate policy measures are not adopted by the government. In other words, energy access is a necessary—but not sufficient—condition to industrial development. The

---

[11] It can either be a gas pipeline or a regasification facility.

[12] The six million people (21% of the population) still lacking energy access would need additional investment in the power grid or off-grid connection.

[13] According to World Bank Data, in 2018, the GDP per capita level in Ghana was still less than 20% of global GDP per capita.

case of Ghana well represents a situation in which limited growth in energy demand does not depend on limited supply but on macro-economic constraints.

As of now, there are only a few countries in the sub-Saharan region where industrial development could drive energy demand growth. South Africa, Kenya and Ethiopia represent different examples of this. South Africa is the most advanced economy in the continent, but its industrial production growth is hindered by frequent electricity shortages. In Kenya, with the blueprint development programme 'Kenya Vision 2030', the government aims to achieve middle-income status by 2030, continuing an extraordinary two-decade-long development trajectory. Ethiopia is one of the fastest growing economies in the world and is embarking on its next phase of economic and social development supported by economic reforms and a bold programme of infrastructure investments.

2.3   The Ongoing Transition

As a consequence of these limitations and due to rapid population growth[14] (United Nations 2019) and uneven progress across the region, the number of energy-poor people is expected to remain unchanged in 2030 (IEA 2018) even in the best-case scenario where all current and announced government policies are implemented. In this context of under-achievement, although a shift to cleaner energy sources is partly taking place, talking of 'transition' may be an overstatement.

Some elements indicate nevertheless that many countries in sub-Saharan Africa are starting to move in the direction prospected by the International Energy Agency, that is, towards a power sector mostly fueled by renewable energy sources and natural gas.

In recent years, many African governments have designed energy strategies aimed at pursuing full energy access with the contribution of regionally abundant renewable energy sources. The number and size of wind and solar photovoltaic (PV) projects grew significantly, reaching a total installed capacity of 6,520 MW at the end of 2018, up from 535 MW in 2013 (Enerdata 2019). Such growth is the result of joint efforts between national governments, development finance institutions and private companies. Even when it comes to gas-to-power projects, many sub-Saharan countries are actively pursuing programmes to grow their gas economies, with gas-fired generation capacity doubled in the past few years, from 10.4 GW in 2010 to 20 GW in 2017 (Enerdata 2019). What Are then the Main Drivers Behind the Move of Sub-Saharan Governments Towards Cleaner Energy Sources?

---

[14] According to the United Nations, sub-Saharan Africa population will be 1.4 bn people in 2030 and will double in the next 30 years, reaching 2.1 bn people in 2050.

2.3.1 Aligning to International Norms to Leverage International Support

One of the drivers of Africa's move towards the energy transition is to abide by the Paris Agreement to which African countries are signatories. Despite sub-Saharan Africa's negligible levels of CO_2 emissions[15] (World Bank 2018a), climate change will impact the continent more than many other geographic areas. This looming scenario makes the region a good candidate to receive 'adaptation support' provided by developed to developing countries in the framework of international climate negotiations.

Prior to the Paris Conference, African countries submitted NDCs outlining how they intended to address climate change and what they would do if adequate financing were available. They perceived COP21 as an opportunity to leverage support from the international community to achieve sustainable development (Africa Union 2015). Adding renewables to their power mix was indeed considered a way to obtain technical and financial support from development finance institutions mandated to support infrastructure enhancement in developing countries and, at the same time, attract investments and capacity building from international energy companies interested in gaining/increasing their foothold in African markets.

2.3.2 Reducing Costs for Energy Access

The second driver behind the move towards cleaner forms of energy is the reduction of costs to access energy. The fact that electricity tariffs often do not cover generation costs, hence putting utilities under financial stress, burdens state budgets and drives African governments to favour the use of cheaper generation sources. At the time of writing, the Levelized Cost Of Electricity (LCOE) of solar PV and wind technology has, in many cases, reached the fossil fuel range (IRENA 2019a, b, c).[16] Furthermore, renewables have enabled the development of off-grid solutions that are increasingly attracting the interest of the most important international energy companies for their ability to provide energy access quickly and cheaply. According to Bilotta and Colantoni, 'off-grid generation has a series of advantages for a large part of the African population, particularly the rural one, as it solves the issue of the dispersion of the population, requires smaller investments and is now affordable even for many of the most fragile consumers'. (2018, p. 4). However, while off-grid generation is, in many cases, the cheapest solution to ensure 'basic' energy access,

---

[15] According to the World Bank, CO_2 emissions in sub-Saharan Africa account for 2.3% of global CO_2 emissions.

[16] Irena reports that the decline between 2010 and 2018 in the global weighted average LCOE is 77% for solar PV and 35% for onshore wind. The global weighted average LCOE of solar PV was USD 0.056/kWh in 2018; the global weighted average LCOE of onshore wind was 0.085/kWh in 2018; the fossil-fuel-fired power generation cost range by country and fuel is estimated to be between USD 0.049 and USD 0.174/kWh.

it is instead largely insufficient to guarantee 'industrial' access, necessary to sustain the pace and size of economic development that the region needs.

2.3.3 Energy Security

Finally, the third driver of the low-carbon transition in Africa is energy security. Reliable supply of energy is one of the most important requirements for significant growth. African governments have started to identify the wind and the sun—abundant and widespread across the continent—as crucial sources of energy and means to increase energy security, through fuel diversification and greater energy independence (IRENA 2019a, b, c). Energy self-sufficiency reduces countries' exposure to the price and supply volatility of importing energy.

2.4 Case Studies

The cases of Kenya, Nigeria and South Africa represent different cases on how natural gas and renewable energy were introduced in the energy system to meet growing electricity demand. Different economic and political contexts in these countries shape the pursuit of the goal.

2.4.1 Kenya

Kenya is one of the most successful cases of renewable energy development in sub-Saharan Africa. Kenya's electricity access rate increased massively from 15% in 2000 to 63% in 2017 (World Bank 2018b) and at the end of 2018 Africa's largest wind power project—Lake Turkana Wind Power project (310 MW)—was completed. However, a slower-than-forecast economic growth led energy demand in the country to be incompatible with the numerous Power Purchase Agreements (PPAs) to which the government had previously committed. As a consequence, despite an incredible rise of renewable energy generation in the country, Kenya is currently facing difficulties in the reduction of electricity cost, hence halting the development of further renewable energy projects.

With a third of the population still lacking access to electricity, a major cause of missing demand is also linked to the current state of the power grid that does not allow power supply to reach final demand. Moreover, the high cost of power generation based on heavy fuel oil drove the government, in the past few years, to attempt shifting to natural gas (i.e. LNG). The shift, however, encountered resistance and did not gather sufficient political consensus. The introduction of natural gas interfered with plans to develop a 1050 MW coal power project.[17]

---

[17]Currently suspended by the National Environment Tribunal.

2.4.2   Nigeria

Nigeria is instead a prime example of infrastructure inadequacy and power system economic unsustainability. Despite having a total installed capacity of 12.5 GW and an energy access rate of 54.4% (World Bank 2018a), the country can deliver only about 5.2 GW of power to its citizens. The grid's inefficiency and non-capillarity and the inability of the national utility to collect cash from its customers are the main causes of under-performance. This leads the government to use pollutants and costly diesel-fuelled generators as backup solutions to face frequent power shortages.

Also, political tensions add up as energy subsidies are the subject of an ongoing challenging negotiation between the government of Nigeria and the World Bank Group for a large financial package in support of the Power Sector Recovery Program.

2.4.3   South Africa

South Africa is a different case, and a peculiar one in the sub-Saharan context, both from an economic development perspective and an energy standpoint. The country's GDP per capita (USD 6,339 in 2018) (World Bank 2019b) is far above the sub-Saharan average (USD 1,573); half of the electricity generated in sub-Saharan Africa is consumed in South Africa; 90% of generation comes from coal. However, the country suffers from power shortages, and its coal fleet is ageing.

> The authorities planned a gas-to-power programme aimed at building and providing 3 GW of gas-fired plants. Gas will be both imported and domestically sourced. Though the government invited expressions of interest for the construction of a 600 MW gas power plant at either the port at Saldanha or at Richards Bay, at the end of 2017, the project was delayed. Later on, in September 2018, the government disclosed the Integrated Resource Plan (IRP) in which it outlined that gas-fired power was going to become more important in the power mix with LNG imports expected to drive the shift initially. However, according to Fulwood, political difficulties associated with the social implications of replacing coal with gas make the project unlikely to be developed before 2025 (2019, p. 16).

In terms of renewables, even though South Africa has been one of the first countries in Africa to develop solar and wind projects with the Renewable Energy Independent Power Procurement Programme (REIPPP), an ongoing political debate on the cost of energy is currently hindering the implementation of several renewable energy projects.

3   Geopolitical Dynamics

Before deep diving into the geopolitical dimensions, it is worth reminding that the literature has so far 'only barely scratched the surface with regard to exploring the potential geopolitical effects of the transition towards more renewable energy sources' (Criekemans 2011, p. 4). Even less with regard to Africa.

In fact, though the energy transition has become a sensational topic of debate, much remains unclear about its geopolitical implications in Africa and globally. Predicting winners and losers in this transition then becomes for some highly uncertain (Hache 2016; Paltsev 2016), due to the increased complexity of energy geopolitics, consequential to a more heterogeneous set of technologies and actors involved. For others (Stang 2016; Huebner), it is more straightforward: winners are those countries with high energy consumption and few own resources like India, China, Mexico, Brazil and Europe. Losers are instead leading oil and gas exporters whose leverage decreases as energy types and suppliers diversify, and the need for long-distance transport of fuels diminishes due to decentral generation and smart grids.

In such blurred situation, a few analytical considerations are due. The first is that the arrival of renewable sources of energy is triggering a 'transposition of the geopolitical logic of oil and gas onto renewables, despite the considerable differences between the energy types and their associated technologies and infrastructure' (Overland 2019, p. 36). This chapter argues that while some dynamics inevitably remain unvaried, namely, dependence on external financing and technology, hence justifying the 'transposition of the geopolitical logic', the peculiar characteristics of renewables[18] will add new dynamics to the geopolitical conundrum. In Africa, this mainly translates in a 'regionalisation of energy relations' (Scholten, p. 23).

Secondly, energy-related issues will partly be 'less about locations and resources, and thus less geopolitical in nature' (Overland 2019, p. 38). Rather, international energy competition 'may shift from control over physical resources and their locations and transportation routes to technology and intellectual property rights' (Overland 2019, p. 38). In other words, we may talk more about tech-politics than geopolitics. A caveat, however, is that natural resources still need to be extracted to produce technology, and Africa currently supplies some critical items. Access to natural 'critical materials' will hence be crucial and is likely to replicate the type of rent-seeking dependent relation with international actors typical of oil and gas.

These initial considerations need to be, however, contextualised in a world where the replacement of fossil fuels cannot realistically happen any time soon. Seemingly then, according to Scholten, 'the fact that both fossil fuels and renewable energy will coexist in the energy mix for the foreseeable future implies that any (practically relevant) understanding of the geopolitics of renewables is in essence about how the energy transition affects fossil fuel dominated interstate energy relations' (Scholten 2018, pp. 10–11).

---

[18]That is, being abundant and intermittent, relying on decentralised generation and usage of rare earth materials in clean tech equipment, mostly electric distribution, requirement of stringent managerial conditions and long-distance losses.

3.1 International Dependence and Regionalised Energy Systems

Even if Africa is becoming the locus for the extraction of 'new' resources needed to manufacture renewable energy technologies, as well as home to a variety of projects aimed at harnessing renewable energy, oil and gas are expected to be around for quite some time still.

On the one hand, there will be (there already are) endogenous drives to diversify economies and reduce reliance on hydrocarbons. On the other hand, global aspirations to increasingly use cleaner energy sources may translate into a decreased relevance of Africa's role in global hydrocarbon fluxes, altering trade relationships and geopolitical connections. According to the IEA New Policies Scenario, global oil demand is expected to peak only in 2040 to then downturn. However, should the pace of the global energy transition unexpectedly accelerate,[19] Sub-Saharan economies—in particular, those heavily reliant on oil revenues[20]—may be profoundly destabilised if diversification strategies are not timely implemented (IRENA, A New World The Geopolitics of the Energy Transformation 2019).

In this context, competition changes. While affordable *access* to oil, coal and gas is the main ground of competition in the fossil fuels world, in a low-carbon world, the struggle will still partly revolve around access to materials, required to manufacture technology, while also partly around how to *finance* infrastructure and control the *technology* needed to harness wind, solar and other renewable power sources. African countries will likely continue relying on external finance and technology to advance.

Nonetheless, these dynamics of dependence are increasingly complemented by dynamics of inter-dependency. While dependence characterises relations between Africa and international actors, interdependence is part of regional integration phenomena, in which the energy sector could act as a catalyst for development (see power pools).

3.1.1 Infrastructure Financing

According to Bilotta and Colantoni, 'the SSA energy sector has suffered from the same chronic difficulty as African infrastructural projects in finding adequate investments, due to the lack of domestic funds as well as to a higher perceived regional risk' (Bilotta and Colantoni 2018, p. 5). Domestic factors of instability, such as political struggles, GDP fluctuations, corruption and lack of transparency, currency risk, led

---

[19] According to the Sustainable Development Scenario of the International Energy Agency, global oil demand falls to 93.9 mb/d in 2025 and 69.9 mb/d in 2040. The Sustainable Development Scenario provides 'an integrated strategy to achieve energy access, air quality and climate goals, with all sectors and low-carbon technologies—including carbon capture, utilisation and storage—contributing to a broad transformation of global energy'.

[20] For instance, according to the World Bank Data, fossil fuel rents amount to over 15% of GDP in Chad, South Sudan, Angola, Gabon, Equatorial Guinea and the Republic of the Congo.

African countries to mostly depend upon external actors to finance the development of energy projects. With renewables, the dynamic seems to be replicated.

Young African entrepreneurs in the energy space face challenges related to access to finance (in addition to lack of technical knowledge). Even though locally bred pay-as-you-go companies, such as M-KOPA and BBOXX, are rising, they face the challenge of having to raise sufficient capital to finance the upfront cost of solar panels due to reluctance from local financial institutions to provide financing. Hence, the need to rely on international investors, meaning that the inherent transaction costs, currency risks and profit expectations will be translated into higher solar home system prices, preventing greater uptake of decentralised off-grid solutions and limiting employment opportunities (IRENA, Renewable Energy and Jobs 2018, p. 23).

As for international investors, the incentives to invest in oil and gas infrastructure differ from those in renewables, including electrification. If in the oil and gas world, incentives are relatively high and resources can be sold on international markets, in the renewables-electric world, the non-exportability of resources and the need to use them in underdeveloped domestic markets make the incentive for the creation of power infrastructure lower. For this reason, domestic and international political support is crucial to guarantee the viability and long-term security of the projects.

Development Finance Institutions (DFIs)[21] are increasingly playing an important role in supporting the energy transition in developing countries.[22] In sub-Saharan Africa, DFIs are involved in infrastructure projects along the entire energy sector value chain. They make use of a varied range of instruments. In the last few years, the most important multilateral DFIs have been enhancing their support to renewable energy projects. Two of the most notable programmes supporting private renewable energy projects in sub-Saharan Africa are the World Bank's Scaling Solar and the European Union's External Investment Plan[23]—the former was launched in 2015, while the latter is more recent.

These two programmes provide technical and financial support to both governments and project developers for the development of power generation projects and mitigation of the off-taker risk through credit enhancement mechanisms (such as partial risk guarantee, which is a core instrument to make power generation projects bankable). However, while these programmes are effective in mobilising private investments in power generation, they are unable to address the deep causes of inefficiency and unsustainability of the sector. They need to be complemented by other initiatives along the electricity value chain for the power system to benefit effectively. Integrated approaches are hence preferred in this context.

The Temane Regional Electricity Project (TREP) in Southern Mozambique well exemplifies how an integrated approach looks like. In 2019, the Board of Directors

---

[21]Development finance institutions are legally independent and government-supported financial institutions with explicit official missions to promote public policy objectives.

[22]According to the OECD, multilateral climate finance from developed countries to developing countries almost doubled in recent years, from USD 15.5 billion in 2013 to USD 27.5 billion in 2017.

[23]The External Investment Plan has identified five priority areas, but we refer to the 'Sustainable Energy and Sustainable Connectivity' investment window.

of the World Bank approved a total USD 420 million of International Development Association (IDA) grants and guarantees to strengthen Mozambique's transmission capacity for domestic and regional markets and increase electricity generation capacity. Developed as an integrated operation including both public and private investments, the TREP entails the construction of a 563-km high-voltage transmission line between Maputo and Temane and a private sector financing of a 400 MW CCGT generation plant in Temane. A USD 300 million grant will be provided for the construction of the transmission line and two IDA payment guarantees to de-risk the sale of electricity and the purchase of gas by the power plant. This approach requires a much more challenging and costly effort by all involved stakeholders. But the joint approach proves very effective as it simultaneously improves supply and demand.

In addition to centralised power generation, the World Bank has started also focusing on energy access programmes through mini-grid and off-grid projects so as to reach sections of the population that are not served by the national grid. This Africa-specific approach is crucial not only to bypass infrastructural deficiencies but also to 'deliver energy in a way that African consumers will be able to afford—even the poorest strata of the population, and rural consumers in particular'. (Bilotta and Colantoni 2018, p. 6).

3.1.2 Control of Technology

The energy transition is above all about technology and innovation. African countries' role in this realm does not differ, so far, from the role played in the oil and gas world. They are suppliers of critical materials needed to manufacture renewable energy technologies but do not hold the leadership in terms of technological innovation or control of technologies' manufacturing. Hence, the relationship of dependence upon external actors is seemingly reproduced.

Suppliers of critical materials
The nascent energy transition and ever-evolving technological advancement have expanded the number of resources considered strategic. Africa is home to some so-defined critical materials, that is, materials 'which most consumer countries are dependent on importing, and whose supply is dominated by one or a few producers' (Overland, 2019, p. 37).

The Democratic Republic of the Congo (DRC), for instance, accounts for most of cobalt[24] world production (more than 60%). Exports are mainly China-bound with the far eastern country being the world's leading consumer in 2018. China primarily uses cobalt in the rechargeable battery industry. Securing access to the resource translated in the acquisition by Chinese companies of eight of the 14 largest cobalt mines in the DRC, accounting for almost half of the country's output. South Africa is

---

[24]Cobalt is a crucial component for the manufacture of lithium-ion batteries.

instead the world's leading producer of manganese[25] (30.6%) followed by Australia (17.2%), Gabon (12.8%), China (10%) and Brazil (6.7%).

Another strategic mineral present in Africa is Nickel. About 23.5% of reserves of this mineral are located in Madagascar and 4% in South Africa (USGS 2019). Other strategic materials such as lithium are overwhelmingly located in South America and are comparatively scarce in Africa in terms of reserves, resources and production.

It is important to stress that, while cobalt, nickel and other minor minerals are extracted from the African soil, their refining and manufacturing take place primarily abroad. In this process, China plays a fundamental role by virtue of its comparative advantage and remarkable refining capacity. In general, the extraction of these minerals presents a series of problems of environmental nature (particularly water-intensive and polluting). On top of that, the issues linked to the highly extractive mining industry (illegal mining activities, conflict subsidisation, low worker protection) have historically been taking a toll on local communities in Africa.

Nonetheless, though these materials are currently crucial, given the fast-pace of technological innovations in the field, it is not improbable that currently used materials may soon be displaced by others. In other words, their geopolitical relevance is vulnerable not because of their geological abundance/scarcity, but because of the rapidity of technological innovation.

Leadership in technological innovation

Advancements in renewable energy technology guarantee the centrality of a country in a wider geopolitical perspective as 'innovation will be a key determinant of the pace of change' (IRENA 2019, p. 27). Though it is clear that leadership in technological innovation mostly does not originate in Africa, Scholten notes that it is, however, unclear 'how developments like great power rivalry between the US and China or the EU and Russia and technical innovations in batteries or ICT will influence the speed and direction of the energy transition and nature of energy systems'. (p. 4).

One way to evaluate countries' approach to technological innovation is to look at how they fare in terms of patents. The foremost leader in terms of renewable energy innovation is by far China, with more than 150.000 patents as of 2016 (IRENA 2019). It is followed by the United States with a little more than 100.000 patents. Japan occupies the third place, soon tailed by the EU, where Germany is in a leading position with 30.000 patents.

As for Africa, the main owner of renewable energy patents is South Africa, with 63% of total African patents, followed by Egypt with 8% (UNEP, EPO 2013). Specifically, South Africa is the leader in the field of mitigation[26] technologies, with 553 patents, accounting for 84.2% of the African total. Northern Africa—represented by

---

[25] Manganese is another important material for batteries and, as of now, it has no satisfactory substitute in its major applications (U.S. Geological Survey, Mineral Commodity Summaries, February 2019).

[26] 'Climate Change Mitigation refers to efforts to reduce or prevent emission of greenhouse gases. Mitigation can mean using new technologies and renewable energies, making older equipment more energy efficient, or changing management practices or consumer behaviour. It can be as complex as a plan for a new city or as a simple as improvements to a cook stove design. Efforts underway

Egypt (18 patents), Algeria (12) and Morocco (11)—follows suit, but lags significantly behind South Africa. Other realities, such as Ghana, Burundi, Mali, Senegal and Zimbabwe, have less than 1% of total African mitigation patents, with only Kenya reaching 1.2%. The rest of Africa possesses 3.9% of patents (25 over 657 in the entire African continent).

The number of African inventions related to renewable energy, as of 2016, was significantly lower than that found in other countries or regions, translating into higher exposure for African states to the shocks brought by the transition. Apart from the economic consequences of these shocks suffered especially by countries highly reliant on fossil fuels, lagging behind in terms of innovation corresponds to decreased political importance.

Nonetheless, African countries still possess a comparative advantage for other types of energy-transition-related technologies. These are Made-in-Africa technologies that facilitate access to energy. Examples are 'smart payment' or 'pay-as-you-go' systems. These smartphone-friendly apps are intended to facilitate payments in particularly remote rural areas where banks are not readily available.

3.1.3 Regionalised Energy Systems/Power Pools

The likely geopolitical implications of increased usage of renewables are first 'a regionalisation of energy relations' (Scholten, p. 23). Renewables intrinsically need to use electricity as a carrier and electricity is currently a regionally traded commodity—rather than internationally traded like oil and gas—due to long-distance losses (IRENA, A New World, p. 47). Second, they imply a 'strategic emphasis on continuity of *service supply* instead of *commodity supply* due to renewables' abundance and stringent managerial conditions' (Scholten, p. 4). As a consequence, a shift to the green economy could increase intra-continental trade while creating important interdependencies.

This adds up to the existing geopolitical conundrums for two reasons. The first refers back to the first two points made about finance and technology. In order to trade electricity regionally, grids need to be improved, and this strongly depends on external finances and technology, which are not always readily available hence making electricity trade less quick or smooth than wished. Secondly, though grids' development is needed in Africa, off-grid systems may develop more quickly and in a more widespread fashion, leading certain areas of the continent to leapfrog the grid system entirely.

Geopolitically, greater cross-border trade in electricity could create geopolitical vulnerabilities for electricity importers, but greater electric interconnection will also increase interdependence among nations, reducing risks of conflict. In fact, Overland challenges the fact that electricity disruptions can be used as a geopolitical weapon inasmuch as 'much of the future international solar and wind power

---

around the world range from high-tech subway systems to bicycling paths and walkways'. (UNEP, Mitigation).

trade will likely involve more symmetrical relationships between different prosumer (producer-consumer) countries than does the unidirectional gas trade (and much past electricity trade)' (2019, p. 38).

In a world in which energy can be produced in various locations, then it is less likely that few actors dominate the scene by controlling routes and chokepoints. However, control over grid infrastructure may become vital. While some argue that countries that dominate electricity grids may exercise undue control over their neighbours and that interstate electricity cut-offs will become an important foreign policy tool, applied strategically in the same way as oil and gas sanctions (O'Sullivan et al. 2017, as cited in IRENA), others note that 'electricity trading tends to be more reciprocal than trade in oil and gas [...] A country that generates solar power may import energy from a neighbouring country when it rains, but export to that neighbour when the sun shines' (2018, 51). As a consequence, 'renewable energy exporters will always be part of a complex web of interdependencies between importers and exporters that would tend to curtail the potential to use renewable electricity as a geopolitical weapon' (IRENA 2018, p. 52). In Africa, however, this may be problematic given the levels of energy infrastructure development across the continent vary widely, potentially either making a country's excess exports virtually impossible (for lack of electricity network or lack of payment capacity), or enhancing the asymmetry between energy exporters (countries with financial capacity to develop the production sector) and importers (countries unable to produce).

In sub-Saharan Africa, there are four power pools that have been 'established to improve generation capacity and transmission infrastructure for greater cross-border trade and ultimately address a cost-effective way of evacuating excess capacity between countries to offset peak demands' (Medinilla et al. 2019).

The Southern African Power Pool (SAPP) was the first electricity regional market created in sub-Saharan countries and it comprises 12 countries—nine of them are already interconnected; the East Africa Power Pool (EAPP) is composed of 11 countries but it is at a very early stage of development since it was established only in 2015; the West African Power Pool (WAPP), created in 1999 by the regional economic union ECOWAS, has ambitious development objectives with a USD 60 bn plan for transmission and distribution investments by 2030; the Central Africa Power Pool (CAPP), established in 2003 and comprising 10 member countries, is the least developed power pool in Africa and the one that would need the major infrastructure investments.

The vast disparities across countries in terms of political economies, governance and infrastructure could hinder efforts to create or further develop 'grid communities' or regional pools. Hence, strategic imperatives for success and growth of power pools in Africa require countries to focus on diversifying power generation sources by taking advantage of the renewable energy potential, which will help mitigate the respective power pools' vulnerability and their impact on regional dynamics. Moreover, power pools will not reach their objectives if member state power utilities do not invest in transmission capacities and maintenance.

Moreover, the challenges to achieving functioning power pools are technical, but also political. Power pooling requires trust and a strong alignment of interests between the region's member states, between the regions (and/or member states) and the national private sector, and between external partners' and member states. Indeed, vulnerabilities (of domestic/regional systems) that hinder the development of regional energy integration could be mitigated if international actors contributed, pragmatically, to fostering 'positive interdependence' through support in developing priority transnational infrastructure.

In this context, it is worth noting that 'increasing electrification of the energy system […] implies the reliance on a single transport modality' (Scholten 2018, 23), hence replicating the same risky dynamic of lack of diversification. At the same time, this would certainly give to landlocked countries (generally disadvantaged in the oil & gas world) better chances of being connected and further their development goals.

3.2   Socio-Economic Implications and Security Risks

Understanding the overall political and economic landscape, in other words, the particular context in which a transition is unfolding is crucial. Most accounts on the shift to renewables tend to see the energy transition as a mere shift in the energy mix from a source to another without accounting for the 'disruptive potential of renewables to redefine energy systems and markets' (Scholten 2018, p. 9) or for the fact that historic contingencies play a crucial role in technological change (Baker et al. 2014, p. 798). At the same time, it is equally important to identify the international macro-economic forces and political actors, institutions and processes that have an impact on how domestic policy choices and debates are shaped, enabled and constrained (Baker et al. 2014, p. 795).

3.2.1   Socio-Economic Growth and Disruptions

The transition to cleaner sources of energy can prove particularly disruptive in countries where fossil fuels provide crucial revenues to the state. While current importers of fossil fuels (or countries where production is for domestic use) are starting to replace fossil fuels for domestic use, many others in Africa are still heavily reliant on these sources. For oil, gas and coal producers, like Nigeria, Angola or South Africa, the decline in revenue generated from fossil fuel energy exports can provide an impetus for political reform and economic diversification. However, a decline in hydrocarbons revenue could also lead to political instability, especially in the short to medium term. These countries, unless they have ambitious strategies of economic diversification, could face some severe challenges in the near future. Indeed, countries in sub-Saharan Africa rank low on the Energy Transition Index designed by the World Economic Forum, signalling their lack of readiness for the energy transition.

However, renewables in Africa could also have a positive impact on economic growth. For instance, off-grid access to electricity can contribute to achieving better levels of 'basic' access across the continent—though 'industrial' access to energy will still be reliant on fossil fuels, hence constrained by the sector's limits. Increased basic access, especially in rural areas, could slow down urbanisation processes, which in turn have a high impact on grid/electricity demand and on air quality (which has worsened significantly across the continent in the past 30 years[27]). Indeed, if citizens were guaranteed a decent and sustainable livelihood in rural areas, the impetus to move to the city might decrease. At the same time, according to IRENA the majority of countries in sub-Saharan Africa 'will benefit from reducing fossil fuel imports and generating renewable energy domestically, because this will boost job creation and economic growth' (IRENA 2019, p. 30), with the exception of Nigeria and Angola, whose dependence on fossil fuels rents will place them at risk.

Competing narratives and vested interests

The potential shift (even before becoming actual) translates into (often competing) narratives being played out, like the 'energy security' and 'sustainability' ones (Jacob 2017, pp. 348–349). Energy security narratives sustain that a country, to develop and industrialise, must be energy-secure and less vulnerable to disruptions to its energy supply. This approach justifies the use of whatever energy resource to support development. The case of South Africa is emblematic: considered by IRENA the leader in renewables development in Africa, the country is also Africa's top coal producer, with the Mpumalanga Province being the second worst sulphur dioxide[28] emission hotspot in the world (Greenpeace 2019). In Tanzania instead, the Minister of Energy and Minerals notes that 'coal to electricity is necessary [...] because its cost will be cheaper for citizens and this electricity will boost industrial growth' (All Africa, 16 January 2016).

The counter-narrative implies instead a strict adherence to the global low-carbon movement. This approach sees clean energy as crucial to reduce dependence on fossil fuels and achieve sustainability and low carbon development. As Jacob notes in relation to Tanzania, 'this narrative is grounded in the claim that ongoing and planned coal investments will create obstacles to Tanzania's efforts to meet its obligations to reduce carbon emissions and mitigate climate change' (p. 350).

Behind these narratives, Okem notes, there is a shift between 'competing industrial sectors and political constituencies' (Asuelime 2018, p. 3). The implications of the shift range from employment concerns, especially in those countries where fossil fuels employ large numbers of the population, to concerns over maintaining loyalty of voters and business circles for political survival (Whitfield et al. 2015). Indeed, if the energy transformation begins to permeate into industrial sectors that have traditionally been dominated by fossil fuel energy, there could potentially be severe social disruption, and we are currently seeing the fear of this disruption playing out in the coal industry.

---

[27] According to UNICEF, between 1990 and 2017 deaths from outdoor air pollution in Africa grew by 60% (2019, p. 4). The figure is expected to grow even further as population growth, industrial growth and consumption growth—all expected in sub-Saharan Africa in the next decades—increase levels of pollution, especially in urban areas.

[28] Sulphur dioxide is a toxic pollutant that causes disastrous air pollution and premature deaths.

Technological and political change is, however, also embedded and affected by broader global processes. In a study on South Africa, Baker, Newell and Phillips (p. 795) argue that global processes have the potential to alter the balance of power within the country between 'entrenched coal-based interests' and 'emerging niches in renewable energy generation' as both levels 'interact with and are backed by' international stakeholders. China, for example, has emerged as the 'leading financier of Africa's coal boom' (Jacob 2017, p. 345).

Employment

Most African countries are expected to benefit, in terms of job creation, from the energy shift (IRENA 2019, p. 30). The sector's workforce—in terms of direct, formal jobs—is already comparable to traditional power grids and utilities in Nigeria and Kenya where the job creation impact is expected to grow in the next few years—by 70% in Kenya and over 100% in Nigeria. However, these are exceptions. Employment of renewable energy still remains limited in Africa as a whole, especially when compared to the growth in Asia (60% growth in 2017 compared to 51% in 2013). Most of Asia's dynamism is based on growing domestic deployment and strong manufacturing capabilities, supported by policies such as feed-in tariffs, auctions, preferential credit and land policies, and local content rules.

In Africa, instead, compared to the oil and gas sector—which is capital intensive but not labour intensive—or mining—which is very labour intensive but localised—employment in the renewables sector is less labour intensive (than oil and gas) but not comprehensively developed along the value chain. In fact, employment is generated especially in the sales and distribution, installation, and operations and maintenance of the supply chain—vis-à-vis manufacturing. As a consequence, as long as there is only a limited domestic capacity to assemble equipment or manufacture products, economic multipliers and the resulting employment and other benefits will accrue elsewhere.

South Africa is the country that employs the largest contingent of workers in the renewable energy sector, close to 35,000, distributed across solar PV, concentrated solar power (CSP) and wind. Important developments are also taking place in the off-grid sector. In Ghana, Africa's largest solar PV project (Nzema plant, generating 155 MW) is likely to induce 2,100 local jobs through subcontracting and demand for goods and services.

However, a global review[29] of skills for green jobs including four countries in Africa (Egypt, Mali, South Africa and Uganda) revealed the existence of a gap between the goals and targets set in environmental policies and the human resources available for their implementation. Skills gaps exist for technical and engineering positions and could grow as the renewable energy sector continues to expand, leading to project delays or cancellations, cost overruns and faulty installations. There are, however, promising experiences. For example, Cape Verde launched a Renewable

---

[29]High-Level Political Forum on Sustainable Development. 'Interlinkages between energy and jobs'. Policy Brief No. 13.

Energy and Industrial Maintenance Center (Cermi), whose main activity is the training of professionals in the areas of design, assembly and maintenance of photovoltaic installations.

3.2.2 Security and Dominance/Cooperation and Conflict

Finally, renewables are likely to reduce conflicts as we know them, but other tensions are likely to arise around cybersecurity and access to important minerals (IRENA, 'A new world', p. 55).

In terms of cybersecurity, the fact that renewables rely on the use of electric grids as the carrier of energy, vis-à-vis pipelines, tankers, rail, road, the sea used to transport coal, gas and oil may reduce risks linked to crucial infrastructure or chokepoints. At the same time, however, grids may be subject to new vulnerabilities. Not because they are in Africa, nor because they are more likely to be cyber-attacked compared to other infrastructure (for instance, oil- and gas-related infrastructure). Rather because, as noted above, 'increasing electrification of the energy system […] implies the reliance on a single transport modality' (Scholten 2018, 23), hence replicating the same risky dynamic of lack of diversification. On the other hand, however, the diffusion of off-grid systems 'may actually make the system more resilient, as many different units will have to be hacked to destabilize the system as a whole' (Overland 2019, p. 38).

In terms of minerals instead, while cartels could develop around materials critical to renewable energy technologies, as noted above the fast-pace of technological innovations makes it likely for many of these materials to be displaced by others rather quickly, hence decreasing the chances of cartels—hard to form and sustain—to emerge (IRENA 2019, p. 54). In other words, their geopolitical relevance is vulnerable not because of their geological abundance/scarcity, but because of the rapidity of technological innovation.

4   Conclusions

Any attempts to generalise sub-Saharan African dynamics, in the energy sector as much as in any other area, risk being over-simplistic given the diversity of structures, ideas, leaderships and international positioning across the sub-continent. While this chapter tried to present specific examples, it also made the effort to identify common traits so to provide a manageable framework to understand the energy transition in sub-Saharan Africa and its geopolitical implications.

Problematising the concept of 'transition' in African contexts, instead of taking for granted that a transition is well received and occurring, served the purpose of placing proper emphasis on Africa's problem number one: access to energy, both 'basic' and 'industrial' access, that is, for households and the industry, respectively. Six out of ten people still lack basic access, and the remaining share is unable to consume as it would like due to non-reliable power networks or inadequate financial means. We then argue that most of Africa has very little to transit away from.

Nonetheless, for many governments, the idea of leapfrogging the conventional path to energy development and shift directly to clean energy is starting to become attractive. This is not only to win the hearts of the electorate—increasingly aware, if not victim, of climate change effects—but also to leverage the ad hoc financial support that international institutions are making available to help Africa achieve sustainable universal energy access.

Despite efforts, however, renewable energy—estimated by the IEA to be a crucial component, together with gas, of Africa's low-carbon energy mix in the next decades—in sub-Saharan Africa represents today a mere 1% reality. This is due to the fact that although Africa remains an important supplier of natural resources (namely, oil and gas but also minerals to manufacture renewables technologies, as well as renewables), it continues to rely on external financial and technological support to develop energy systems. Nonetheless, the intrinsic features of gas and renewables are shaping the creation of regional energy markets. These will inevitably imply increased cooperation and coordination between neighbours across the region.

As a fact, looking at the energy transition in Africa will not merely imply monitoring the evolution of Africa's dependence upon external actors, but will require a more alert acknowledgement that regional interdepencies are ever growing.

References

African Energy Live Data (2019). Retrieved from https://www.africa-energy.com/database
African Union (2015) Press release, Africa calls for a fair, equitable and legally binding agreement during the COP 21. Retrieved from https://au.int/en/newsevents/16848/africa-calls-fair-equitable-and-legally-binding-agreement-during-cop-21
Asuelime L (2018) Expanding the frontiers of the political economy of sub-Saharan Africa's energy. In: Asuelime, Okem (eds) The political economy of energy in sub-Saharan Africa, Routledge
Baker L, Newell P, Phillips J (2014) The political economy of energy transitions: the case of South Africa. New Polit Econ 19(6):791–818. https://doi.org/10.1080/13563467.2013.849674
Bilotta N, Colantoni L (2018) Financing energy access in sub-Saharan Africa. Istituto di Affari Internazionali. IAI Papers (18): 1–21. Retrieved from https://www.iai.it/it/pubblicazioni/financing-energy-access-sub-saharan-africa
Criekemans D (2011) The geopolitics of renewable energy: different or similar to the geopolitics of conventional energy? Conference paper presented at global governance: political authority in transition, ISA annual convention 2011, Montréal, Québec, Canada
Enerdata (2019) Energy mix, electricity production
EPO (2013) Patents and clean energy technologies in Africa. https://www.epo.org/news-issues/technology/sustainable-technologies/clean-energy/patents-africa.html
Fulwood M (2019) Opportunities for gas in Sub-Saharan Africa. Oxf Energy Insight (44): 1–24. Retrieved from https://www.oxfordenergy.org/wpcms/wp-content/uploads/2019/01/Opportunities-for-Gas-in-Sub-Saharan-Africa-Insight-44.pdf
Global Commission on the Geopolitics of Energy Transformation & IRENA (2019) A new world: the geopolitics of the energy transformation. Retrieved from https://geopoliticsofrenewables.org/assets/geopolitics/Reports/wp-content/uploads/2019/01/Global_commission_renewable_energy_2019.pdf

Greenpeace (2019) Global air pollution map: ranking the World's worst so2 and no2 emission hotspots.https://storage.googleapis.com/planet4-africastateless/2019/03/625c2655-ranking-so2-and-no2-hotspots_19-march-2019.pdf

Hache E (2016) La géopolitique des énergies renouvelables: amélioration de la sécurité énergétique et / ou nouvelles dépendances? (The geopolitics of renewables: ddoes more energy security come with more energy dependencies?), Revue Internationale et Stratégique 1(101):36–46

International Energy Agency (2018) World energy outlook 2018. Retrieved at https://www.iea.org/weo2018/

International Renewable Energy Agency (2019a) Tracking SDG 7: the energy progress report 2019. Retrieved from https://irena.org/publications/2019/May/Tracking-SDG7-The-Energy-Progress-Report-2019

International Renewable Energy Agency (2019b) Renewable power generation costs in 2018. Retrieved from https://www.irena.org/media/Files/IRENA/Agency/Publication/2019/May/IRENA_Renewable-Power-Generations-Costs-in-2018.pdf

International Renewable Energy Agency (2019c) Scaling up renewable energy deployment in Africa. Retrieved from https://www.irena.org/-/media/Files/IRENA/Agency/Regional-Group/Africa/IRENA_Africa_impact_2019.pdf?la=en&hash=EECD0F6E8195698842965E63841284997097D9AA

Jacob T (2017) Competing energy narrative in Tanzania: towards the political economy of coal. Afr Aff 116(463):341–353. https://doi.org/10.1093/afraf/adx002

Kojima M, Trimble C (2016) Making power affordable for Africa and viable for its utilities. Retrieved from http://documents.worldbank.org/curated/en/293531475067040608/Making-power-affordable-for-Africa-and-viable-for-its-utilities

Medinilla A, Byiers B, Karaki K (2019) African power pools

O'Sullivan ML, Overland I, Sandalow D (2017) The geopolitics of renewable energy. Columbia Centre on Global Energy Policy, Working Paper. Retrieved from https://energypolicy.columbia.edu/sites/default/files/CGEPTheGeopoliticsOfRenewables.pdf

Overland I (2019) The geopolitics of renewable energy: debunking four emerging myths. Energy Res Soc Sci 49:36–40. https://doi.org/10.1016/j.erss.2018.10.018

Paltsev S (2016) The complicated geopolitics of renewable energy. Bull Atom Sci 72:390–395. https://doi.org/10.1080/00963402.2016.1240476

Scholten D (ed) (2018) The geopolitics of renewables. Springer, New York

Stang G (2016) Shaping the future of energy. European Union Institute for Security Studies (EUISS), Brief Issue, 24

Stratfor (2018) How renewable energy will change geopolitics. Retrieved from https://worldview.stratfor.com/article/how-renewable-energy-will-change-geopolitics

United Nations Conference on Trade and Development (UNCTAD) (2019) Economic development in Africa Report 2019, made in Africa—rules of origin for enhanced Intra-African trade, New York

Whitfield L, Therkildsen O, Buur L, Kjær AM (2015) The politics of African industrial policy: a comparative perspective. Cambridge University Press, New York

World Bank (2018a) Open data: Access to electricity (% of population). Retrieved from https://data.worldbank.org/indicator/EG.ELC.ACCS.ZS

World Bank (2018b) Open data: CO_2 emissions (kt). Retrieved from https://data.worldbank.org/indicator/EN.ATM.CO2E.KT?locations=ZG-1W

World Bank (2019a) Open data: Population, total. Retrieved from https://data.worldbank.org/indicator/SP.POP.TOTL

World Bank (2019b) Open data: GDP per capita (current US$). Retrieved from https://data.worldbank.org/indicator/NY.GDP.PCAP.CD

In-Depth Focus on Selected Issues

Technologies for the Global Energy Transition

Manfred Hafner and Michel Noussan

1 Introduction

The United Nations Intergovernmental Panel on Climate Change 2018 Report stated, "Limiting global warming to 1.5 °C would require rapid, far-reaching and unprecedented changes in all aspects of society." In recent years, it has become clear that that scenario would require not only a transformation of our energy system in order to meet our global emissions targets, but also a rethinking of the way we control the temperature of our homes, travel around our planet, and manufacture our goods.

In order to meet this transformation by mid-century, scientists, engineers, and technical experts are needed in the crucial role of designing pathways for the decarbonization process of specific, energy-intensive sectors, notably power, industry, transport, and buildings. Fondazione Eni Enrico Mattei (FEEM) and the Sustainable Development Solutions Network (SDSN) invited more than 60 technical experts from around the world to gather in Milan in April 2019 to discuss the state of decarbonization technologies that can accelerate the global shift toward decarbonization. The outcome of that workshop, followed by an extensive external consultation and review process by a large number of scholars and stakeholders (from international agencies, academia, research centers, think tanks, non-governmental organizations, public institutions, and the private sector), was the basis of the "Roadmap to 2050: A Manual for Nations to Decarbonization by Mid-Century" (Carnevale and Sachs 2019). Some of the contents of this chapter are partially based on the work of this Roadmap to 2050.

In order to decarbonize the global economy, energy demand growth needs to be uncoupled from economic growth, and then the remaining energy demand needs to be decarbonized. This chapter provides an overview of the latest decarbonization technologies available for national governments to develop their low-emission

M. Hafner (✉) · M. Noussan
Fondazione Eni Enrico Mattei (FEEM), Milan, Italy
e-mail: manfred.hafner@feem.it

© The Author(s) 2020   177
M. Hafner and S. Tagliapietra (eds.), *The Geopolitics of the Global Energy Transition*,
Lecture Notes in Energy 73, https://doi.org/10.1007/978-3-030-39066-2_8

development strategies as outlined in the Paris Agreement. Following the Paris Climate Agreement's aim to strengthen the global response to the climate crisis "in the context of sustainable development and efforts to eradicate poverty," the chapter is conceived on a "systems approach," aspiring to simultaneously address multiple objectives and promote policy instruments and technological solutions that can be used across sectors. The multiple objectives span decarbonization and environmental sustainability, economic prosperity (including poverty reduction), and social inclusion that leave no one behind. Needed policy instruments include public investments, phase out of subsidies to fossil fuels, market mechanisms, regulatory framework, and regulations on land use, while technological solutions address a wide range of current and emerging solutions, from smart power grids to synthetic fuels.

According to IEA (2019a), global CO_2 emissions caused by fossil-fuel combustion for energy production totaled 32.3 Gt in 2016. The first responsible has been the heat and electricity generation, with 13.4 Gt (41%), followed by transport (7.9 Gt, 24%), industry (6.1 Gt, 19%), and buildings (2.7 Gt, 8%). However, if electricity and heat generation are allocated to the relevant final sectors, industry takes the lead with 11.8 Gt (36%), while buildings and transport are comparable, with 8.6 and 8.1 Gt, respectively (i.e., 27% and 25%). The remaining emissions are related to other sectors (including agriculture/forestry, fishing, and other non-specified).

The chapter discusses the main technology options for decarbonization both of the power sector and of the three final sectors: industry, transport, and buildings. At the end of the chapter, we provide strategy and policy recommendations from a technology point of view on how to decarbonize these sectors by mid-century and of the necessity to take a systems approach.

2   The Power Sector

Electricity is the fastest growing energy vector, its demand has increased by a 3% annual growth rate since 2000 (IEA 2018), around two-thirds faster than the total final energy consumption at global scale. Worldwide electricity demand in 2017 reached 22,200 TWh according to the latest IEA's World Energy Outlook, and the future demand is expected to increase in all the different scenarios. The main reasons are an increasing access to electricity in developing countries and an increasing consumption of current users due to increased well-being. Although 840 million people still lack access to electricity as of 2017 (World Bank Group 2019), this number decreased from 1.2 billion in 2010 and is expected to further decrease.

Moreover, electrification is seen as a strong tool for decarbonization: the lower the emissions target, the higher the electricity consumption that is required. However, electrification itself is not sufficient to push toward decarbonization, since the development of low-carbon generation sources need to keep the pace with the increase of demand, which is a rather challenging task, especially in developing countries. While wind and solar have significantly ramped up, they are still representing only 6% of the total electricity generation worldwide, which remains largely dependent on fossil fuels that provide around two-thirds of the total generation, with coal holding the lion's share with 40% of the total generation. As a result, in 2017, each kWh of electricity consumed in the world resulted in 480 g of CO_2 emissions (IEA 2018).

The decarbonization of the power sector is generally associated with increasing penetration of renewable energy sources. However, especially for non-dispatchable renewables such as solar and wind, they need to be associated with other flexibility options, such as dispatchable power plants, electricity storage, grid interconnections, demand-side management, and sector coupling. Moreover, other low-carbon sources may be needed, including nuclear and fossil fuels coupled with carbon capture, utilization, and storage. The generation mix is strongly country-specific, and this will be reflected in the technological choices that will be performed to address the decarbonization targets.

2.1 Renewable Energy Sources

The main strength of renewables is to provide a solution to generate electricity with an alternative to the traditional combustion of fossil fuels, which causes significant CO_2 emissions. Different sources are available, including hydro, solar, wind, bioenergy, geothermal, tidal, and waves. The technologies used for electricity generation are at different levels of maturity, but many are showing interesting potential.

Hydropower is currently the most important renewable source for electricity generation, with more than 1 TW installed worldwide in 2016 and roughly 70% of the electricity generation from renewables (World Energy Council 2019). Hydropower includes a number of applications, ranging from base load to peak power matched with reservoir or pumped storage plants. There is still significant potential for hydro generation worldwide, although concentrated in a relatively limited number of countries with favorable morphological conditions.

This limitation is not applied to wind, which is virtually available everywhere, although with strong variations in wind consistency and strength. Wind power has shown a remarkable growth in last decades, reaching 4% of the world power generation in 2017 (IEA 2018). While the current wind farms are mostly onshore, offshore facilities are spreading, supported by the lower environmental impacts related to noise and visual impacts, together with higher capacity factors and the possibility of exploiting larger areas. A potential breakthrough technology is floating offshore wind, which is currently limited to some pilot plants in different countries but holds the potential of unlocking even larger areas.

The other source that showed a large increase in the last decade is solar energy, mainly from PV technology. The strong decrease of installation costs, driven by subsidies and manufacturing upscaling, has allowed the spread of plants both at utility scale and for final users, unlocking the paradigm of distributed generation. As of 2017, PV generated 435 TWh, with a 2% share worldwide (IEA 2018). Further cost decreases are expected (especially for inverters and balance of system, rather than PV panels themselves). A marginal share is represented by concentrating solar power (CSP), which may increase its potential thanks to higher capacity factors and the possibility of exploiting a larger flexibility in comparison with PV systems.

The main limitation of wind and solar plants is their variability, resulting in the impossibility of dispatching them when needed. In contrast, this is one of the advantages of bioenergy, which is used in thermoelectric power plants just like fossil fuels, but with a closed CO_2 cycle, meaning that the combustion emissions are absorbed by the biomass during its life. Bioenergy has traditionally been considered as carbon-neutral, but some studies are highlighting that this is not always the case, since attention must be paid on land use issues, as well as on the effect of a slow capture compared with an instantaneous emission (Hausfather 2018). Although biomass is not primarily used for power generation, it totaled 623 TWh in 2017, including solid, liquid, and gaseous bioenergy. Future developments may be focused on crops that can be cultivated in marginal areas with a specific attention on the sustainability of the entire supply chain, as well as from emerging technology like microalgae.

Other RES that have currently a marginal role in power generation include geothermal, waves, and tides. While the former has a long history and shows a narrow potential due to the limited sites available worldwide, marine energy may play a role in the decarbonization of the power sector thanks to its large potential, although the technology is currently at early stage of development.

2.2 Energy Storage and Other Flexibility Solutions

The higher the penetration of renewables, especially non-dispatchable such as solar and wind, the higher the need of technology solutions supporting the flexibility of the power system. A number of options are available, including batteries, sector coupling, networks interconnections, demand response, and traditional dispatchable power plants.

The power network balance has traditionally been obtained by dispatchable power plants, mainly based on fossil fuels. Gas-powered units have generally a higher degree of flexibility, providing higher speed and ramps. Another common option is hydro, both dam and pumped hydro, although the deployment of this option is heavily limited by the geographical and morphological conditions. The predominant role of dispatchable power plants is expected to make way for other solutions.

Among those, electricity storage through batteries is gaining momentum, both with utility-scale and behind-the-meter installations. The current upscale of lithium batteries for electric vehicles is leading to a strong cost decrease, resulting in competitive solutions in multiple countries. Electric batteries provide solutions for short-term electricity storage, i.e., up to some days, while for applications on wider time frames other solutions are required (such as the production of synthetic fuels).

An additional flexibility option that is already in operation in different countries is demand response, both in industry and buildings, which can be fostered by time of use electricity tariffs or coordinated by virtual aggregators that have the control over a large number of users. In this perspective, demand response is tightly related with a growing digitalization of the power networks, which involve the implementation of smart grids, where the increasing role of distributed generation leads to the shift from consumers to prosumers.

Sector coupling, also referred to as P2X (power-to-everything), represents the idea of coupling the electricity system with other sectors (heating and cooling, mobility, desalination) or to generate other energy carriers (gases and liquids). This option is increasingly seen as an opportunity to accommodate the excess of power generation from RES, especially in countries with high penetrations of wind generation. In some cases, sector coupling can also be operated as electricity storage, if the technology allows a bi-directional operation, such as power-to-gas coupled to fuel cells or vehicles-to-grid.

Finally, network interconnections should not be overlooked, since they are an important flexibility solution, both at local and international levels. The connection of a larger pool of both users and generation units can support a better matching of demand and supply, by decreasing the limitations and the bottlenecks that may be related to the transmission and distribution infrastructure.

2.3 Other Generation Sources

While much attention is put on renewables, which are necessary to reach a fully sustainable power generation in the long term, other options may be needed in the short term to reach a faster decarbonization of the sector. In particular, natural gas is seen as a promising solution to facilitate the phase out of coal power plants, especially in developing countries. However, some experts fear a potential lock-in of fossil-fuel technologies. Another source that is not yet object of consensus is nuclear, which allows generating low-carbon electricity but with other potential environmental impacts.

Natural gas is seen by many as a potential bridge fuel by temporarily offsetting the decline in coal use. Others have contended that such option is incompatible with the current climate targets and that methane leakage from natural gas systems (especially upstream) may eliminate any advantage that natural gas has over coal (Levi 2013). A transition from coal to natural gas power generation has already happened in the US, mainly due to the rise of shale gas production, but with the effect of lowering the international price of coal and shifting its use to other countries. China has partially seen a similar shift, mainly caused by the need to limit the local pollution in large cities. However, natural gas will need to face a strong economic competition with both coal and renewables, and the investment in plants and networks that will operate for a limited time lead to the fear of stranded assets for investors.

Nuclear power has not seen significant improvements in the last two decades, resulting in a decrease of its share in power generation from 17% in 2000 to 10% in 2017 (IEA 2018), also due to the worldwide economic recession and to the concerns raised by the Fukushima Daiichi nuclear disaster that had consequences on the nuclear policies of several countries (including Japan, Germany, and Italy). However, most countries with nuclear power or with plans to add nuclear power to their energy mix have maintained an interest in developing the technology. In several cases, visible delays in nuclear implementation have resulted from safety reviews and resultant required actions (Nuclear Energy Agency 2017).

In addition to current nuclear fission technology, potential developments are expected from small nuclear power reactors, driven by the desire of reducing investment costs as well as decreasing the importance of centralized power generation at large sites. There seems to be a renewed interest in nuclear fusion, which is attracting private investors through different technological options (The Economist 2019). However, all of these options are still at very early stages, since no solution has yet reached a net energy gain (i.e., they have to produce more energy than they consume), and therefore any possible success will have consequences on a long-term horizon.

2.4   Carbon Capture, Utilization, and Storage

Another technological solution that may play a significant role in the decarbonization of the power sector is the carbon capture, utilization and storage (CCUS), aiming at complementing power plants based on combustion, both for fossil fuels and eventually for bioenergy (BECCUS), to reach net negative emissions. Different technologies and concepts are available.

The first projects related to CCS were aimed at exploiting the carbon dioxide to enhance oil recovery in depleted reservoirs. Since the first large-scale facility, which dates back to 1972, almost one hundred CCS facilities and nine test centers worldwide have started up or begun construction (Global CCS Institute 2018). These projects generally involve the post-combustion separation of the CO_2, with removal efficiencies around 90%, and require additional power consumption and a dedicated infrastructure for the transport of the gas to the storage site. Transport and storage risks are among the causes leading to a low public acceptance of these projects, which appears to be higher when carbon dioxide is reused instead of sequestrated (Arning et al. 2019).

Many alternative solutions are being proposed for CO_2 utilization. Sometimes referred to as carbon-to-value, CCU includes the multiple technologies that allow to recycle the carbon dioxide stream obtained by flue gases or air to manufacture a range of products, including cement, carbonates, chemicals, plastics, and synthetic fuels. The aim of these processes is to find an alternative to CCS, by overcoming the concept of burying carbon emissions underground and providing effective value by creating market products that would have consumed other resources for their production. In some cases, e.g., when producing fuels, the carbon dioxide is released again into the atmosphere, but the entire cycle is (almost) carbon-neutral. There is a growing interest in CCU applications worldwide, and some companies are already providing commercially competitive solutions, although often at limited scale.

While the previous concepts are generally coupled to the combustion flue gases of thermoelectric power plants, another technology still in early maturity is gaining interest in the scientific community: direct air capture (DAC). The idea of DAC is to use specific solutions, including membranes, to capture the CO_2 directly from the concentration in the air. This would allow a broader flexibility in locating the facility. One of its main challenges is related to its significant energy consumption, although experts are confident it can become cost-competitive with other CO_2 capture technologies if massively deployed (Fasihi et al. 2019). Another limitation appears to be the rate at which this technology can be scaled up (Realmonte et al. 2019).

3   The Industry Sector

The industrial sector is composed of a large variety of activities, which have different purposes and characteristics, and for which specific decarbonization solutions and technologies are required. If heat and electricity emissions are allocated to the relevant final sectors, industry is the single final sector with higher CO_2 emissions, with 36% of the 32.3 Gt of emissions estimated for 2016 (IEA 2019a). The emissions in industrial applications include both the direct emissions during the processes, and the indirect emissions caused by the fuel combustion to provide the energy required by the processes.

Decarbonization strategies in industries include three main areas: actions on the demand/reuse of products and materials, energy efficiency in the industrial processes, and different sources in the energy supply. The first area includes both actions devoted to decrease products demand and increase recycling both for industrial stakeholders and for final consumers. The energy efficiency measures in industrial processes involve a number of technologies that are already available, but that are not economically viable due to the absence of specific incentives to support low-carbon solutions (e.g., carbon tax). Finally, the use of different sources for energy supply in industry may include electricity and hydrogen produced from RES, sustainable biomass, or fuel combustion coupled with carbon capture systems.

A general concept for different industrial sectors is that there are currently no purely technological limitations blocking major decarbonization routes. The barriers are economic and not technological: for the most part, we have the technologies today, but they are expensive. Future technological advancements might very well reduce those economic barriers (Carnevale and Sachs 2019). An additional aspect related to economics is represented by the high cost of production plants, resulting in long average lifetimes (in some cases up to 50 years). Therefore, since the turnover is limited, the implementation of new systems would require some time in existing plants and would need to be economically competitive for new plants. Moreover, large industrial facilities are heavily integrated, and therefore a retrofit of a part may require the adaptation of the other units, resulting in the need of a systemic approach.

This section will mainly be devoted to three main industrial applications, due to their major contribution for carbon emissions: cement, steel, and chemicals. Some of the solutions that will be presented can also be applied to other industries. A final paragraph will be devoted to the technologies related to Internet services: although they are not usually considered as an industrial sector, their dramatic growing in the last decades calls for attention on the increasing energy consumption they require.

3.1 Cement

The share of global CO_2 emissions deriving from the cement industry is about 5%. More than 50% of these are process-related and cannot be avoided (Markewitz et al. 2019). These emissions are caused by the manufacturing of cement from limestone, since the heating of limestone ($CaCO_3$) releases carbon dioxide to produce CaO, the primary component of Portland cement. These emissions are currently dispersed into the atmosphere, since there is no incentive or regulation to support alternative solutions. Since the process emissions represent the most significant share in the cement industry, substitution of clinker with other materials for cement production (blended cement) reduces carbon dioxide emission significantly (Nidheesh and Kumar 2019). Another option, without modifying the chemical process, would be to install carbon capture systems to avoid those emissions, although some technical challenges are related to the very high temperatures at which those processes occur.

Other actions that can support a reduction of the CO_2 emissions include the use of dry kiln instead of wet kiln, efficient kiln drives, low-pressure drop cyclones for suspension preheaters, heat recovery for power generation, kiln shell heat loss reduction, kiln combustion system improvements, seal replacement, oxygen enrichment, conversion to reciprocating grate cooler for clinker making in rotary kilns, adjustable speed drive for kiln fan for clinker making in all kilns, indirect firing for clinker making in rotary kilns, modern power management systems, and use of modern clinker coolers (Fellaou and Bounahmidi 2017). Moreover, indirect emissions are caused by fossil-fuel combustion for heating purposes, which can be substituted by other low-carbon sources, mainly biomass.

3.2 Steel

The production of iron and steel is not only associated with fossil-fuel combustion CO_2 emissions, but it includes also process emissions. Iron and steel products are basic materials at the core of modern industrial systems, additionally being essential also for other decarbonization options like hydro and wind power (Mayer et al. 2019). Iron and steel production is estimated to cause 25% of the global CO_2 emissions from the industrial sector (Serrenho et al. 2016). While continuous process improvements and retrofitting measures have led to a relative decoupling of emissions from fuel combustion in last decades, especially in Europe, process emissions are essentially unavoidable under current conventional best-available technologies (Mayer et al. 2019).

To reach long-term decarbonization targets, multiple studies have concluded that a strong decline in CO_2 emissions is achievable only by a combination of BATs and CCS, or with a major decline of the sector's output. This latter solution appears unlikely in the medium term, unless alternative materials become viable, such as polymers for the automotive applications or wood for construction. Another possibility would be the scale-up of steel scrap recycling, although it is not expected before some decades and the quality of the final product may not reach the required standards.

The current most diffused iron and steel production routes are the blast-furnace basic-oxygen-furnace route and the route of carbon-based direct reduced iron (which is fed into an electric arc furnace), representing 71.6% and 28.0% of global steel production, respectively, in 2017 (World Steel Association 2018). Two promising emissions-free breakthrough alternatives are the route of hydrogen-based direct reduced iron (fed into an EAF) and the plasma-direct steel production route. However, since they both rely on electricity consumption, a low-carbon power generation mix is essential to limit the indirect emissions.

3.3   Chemicals

Within the industrial sector, the chemical industry is one of the largest energy users, accounting for 12% of global industrial energy use (Sendich 2019). The chemical industry is usually divided into basic chemicals that are the basis for other products, and specialized chemicals, including medicine, soap, and paints. Basic chemicals, or commodity chemicals, generally require significant energy for their production, but due to their large-scale production they are sold at low prices. They include raw material gases, pigments, fertilizers, plastics, and rubber. Fossil products (mainly natural gas and oil) are used to produce chemicals both as fuels and feedstocks. The energy production is mainly necessary for process steam and for equipment (e.g., pumping), and the largest feedstock use is required by the petrochemical industries.

The top five commodity chemicals with both the largest production volume and energy consumption worldwide are ammonia, ethylene, propylene, methanol, and benzene/toluene/xylene (BTX). All of these chemicals require energy for their synthesis, and in some cases also hydrogen is involved as a feedstock. While current hydrogen production is mostly performed through natural gas steam reforming, mainly for economic reasons, water electrolysis is a mature technology, and may be adopted if supported by carbon pricing policies. Some of these chemicals, particularly methanol and ammonia, may be used as synthetic fuels in a low-carbon energy system if produced by renewable electricity, leading to a closed carbon cycle.

Some technological solutions related to the chemical processes include the possibility of exploring electrochemistry, since there are many possible routes for electricity to drive a chemical reaction (Schiffer and Manthiram 2017). Electrification may allow chemical reactions at lower temperatures, supporting the development of smaller units distributed in locations with high availability of renewables for power generation. An additional advantage would be the decreasing distribution costs. Finally, electrochemical reactions facilitate the products separation, which can be an energy-intensive step in the current technological processes.

Electrification of the chemical industry may be integrated into broader trends in modular and local manufacturing that have been enabled by robotic automation and additive manufacturing methods to support a new paradigm of a fully integrated, decarbonized, local manufacturing that starts with renewable resources and ends with desired commercial products (Schiffer and Manthiram 2017).

3.4 Information and Communication Technologies

Although not traditionally included into the industry sector, the potential growing rates of the power consumption of the information and communication technologies (ICT) sector may reach important shares, and therefore it should be considered into the analysis of future energy systems. The energy consumption of the sector is limited to electricity, thus more easily manageable in the hypothesis of a low-carbon power mix, but dramatic rises of power demand with respect to the forecasted scenarios may have an impact on the deployment of RES power plants.

Andrae and Edler (2015) calculate that the ICT sector will represent 21% of global electricity consumption by 2030, reaching 8,000 TWh from a base of around 2,000 TWh in 2010. Two additional scenarios are presented by the authors, with relative power consumption ranging from 8 to 50% of global electricity use by 2030. The International Energy Agency (IEA 2017), provides some figures for the power consumption of communication networks (185 TWh in 2015) and data centers (at 194 TWh in 2014), which together represent around 2% of the global electricity consumption. A moderate growth in the energy consumption of data centers of 3% by 2020 is expected, but there is a greater uncertainty for the estimation of future consumption for networks, with scenarios varying between growth of 70% and a decline of 15% by 2021 depending on trends in energy efficiency.

In general, it is not clear if the current positive trends in energy efficiency, particularly for data centers, will be able to compensate the dramatic increase of data demand for the users (Morley et al. 2018), in particular, if future technologies will be massively adopted, including smart devices and automated vehicles. The growth of Internet traffic is the combined results of multiple phenomena: the increase of Internet users, the rise of the average devices per user (there will be 3.6 networked devices per capita by 2022, up from 2.4 networked devices per capita in 2017, Cisco 2019), as well as the constantly increasing speed and contents that are available for the users. These aspects result in an exponential growth of Internet traffic, with global traffic flows rising from 100 GB per second in 2002 to 26,600 GB per second in 2016, and the volume of traffic is expected to nearly triple within the next 5 years (Cisco 2019).

Finally, an aspect that deserves attention is the daily pattern of ICT consumption, since the trends suggest that the increase of Internet use during peak hours is rising even faster, driven by the video streaming (Morley et al. 2018). For this reason, particular attention should be paid to the management of peak electricity demand, which may become most critical than in current electricity networks. Decarbonization strategies should deal with Internet-related energy demand as it develops, rather than allowing it to become a "problem" that will be harder to tackle once data-intensive services are more thoroughly embedded in normal, everyday life and thereby "locked in" (Morley et al. 2018).

4   The Transport Sector

The total final energy consumption of the transport sector reached 2.8 Gtoe in 2017 (IEA 2018), almost 29% of the total, and 92% is represented by oil products, although they account for less than 50% of the growth in demand over the previous year. Mobility demand is showing a constant increase at global level, which is expected to continue for the next three decades. Both passenger and freight transport is expected to increase nearly threefold between 2015 and 2050 (ITF 2019), based on the current path. The reduction of CO_2 emissions in transport needs a combined approach, tackling both a limitation of the demand and the deployment of low-carbon alternative technologies as well as compensation measures.

This section will present the main technological options that are available in the four main segments of passenger and freight transport, i.e., road, rail, aviation, and shipping. Currently, three-quarters of the final energy consumption in transport is due to the road segment, with roughly 10% each to aviation and shipping and just 2% to rail, also thanks to its higher energy efficiency compared to other modes (International Energy Agency 2016). These shares have remained rather unchanged in the last three decades, although the total transport energy demand is more than doubled.

4.1   Road Transport

Road transport is probably the most various segments of the transport sector, ranging from passenger cars, buses, and two-wheelers to heavy trucks for freight. The current situation is seeing a predominance of internal combustion engine (ICE) technology, based on oil products and on a limited share of alternative fuels, including biofuels and natural gas. The two main potential alternative powertrains are based on electricity, either by its direct use through its storage supported by on-board batteries, or through its conversion from hydrogen thanks to a fuel cell. The opportunities and challenges of these technologies are related to a number of aspects, including cost, range, flexibility, reliability, performance, and charging time.

There is an increasing interest in electric vehicles (EVs) worldwide, especially for passenger light-duty vehicles (LDVs). EVs generally include multiple technologies, from hybrid EVs (HEVs), which have both a traditional ICE powertrain and an electric engine (but usually no possibility of directly charging from the external grid), to battery EVs, which are fully electric. An intermediate technology is the plug-in hybrid (PHEV), which is a hybrid vehicle with a larger battery and the possibility of connecting to an external power source. Finally, fuel-cell-powered EVs (FCEVs) are often grouped in the category of EVs, although they are basically running on hydrogen to supply the electricity needed by the vehicle.

The penetration of EVs in the global market currently remains marginal, although significant improvements are being made in the last years. The global EVs car fleet reached 5.1 million in 2018, almost doubling the number of new EVs sales (IEA 2019b), but compared to a total car fleet around 1 billion. The world largest market remains China, followed by the US and Europe, and the largest share of EVs market share is in Norway, where in 2018 EVs reached 46% of the new vehicles sales. However, electricity penetration in passenger transport is going beyond cars, especially in China. The country hosts the vast majority of the estimated 300 million of electric two-wheelers in the world (IEA 2019b), while electric buses reached a world fleet of around 460,000 vehicles, with 100,000 new sales in 2018.

The deployment of EVs requires a parallel development of a proper charging infrastructure, whose capillarity is essential to support the use of EVs. In particular, a total of 5.2 million charging points for LDVs are estimated worldwide, mostly slow chargers installed at private houses and workplaces, including an estimated 540,000 public chargers, of which 150,000 fast chargers (78% in China).

The limitations of EVs deployment are mainly related to the limited available range due to the low energy density of batteries, together with the high duration of charging and the current limited availability of charging points. The investment costs are currently higher than for ICE cars, but a massive upscaling of EVs manufacturing may lead its future cost to be in line with traditional vehicles. Moreover, expected improvements in batteries technology may also increase the performance of EVs. The chicken-and-egg problem of vehicles and charging infrastructure deployment may be currently slowing down the adoption of EVs, especially in Europe.

The limitations related to BEVs are among the reasons that may lead to the adoption of hydrogen-powered vehicles, which promise longest ranges and shorter charging times. However, the generation, compression, and use of hydrogen lead to an increased energy chain leading to a lower system efficiency when compared to batteries. This aspect may not be too critical in terms of emissions as long as the electricity is produced by RES, but the need of additional power generation should be carefully taken into account. Other issues related to hydrogen are the transportation and the installation of a proper refueling infrastructure.

An alternative pathway for road transport decarbonization is related to biofuels (e.g., biodiesel, biomethane) or to synthetic fuels (e.g., methanol) supported by electricity generation from RES. These solutions have the advantage of exploiting the existing powertrains as well as distribution infrastructures, although their limited availability would suggest that their use may be prioritized to transport segments that are harder to electrify (i.e., aviation and shipping).

Finally, heavy trucks for freight transport have different requirements than LDVs, since their higher size and required ranges may deter the use of electric batteries, although some companies are currently evaluating their technical feasibility. The most probable alternatives include hydrogen-powered trucks or electric road systems, in which the vehicles are constantly supplied with the required electricity during the travel on highways. These systems may be integrated with different powertrains (e.g., batteries, hybrid, or hydrogen) to enhance their flexibility and limit the need of a capillary power infrastructure outside of the main roads.

4.2   Rail Transport

Rail networks carry around 8% of the world's motorized passenger movements and 7% of freight transport, but account for only 2% of energy use in the transport sector, thanks to their high efficiency (IEA 2019c). Rail transport includes conventional railways around and between cities, high-speed railways and urban networks (including subways and tram). The majority of passenger transport on conventional railways is located in Asia, with India accounting for 39% of the total, followed by China with 27%, and Japan with 11%. Today China accounts for about two-thirds of high-speed rail activity, having overtaken both Japan (17%) and the European Union (12%). The regional distribution of urban rail activity is more even; China, European Union, and Japan each have around one-fifth of urban passenger rail activity.

Rail transport is currently the most electrified transport mode worldwide, although with different levels depending on the area. Considering conventional railways, three-quarters of passenger rail transport and almost half of all freight rail are electric (IEA 2019c), the remainder being powered by diesel trains. High-speed rail and urban rail are completely powered by electricity.

As a result, the challenges for rail transport decarbonization appear lower than for other modes, as the options are clear and already available. Efforts are needed to further improve the power infrastructure in some railways that are not yet electrified, together with the extra-sector improvement of the share of power generation from renewables. However, the electrification of railway lines with low utilization factors is often not economically affordable, and therefore other technologies may play a role, including battery-electric trains and hydrogen fuel cell trains.

Potential alternatives to high-speed railways include maglev (magnetic levitations) and Hyperloop concepts. Maglev trains are already in operation in six locations in Asia, but the only train that is operating at a speed higher than the normal high-speed trains is the one connecting Shanghai City center with the airport, reaching a top speed of 430 km/h over the 30 km of its length. Another project currently under construction in Japan plans to connect Tokyo and Nagoya, but the benefits provided by increased speed come at the cost of a four to five times higher energy consumption in comparison with the current high-speed train connecting those cities. Another alternative technology, the Hyperloop, is based on a low-pressure tube in which a passenger or cargo pod is operated through an electromagnetic propulsion system.

According to some feasibility studies, this technology could be more efficient than the current high-speed trains, but there are not yet any real figures from services in commercial operation.

Further actions are possible in the optimization of the demand and logistics, especially in freight, and the improvement of the performance and energy efficiency of railways. These solutions may include the implementation of on-board energy recovery devices (including regenerative braking and energy storage), as well as the use of lighter materials and a decreased use of energy-intensive power electronics.

4.3   Aviation

Aviation is among the most critical transport segments, due to its constantly increasing passenger demand, especially for long-haul flights, and the high energy density that is required. Demand for domestic and international air transport combined will rise from 7 trillion passenger-kilometers in 2015 to 22 trillion in 2050, according to (ITF 2019). Moreover, since air travel is a highly regulated environment, the access to innovative technologies is strongly related to the policies implementation.

The most promising pathway for air transport decarbonization is the development of advanced sustainable jet fuels, either by incorporating biofuels, or by the use of synthetic fuels based on power generation from RES. The blending of biofuels with the current fossil-based products may provide the possibility of gradually integrating low-carbon solution in the existing system, without the need of major changes in the current fleet, which usually has a renewal rate around 30 years. However, to match the requirements of existing certifications, the fuels need to provide high energy density and low freezing temperatures, which is rarely the case for available biofuels. Thus, a blending with fossil fuels is currently required to meet those strict standards.

The electrification of short- and medium-haul flights is being evaluated by different manufacturers, but major hurdles remain related to the energy density provided by the current batteries. Moreover, the largest part of air travel demand worldwide is related to long-haul flights, since travels shorter than 600 nautical miles (that would be the target of potential electric aircrafts) currently represent half of the departures but only 15% of the total fuel use (Schäfer et al. 2019). Hybrid technology may enter the market soon if it proves to bring economic advantages, and the combination with biofuels may help in reaching the decarbonization goals.

Lastly, energy efficiency measures are necessary to compensate the expected increase in demand. These solutions include the use of innovative materials such as advanced composites and airframe metal alloys, with lower weight and improved performance, and the development of new plane designs. As in any sector, energy efficiency is usually the most effective measure to support decarbonization, and available solutions should be always prioritized before thinking of a sustainable energy supply.

4.4 Shipping

The other transport segment that shows severe challenges for decarbonization is maritime shipping, especially for long-haul freight transport. Solutions and opportunities differ depending on the range and the purpose of the trip.

Short-haul naval shipping for freight and passenger transport, especially in inland waterways, is already seeing an evolution toward electrification in some countries. The ships operation on fixed routes allows a better planning of the battery size as well as the management of charging during the load and unload of the ship. Electric ferries are already in operation in Denmark and Norway, and different technologies are under evaluation, including hydrogen fuel cells (Norled 2019) and flow batteries (Valentine 2018).

For long-haul freight maritime transport, which represents the largest share of fuel consumption in the sector and totally relies on oil-based bunker fuel and diesel, some potential alternative fuels are under evaluation. The current regulations limiting pollutant emissions in some areas are pushing toward cleaner fuels, including liquified natural gas (LNG). Other low-carbon options, although not yet commercially available, are hydrogen or ammonia produced through electricity from RES. Ammonia has higher volumetric energy density than hydrogen and more practical storage temperatures and pressures, and its production requires less energy than other synthetic fuels like methanol or ethanol (Laursen 2018). However, for the development of all these alternative fuels, a dedicated infrastructure is required, both for their production and distribution and for their supply in the ports worldwide, since a diffused availability of fuel supply is at the basis of its use for long-haul freight.

Besides the supply technologies, energy efficiency measures can be implemented to decrease the energy demand of shipping. These include the use of lighter materials, the slender design, the decrease of friction (air lubrication, hull chemical coating), on-board waste heat recovery, wind assist, or exploitation of renewables sources with on-board devices such as kites (Traut et al. 2014). Moreover, a better electrification of ports and the availability of power supply for boats would allow the combustion of fuels for on-board power demand during the loading/unloading operations.

5 The Buildings Sector

Buildings floor area in the world is currently estimated at around 223 billion m^2, and it is expected to grow to 415 billion m^2 by 2050 (Dean et al. 2016). The largest increase is expected from Africa, in line with the expected increase of population which may reach a total of 2.5 billion people in the continent by 2050 (United Nations 2019). Buildings are the final sector with highest final energy consumption at global scale, with 3.05 Gtoe as of 2017, although energy efficiency measures have limited its energy demand growth in last decades (IEA 2018). Moreover, an additional impact is related to the embodied energy, i.e., the energy required for materials and construction, although these impacts are generally included into the industry sector.

The strong push from growing population and increasing income levels in emerging economies and developing countries represent the main drivers for building stock rise, leading to a potential increase of energy consumption in the sector reaching 50% by 2050 if no action is taken (Dean et al. 2016). In the 2010–2017 period, energy consumption increase (+5%) has been lower than the increase of floor area (+17%), thanks to the improved energy efficiency in new buildings and in renovations. Electricity (+15%) and renewable energy sources (+14%) contributed more than natural gas (+5%) to the substitution in final energy use of less-efficient coal-based technologies (-8%), while other fuels (oil and biomass) remained almost stable. Natural gas and electricity constitute the main energy source in OECD countries, while non-OECD countries still mainly rely on biomass and coal, with slow shifting to electricity and gas (Carnevale and Sachs 2019).

Energy consumption in buildings is related to multiple aspects and services, including heating and cooling, lighting and appliances. Moreover, although the majority of buildings are related to housing purposes, other applications with specific energy demand needs and patterns include shops, offices, schools, hospitals, etc. The amount and type of energy demand is related to the specific activity and the occupancy schedule and density for each building. Some services are also strongly dependent on external weather conditions, both for temperature and solar radiation, and strong variations occur on a geographical and chronological basis.

This section will present the main decarbonization options for buildings, by considering heating, cooling, and electricity demand. In comparison with other sectors, energy efficiency measures are of utmost importance, and although many actions have already been undertaken, especially in developed countries, there is still a huge potential for energy savings and rational energy use in the building sector.

5.1 Space and Water Heating

Space heating is particularly significant in temperate and cold climate regions, with a strong seasonality imbalance between winter and summer. Also domestic hot water production is generally higher in those regions, although it shows a more evenly distribution both across the world regions and the months of the year. Fuel consumption represents roughly two-thirds of the total final consumption in buildings as of 2017 (IEA 2018), of which less than half is from renewables (mainly traditional biomass used for heating and cooking), and therefore significant measures are required to address this aspect.

The most impactful actions for space heating decarbonization are related to buildings insulation, both for building envelope and for windows. An increasing number of countries worldwide is adopting buildings energy performance regulations, with stringent limits for both new buildings and renovations. However, it has to be reminded that the current renovation rate of buildings worldwide is around 1%, but to reach a total decarbonization by 2050 an increase up to a 3% rate would be necessary (Dean et al. 2016).

Considering energy supply, the most promising solution to substitute fossil fuels is the switch toward heat pumps (HPs), which are a mature and efficient technology for space heating, especially coupled with low-temperature heating systems (such as radiant floors or low-temperature radiators). Heat pumps are based on the principle absorbing heat to a low-temperature heat source (which is usually the outdoor air or the ground) and supplying heat at a higher temperature, thanks to an additional energy input, usually electricity (although gas- or heat-powered HPs are available). The coefficient of performance (COP) of the current heat pumps, i.e., the ratio between the useful heat supply and the electricity consumption, is generally between 3 and 5, depending on the working conditions. Ground-source heat pumps have generally a better performance thanks to the higher temperatures of the ground compared to outdoor air, especially in winter.

While heat pumps are an interesting and promising solution, it has to be observed that the electrification of space heating should be deployed in parallel with the decarbonization of the power sector. The strong seasonality of heat demand may become an issue if the future power mix would be strongly based on solar energy, since the generation pattern would not be well-matched with the heat demand profile (Jarre et al. 2018). Therefore, proper storage strategies would be needed, both on a day/night basis and on a weekly or seasonal basis to compensate the weather fluctuations (both programmed and unexpected).

An alternative solution for the heating sector decarbonization is district heating (DH), which is a mature technology for the energy supply to buildings in dense areas. DH systems are based on a centralized heat generation facility, from which heat is supplied to the users through a network of insulated pipes. While traditional DH systems are based on fossil-fired cogeneration units, the potential upgrading toward low-temperature RES-based DH systems can be obtained through the integration of solar energy, heat pumps, waste heat, and biomass. Some technical challenges remain, but DH can play a significant role in densely populated areas, also integrating distributed heat sources, developing smart thermal networks. Moreover, DH systems can be integrated with power networks through sector coupling, to exploit the availability of excess electricity from RES to generate heat to be stored or directly supplied to final users.

Finally, a limited role may be played by other renewable sources, mostly local wood biomass and solar thermal. Rural areas with a low population density and severe climate conditions may benefit from the use of local wood biomass, which can be used for high-temperature heat generation. Wood biomass is already used in many world regions, although in developing countries it is often used for cooking in low-quality appliances, leading to severe problems for safety and pollutants emissions. On the other hand, the use of local biomass in modern stoves is already a sustainable and low-carbon solution to substitute the use of fossil fuels in many rural areas. Space heating can also be integrated with solar collectors, especially in middle seasons and in climates that are not showing extreme conditions. However, solar thermal is usually mostly used to provide domestic hot water, especially in summer. Among its main advantages, there are the low cost, the simple system configuration, and the relatively high efficiency (up to 70–75% of the solar radiation can be converted into useful heat).

5.2 Space Cooling

While space heating has traditionally been a significant cause of energy consumption in buildings, the role of cooling is progressively increasing worldwide. This growing trend is expected to continue, sustained by the climate change and the growing per-capita income in developing countries. Current cooling technologies are relatively limited, and the large majority of cooling worldwide is supplied by distributed electricity-powered chillers, with some few exceptions including district cooling networks or solar cooling units.

The most impactful actions to limit energy consumption would be the limitation of the rising cooling demand. This could be obtained through building design strategies aimed at minimizing the cooling needs (cool roofs, shading systems, night ventilation), integrated with solutions for the free cooling when the outdoor conditions are favorable. These solutions would be even more impactful in hot climate countries, where cooling demand is particularly high. Non-residential buildings show a significant potential for energy savings through a proper energy management, since they often operate cooling equipment with very low set point temperatures and without a proper attention to limit the flow of cooled air to the outside.

Considering cooling supply, the installation of high-efficiency chillers and their proper operation could lead to significant energy savings. Attention must be paid on maintenance operations, especially for air filters, to avoid unnecessary additional energy consumption. In specific contexts, the use of high-efficiency district cooling networks or solar cooling units may provide performance improvements with respect to common solutions.

5.3 Lighting, Appliances, and Cooking

Besides heating and cooling, there is an increasing energy demand in buildings related to power supply for lighting and appliances, mainly driven by the diffusion of technologies providing multiple services to the users. As long as electricity consumption is concerned, there is a huge difference across developed and developing countries. In the former, the increased integration of houses with digital technologies and the Internet of Things (IoT) may lead to an optimized supply and use of services, but dramatically increasing the power demand of households. At the same time, developing countries are still facing severe problems of energy access, especially in Africa, where many buildings are able to power any appliance or electric lighting system. Universal energy access will increase the living conditions of million of people, and at the same time boost the installation of appliances that will lead to higher power consumption also in existing buildings. For this reason, it is important to ensure regulations that require high standards in energy efficiency both for lights and other appliances.

Another cause of energy consumption in residential buildings is cooking. While in developed countries most houses have access to gas- or electricity-fired cooking systems, the majority of the world is still relying on biomass or coal. While burning local biomass has a limited impact on CO_2 emissions, thanks to the closed CO_2 cycle, it has severe consequences for the indoor air quality and for safety. For this reason, it is important to support the deployment of clean and efficient cooking technologies. Induction cooking is providing higher efficiencies than traditional electric and gas cooking, and its deployment may play an important role toward decarbonization.

6 Conclusions: Strategies and Policy Recommendations

Following the Paris Climate Agreement's aim to strengthen the global response to the climate crisis "in the context of sustainable development and efforts to eradicate poverty," it is necessary to use a "systems approach," aspiring to simultaneously address multiple objectives and promote policy instruments and technological solutions that can be used across sectors.

In this section, we synthesize some of the most important strategies and policy recommendations distilled from the earlier sections of this chapter and from the "Roadmap to 2050: A Manual for Nations to Decarbonize by Mid-Century" (Carnevale and Sachs 2019) to which the authors of this chapter have actively contributed.[1]

There is a broad consensus on technology pathways to decarbonization which points to six main technological pillars: (1) **Zero-carbon electricity**: a shift toward zero-carbon electricity mix; (2) **Electrification of end uses**: the penetration of electricity, built on existing technologies, can enable a green conversion for the sectors currently using fossil-fuel energy; (3) **Green synthetic fuels**: deployment of a wide range of potential synthetic fuels, including hydrogen, synthetic methane, synthetic methanol, and synthetic liquid hydrocarbons applicable for harder to abate sectors; (4) **Smart power grids**: systems able to shift among multiple sources of power generation and various end uses to provide efficient, reliable, and low-cost systems operations, despite the variability of renewable energy; (5) **Materials efficiency**: improved material choices and material flows, such as reduce, reuse, and recycle to significantly improve materials efficiency; (6) **Sustainable land use**: mainly involving the agriculture sector, as it contributes up to a quarter of all greenhouse gas

[1] Other authors who contributed to this "Strategies and Policy Recomendations" section are Niccolò Aste, Marco Bocciolone Dimitri Bogdanov, Ed Brost, Christian Breyer, Victoria Burrows, Carlos Calvo Ambel, Emanuela Colombo, Claudio del Pero, Thomas Earl, Fabrizio Leonforte, Maurizio Masi, Renato Mazzoncini, and Alessandro Miglioli.

emissions from deforestation, industrial fertilizers, livestock, and direct and indirect fossil-fuel uses.

The present chapter has discussed the main technological decarbonization strategies available to decarbonize our energy systems for the power sector as well as the three final energy consumption sectors: industry, transport, and buildings. Here below, we present the main strategies and policy recommendations for each sector.

6.1 Strategies to Decarbonize the Power Sector

The power sector is already undergoing a decarbonization process in multiple countries around the world. The traditional centralized organization of the power system is now facing a paradigm shift to distributed and renewable generation. This new model is closely related tot he implementation of smart grids, where the end users act as prosumers. Digital technologies will be at the center of this revolution, unlocking the potential of different business models like virtual aggregators and peer-to-peer energy trading.

The current technologies supporting this transition can be classified into four main groups: (i) low-carbon energy sources (on- and offshore wind, solar PV and concentrated solar power, hydropower, biomass, nuclear, and geothermal); (ii) short-term and long-term electricity storage solutions; (iii) other flexible options such as network interconnections, sector coupling, supply response (hydro reservoirs, bioenergy) and demand-side management (DSM); (iv) carbon capture, utilization, and storage (CCUS), and variants including bioenergy CCUS and direct air capture. While many of these technologies are already cost-competitive and may offer even lower costs in the future, others (e.g., electricity storage and carbon capture) require future technological developments and/or increased economies of scale to support their effective deployment at the levels needed to reach a full decarbonization of the power sector.

The total decarbonization target will require a combination of multiple technologies. Depending on local conditions, the mix of available power options will vary from one country to another, and thus there will be no one-size-fits-all solution. The implementation of transition technologies may also be required. Coal should be phased out earliest given its high carbon content and its contributions to air pollution. While natural gas may play a crucial role during a transition period, it will also need to be either decarbonized or progressively phased out. To allow for unanticipated technological breakthroughs and cost reductions, energy policies need to be flexible, to be regularly assessed, and be adaptive to ongoing technology advances in order to allow each potential low-carbon solution to be supported and deployed.

While many national and international policies are heavily oriented toward the electrification of energy systems, electrification must proceed alongside decarbonization and uncoupling of energy demand from economic growth in order to fight climate change. Also, the energy efficiency potential along the whole electricity chain should not be underestimated. Moreover, a strongly integrated approach across sectors and energy pathways is essential for addressing climate change issues. Finally, secondary effects and a holistic perspective on the entire lifecycle of technological solutions should be considered to avoid potential rebound effects from specific technology choices.

6.2   Strategies to Decarbonize the Industry Sector

Heavy industry emits a large share of global greenhouse gas emissions, because industrial processes employ high temperatures and depend on high energy densities to enable the chemical processes involved. The industrial sector of the worldwide economy consumed more than half (55%) of all delivered energy in 2018. Within the industrial sector, the chemicals industry is one of the largest energy users, accounting for 12% of global industrial energy use.

Three energy-intensive sectors have been considered: cement, iron and steel, and petrochemicals (plastics, solvents, industrial chemicals). Fully decarbonizing such complicated and integrated industrial environments requires a multidimensional approach. Strategies include (i) reducing demand for carbon-intensive products and services; (ii) improving energy efficiency in current production processes; (iii) deploying decarbonization technologies across all industries, which in turn can be split between four supply-side decarbonization routes: electrification, use of biomass, use of hydrogen and synthetic fuels, and use of carbon capture technology.

Some material efficiency options for the three industry sectors analyzed include (i) for cement: building design optimization, concrete reuse, materials substitution; (ii) for iron and steel: optimization of scrap recycling, product design for efficiency, more intensive use of products; (iii) for petrochemicals: chemical and mechanical recycling, plastic demand behavior change, use of renewable feedstocks, and product eco-design to better enable recycling. For these industries, improvements in energy efficiency should run in parallel with material efficiency and demand reduction.

Appropriate technology for energy efficiency exists today and it can be applied in any country. Some of the key solutions for energy efficiency improvement include (i) for cement: switch to dry kilns, multistage cyclone heaters; (ii) for iron and steel: reuse of high-pressure gas for power, coke dry quenching; (iii) for petrochemicals: energy efficiency in monomer production and naphtha catalytic cracking.

There are currently no pure technological limitations blocking major decarbonization routes across any industrial sector. The barriers are economic and not technological; we have the technologies today but they are expensive.

Of course, also geographical contexts will impact technology decision-making. Countries investing in new plants should go for zero-carbon technology rather than investing in energy efficiency improvements in plants at the end of their life. In contrast, countries where legacy plants and facilities will continue to operate for years to come should invest in energy conservation and energy efficiency improvements for existing processes. Additionally, the possibility of combining more of these solutions in a given country or facility will vary and depend on the geographical distribution of resources and social acceptability of specific technologies.

6.3 Strategies to Decarbonize the Transport Sector

The transport sector requires deploying a diverse mix of decarbonization solutions to meet the challenges within each of its four main segments: roadways, railways, aviation, and navigation. Effective decarbonization pathways in transport rely mostly on technological solutions, new sustainable fuel development and fuel shifts, and are complemented by demand reduction and modal shift strategies.

Direct electricity usage (through either batteries or electrified railways and electric road systems), hydrogen, synthetic fuels, and sustainable biofuels (properly allocated to hard-to-decarbonized modes) will all be important for transport decarbonization. Strategies include (i) in the road segment, CO_2 emissions are easier to abate due to electric vehicles and fuel-cell electric vehicles for short-to-medium haul (freight, passenger, light-duty, or heavy-duty categories); (ii) the pathways for railway decarbonization are mostly based on fuel shifts from diesel to electricity or hydrogen; (iii) in aviation, advanced jet fuels (such as synthetic fuels) are the only way to decarbonize the current fleet and the relevant one in the near future. Modal shift from air to land could be enhanced with innovative alternatives such as ultra-high-speed trains with the right policies in place; (iv) long-haul navigation is hard to abate while short-haul navigation can be supplied by electricity or hydrogen technologies. Ammonia and hydrogen are currently being investigated in long-haul navigation.

The use of biofuels and the sustainability of biomass for biofuels need to be carefully assessed so as to avoid: competition with food production, deforestation or loss of biodiversity in natural regions, and competition with industries that currently use the biomass for higher value products or uses. As sustainable biofuels will only be available in limited volumes, its use should be prioritized in hard-to-abate modes like aviation.

Regulatory frameworks need to be technology agnostic to create a fertile environment for innovation, unleashing the potential of the research while fostering virtuous behaviors of citizens in all transport modes.

6.4 Strategies to Decarbonize the Buildings Sector

Buildings represent an estimated 36% of global final energy consumption and 39% of the global energy-related greenhouse gas (GHG) emissions. The goal of total decarbonization in the buildings sector includes the construction of new buildings and districts with zero or almost zero energy consumption from fossil fuels and the total renovation of existing buildings with the same net zero-carbon standards. Current renovation rates account for about 1% of existing building stock each year, while to achieve 100% zero-carbon goal by 2050 it is necessary to ensure a renovation rate higher than 3%. It should be noted that the CO_2 emissions resulting from material use in buildings represent almost one-third of building-related emissions: the construction industry must radically change its manufacturing structure in order to abate this increasing embodied energy.

In general, using a combination of readily available technologies and approaches, and performance-based design metrics, net zero-carbon buildings and districts can be achieved today by (i) maximizing the buildings energy efficiency mainly through passive and low embodied carbon solutions; (ii) adopting high-efficiency technical systems and advanced control/management strategies: phasing out inefficient solutions, encouraging of low-carbon systems such as heat pumps and district heating and the adoption of advanced control/management strategies; (iii) maximizing on-site or nearby renewable energy production and self-consumption while electrifying the buildings sector, to completely cover or exceed the total energy demand of each building with the minimum exchange of energy with the grid (thus stimulating energy management, storage, and exchange at district level).

In order to achieve the overall decarbonization of the buildings sector, energy consumption related to building codes be addressed. The strategies include (i) establishing advanced building energy codes with mandatory performance standards and setting minimum energy performance levels for existing buildings. Also, policies and subsidies to favor the retrofit of existing buildings rather than new constructions are absolutely necessary; (ii) achieving high-efficiency building envelopes at negative life cycle cost, mandating energy performance standards for envelope components and work with industry to deliver non-invasive and whole-building retrofit packages. Policy-makers should develop strategic frameworks to create the adequate market conditions for low-carbon technologies, guiding building owners and designers in making the correct choices; (iii) mandating minimum energy performance standards for stand-alone heating equipment, prevent expansion of fossil-fuel heating, and pursue strategy to shift demand to high-efficiency and integrated energy solutions with net zero emissions; (iv) pursuing low-cost solar cooling technologies such as high-efficiency and renewable district cooling where appropriate. Mandating the use of waste heat from large-scale cooling for heating and hot water use on-site or via district systems, local governments are uniquely positioned to advance district energy systems in their various capacities; (v) implementing regulations and measures obstructing energy self-consumption such as specific additional taxes or levies should be lifted and administrative procedures to allow self-consumption should be user-friendly; (vi) achieving affordable thermal storage and low-cost solar thermal systems (for low-income countries only); (vii) implementing training and capacity building activities for the construction sector must be adequately promoted, while also pushing the development of specific DSS (decision support system) or design-aid tools to strongly increase the application of climate-responsive and integrated building design.

6.5 An Integrated Systems Perspective Needed

Due to the complexity of the decarbonization process across the whole energy system, it is important to adopt an integrated system approach. A systems perspective recognizes the interconnectivity of actions toward any one or more of these objectives, using any one or more of the mentioned policy instruments or technological solutions. An action in one can be detrimental to another, while some combined efforts could amplify their cumulative effects and achieve multiple objectives. For example, the power grid itself represents a complex system that must continue to operate reliably and efficiently even as it undertakes the deepest transformation in its history. No single policy or technology can achieve decarbonization by itself or be implemented without due consideration to its ripple effects, or to the delicate state of the current, broader system.

In taking a systems approach, many complementarities need to be considered for managing the complexity of the energy system: (i) complementarities of variable renewable energy sources. Wind, solar, and hydropower vary by the minute, day, season, and year. Digital systems will play a large role in coordinating the augmented grid complexity and the required flexibility; (ii) complementarities among zero-carbon technologies. As one obvious example, zero-emission vehicles depend on complementary zero-carbon energy sources and the infrastructure to fuel them; (iii) complementarities of public and private investments. Parts of the energy system are in private, for-profit hands, and parts are publicly owned. It will take significant effort and analysis to harmonize public and private investments, to recognize the diverse role they can play, and the synergies their joint action can create; (iv) complementarities of natural and engineered systems. Achieving net negative emissions would require biological storage of carbon dioxide (CO_2) in vegetation and soils via preservation of existing forests, restoration of degraded habitats, and reforestation to increase natural carbon sinks. Energy strategies that amplify land use degradation must be ruled out; (v) complementarities of mitigation and adaptation. Adaptation measures can also contribute to mitigation strategies. Forest restoration and protection of coastal wetlands would help resist storm surges from rising sea levels, promote resilient food production, and secure carbon, thereby serving both adaptation and mitigation purposes; (vi) complementarities of centralized and decentralized solutions. Renewable energy resources are by nature different from one place to another and restriction on land availability and use may require different power configurations; (vii) complementarities of actions and strategies in different geographies. Efforts to address decarbonization might be similar for big cities in North America and in Europe, but they would not apply to sub-Saharan Africa. Urban areas are also different from rural areas where the fight to bring access to energy and other services to all is still a challenge. Trying to impose the same pathway in different contexts can lead to failure and to the continuation of business-as-usual scenarios; (viii) complementarities of R&D activities supported by research institutions and academia, funded by public and private sectors. These activities should aim at promoting breakthrough innovation to feed continuously the process of decarbonization and keep under control any risk of lock-into solutions that may fail to contribute to total decarbonization in the long run.

References

Andrae A, Edler T (2015) On global electricity usage of communication technology: trends to 2030. Challenges 6:117–157. https://doi.org/10.3390/challe6010117

Arning K, Offermann-van Heek J, Linzenich A, Kaetelhoen A, Sternberg A, Bardow A, Ziefle M (2019) Same or different? Insights on public perception and acceptance of carbon capture and storage or utilization in Germany. Energy Policy 125:235–249. https://doi.org/10.1016/J.ENPOL. 2018.10.039

Carnevale P, Sachs J (2019) Roadmap to 2050: a manual for nations to decarbonize by mid-century

Cisco (2019) Cisco visual networking index: forecast and trends, 2017–2022 White Paper [WWW Document]. https://www.cisco.com/c/en/us/solutions/collateral/service-provider/visual-networking-index-vni/white-paper-c11-741490.html. Accessed 8 Sept 2019

Dean B, Dulac J, Petrichenko K, Graham P (2016) Towards zero-emission efficient and resilient buildings. Global Status Report. Global Alliance for Buildings and Construction (GABC)

Fasihi M, Efimova O, Breyer C (2019) Techno-economic assessment of CO_2 direct air capture plants. J Clean Prod 224:957–980. https://doi.org/10.1016/J.JCLEPRO.2019.03.086

Fellaou S, Bounahmidi T (2017) Evaluation of energy efficiency opportunities of a typical Moroccan cement plant: part I energy analysis. Appl Therm Eng 115:1161–1172. https://doi.org/10.1016/J.APPLTHERMALENG.2017.01.010

Global CCS Institute (2018) CCS Global Status 2018 84

Hausfather Z (2018) Not carbon neutral: assessing the net emissions impact of residues burned for bioenergy. Environ Res Lett 13:035001. https://doi.org/10.1088/1748-9326/aaac88

IEA (2019a) CO_2 Emissions statistics [WWW Document]. https://www.iea.org/statistics/co2emissions/. Accessed 8 Aug 2019

IEA (2019b) Global EV outlook 2019

IEA (2019c) The future of rail

IEA (2018) World energy outlook 2018: the future is electrifying

IEA (2017) Digitalization and energy

International Energy Agency (2016) World energy balances. https://doi.org/10.1787/data-00512-en

ITF (2019) ITF transport outlook 2019. https://doi.org/10.1787/9789282108000-en

Jarre M, Noussan M, Simonetti M (2018) Primary energy consumption of heat pumps in high renewable share electricity mixes. Energy Convers Manag 171:1339–1351. https://doi.org/10.1016/j.enconman.2018.06.067

Laursen W (2018) With ammonia, there's no "chicken or egg" dilemma. Marit Exec

Levi M (2013) Climate consequences of natural gas as a bridge fuel. Clim Change 118:609–623. https://doi.org/10.1007/s10584-012-0658-3

Markewitz Z, Ryssel M, Wang S, Robinius S (2019) Carbon capture for CO_2 emission reduction in the cement industry in Germany. Energies 12:2432. https://doi.org/10.3390/en12122432

Mayer J, Bachner G, Steininger KW (2019) Macroeconomic implications of switching to process-emission-free iron and steel production in Europe. J Clean Prod 210:1517–1533. https://doi.org/10.1016/J.JCLEPRO.2018.11.118

Morley J, Widdicks K, Hazas M (2018) Digitalisation, energy and data demand: the impact of Internet traffic on overall and peak electricity consumption. Energy Res Soc Sci 38:128–137. https://doi.org/10.1016/J.ERSS.2018.01.018

Nidheesh PV, Kumar MS (2019) An overview of environmental sustainability in cement and steel production. J Clean Prod 231:856–871. https://doi.org/10.1016/J.JCLEPRO.2019.05.251

Norled AS (2019) Norled to build world's first hydrogen ferry, to enter service in 2021. Fuel Cells Bull 2019:6. https://doi.org/10.1016/S1464-2859(19)30054-9

Nuclear Energy Agency (2017) Impacts of the Fukushima Daiichi accident on nuclear development policies. Nucl Energy Agency 1–67. https://doi.org/10.1787/9789264276192-en

Realmonte G, Drouet L, Gambhir A, Glynn J, Hawkes A, Köberle AC, Tavoni M (2019) An inter-model assessment of the role of direct air capture in deep mitigation pathways. Nat Commun 10:3277. https://doi.org/10.1038/s41467-019-10842-5

Schäfer AW, Barrett SRH, Doyme K, Dray LM, Gnadt AR, Self R, O'Sullivan A, Synodinos AP, Torija AJ (2019) Technological, economic and environmental prospects of all-electric aircraft. Nat Energy 4:160–166. https://doi.org/10.1038/s41560-018-0294-x

Schiffer ZJ, Manthiram K (2017) Electrification and decarbonization of the chemical industry. Joule 1:10–14. https://doi.org/10.1016/J.JOULE.2017.07.008

Sendich E (2019) Energy products are key inputs to global chemicals industry. Today Energy, EIA

Serrenho AC, Mourão ZS, Norman J, Cullen JM, Allwood JM (2016) The influence of UK emissions reduction targets on the emissions of the global steel industry. Resour Conserv Recycl 107:174–184. https://doi.org/10.1016/J.RESCONREC.2016.01.001

The Economist (2019) Fusion power is attracting private-sector interest. Econ

Traut M, Gilbert P, Walsh C, Bows A, Filippone A, Stansby P, Wood R (2014) Propulsive power contribution of a kite and a Flettner rotor on selected shipping routes. Appl Energy 113:362–372. https://doi.org/10.1016/J.APENERGY.2013.07.026

United Nations (2019) World population prospects 2019 [WWW Document]. https://population.un.org/wpp. Accessed 9 Aug 2019

Valentine H (2018) Competition increases in maritime battery technology. Marit Exec

World Bank Group (2019) The energy progress report 2019, Tracking SDG7

World Energy Council (2019) World energy council database [WWW Document]. https://www.worldenergy.org/data/resources. Accessed 8 Jan 2019

World Steel Association (2018) Steel statistical yearbook 2018

Policy and Regulation of Energy Transition

Karolina Daszkiewicz

1 Overview and Background

The world is at a turning point for what concerns energy trends. At a first glance, one can say that almost nothing happened over the last two decades: the share of fossil fuels in the overall global energy mix remained constant, at around 80%. However, this global number hides many different important energy trends: the rise of China, to become the largest energy consumer of the world, mainly fuelled by coal and oil growth; shale oil and gas reshaping not only the consumption in the United States, but also shifting historical importers/exporter balance; the start of the next new energy giants: India, South East Asia and Africa.

On the other side, we have been seeing clear signals of the will of many governments to steer away from fossil fuels and to move towards a clean energy future. The implementation of vehicles standards for cars, energy efficiency labelling for appliances and support for renewables technologies are strong examples of pivotal policies that have been implemented over the recent decades with climate and sustainability angles, changing the way that we consume and produce energy. This resulted in the stall of the growth of energy-related CO_2 emissions for several years.

But despite these encouraging signals, the CO_2 emissions grew again in 2018, and we are far from achieving, and even being on the right track with the goals that most governments of the world agreed upon at the UNFCCC's conference of parties 21 (COP21), held in Paris in 2015. COP21 was a turning point, but countries' actions need to be stepped-up. As shown in the International Energy Agency's World Energy Outlook 2019, the gap between the efforts currently envisaged and the Paris Agreement pledges is huge. This can—and must—be filled by a series of actions that requires a series of actions and policies to be put in place.

K. Daszkiewicz (✉)
Energy Policies and Markets Expert, Paris, France
e-mail: daszkiewicz.karolina@gmail.com

© The Author(s) 2020
M. Hafner and S. Tagliapietra (eds.), *The Geopolitics of the Global Energy Transition*,
Lecture Notes in Energy 73, https://doi.org/10.1007/978-3-030-39066-2_9

Many technologies and decarbonisation options are available, and it is up to each country to choose the best combination for them. Nevertheless, many areas still lag behind: removing the barriers that prevent the realisation of the huge energy efficiency potential, continuing the increasing deployment of renewable technologies in all sectors, supporting carbon, capture, utilisation and storage (CCUS) and place of nuclear technologies in the mix, creating the right conditions for these investments to be forthcoming are among some of the key policies that will need to be stepped-up in ambition over the next years.

This chapter is to outline the evolution of policies and regulations driving the energy transition with a focus on renewable energy technologies and energy efficiency.

2 Policy Classification

Countries have a range of policies and measures at their disposal to influence the deployment of renewables and energy efficiency improvements. These policies are multiple, ranging from tax benefits or waivers to capital grants, measures rewarding heat or power generation, self-consumption, or energy efficiency codes and mandates and many other tools.

Various energy strategies, targets and the majority of the policies aiming at decreasing investment costs can be used to trigger the deployment of renewables in all sectors—electricity, transport or heating and cooling. However, these policy tools can be adjusted and applied to incentivise improvements in the energy efficiency area. Price-finding mechanisms (e.g. auctions or administratively set tariffs) are mostly applicable to power generating renewables.

The below Table 1 attempts to group these policy tools and measures into overarching policy categories, enlists policy types, and track, if adjusted accordingly, the applicability of these tools across renewable sectors and the energy efficiency, provided necessary adjustments were implemented. Due to the breadth and depth of policy forms available, the below classification is not exhaustive.

Other approaches to classification of the policy types can be applied by sorting them from the angle of project size or purpose (electricity sector). Categorising policies per their exposure to market forces could be another approach (IRENA, IEA and REN21 2018).

In this chapter, only a few of the above-enlisted policy types will be explained in detail, in particular the ones that are responsible for bringing online the most significant shares of new capacities or particular relevance for either renewables or energy efficiency.

Table 1 Policy classification and applicability to renewables and energy efficiency areas

Policy category	Policy type	Applicability			
		Renewables			Energy efficiency
		Power	H&C	T	
Targets setting and strategic planning	Energy strategies	✓	✓	✓	✓
	Action plans	✓	✓	✓	✓
	Targets	✓	✓	✓	✓
Policies targeting upfront investment costs	Grants	✓	✓		✓
	Rebates	✓	✓		✓
	Soft loans	✓	✓		✓
	Tax benefits	✓	✓	✓	✓
	Tax waivers	✓	✓	✓	✓
	Depreciation tax benefit	✓			
	Generation tax benefit	✓			
Policies targeting energy generation	Feed-in tariffs (FITs)	✓	✓		
	Feed-in premiums (FIPs)	✓	✓		
	Auctions	✓			
	Tenders	✓			
	Contract for difference (CfD)	✓			
	Certificates	✓	✓		✓
	Policies for self-generation, self-consumption and sell of electricity to the grid	✓			
Regulatory	Rules on connection and dispatch	✓			
	Mandates and obligations	✓	✓	✓	✓
	Standards		✓	✓	✓
	Labels				✓
	Portfolio Standards	✓	✓		✓
	Regulatory environment not prohibiting or permitting corporate PPAs	✓			

(continued)

Table 1 (continued)

Policy category	Policy type	Applicability			
		Renewables			Energy efficiency
		Power	H&C	T	
Other policies	Education and information dissemination	√	√	√	√
	Training	√	√	√	√
	Research, Development and Deployment programmes	√	√	√	√

Note: H&C = Heating and Cooling, T = Transport sectors

3 Renewable Energy

3.1 Renewable Energy Policy Evolution and Geographical Spread

The deployment of wind and solar photovoltaics (PV) technologies progressed at a rapid pace over the past two decades growing from a nascent level and from few locations. While renewables in power sector made an impressive progression, renewables in transport and heating sectors developed at a slower pace lagging behind which is closely related to the level and effectiveness of corresponding policy adoption.

In 2000, only few countries had in place policies directly targeting renewable power technologies, with various rates of success. In 2010, already 45 countries had a renewable target in place and around 60 countries had some form of a measure remunerating directly renewable power generation gaining experience with feed-in tariffs. By 2017, the number of countries that adopted renewables target had grown to almost 180, of which two-thirds had in place policies incentivising renewable power generation. Robust policy adoption corresponded directly with increasing annual net additions commissioned in countries across all regions.

Countries aiming to support renewables in transport sector mainly use various forms of biofuel blending mandates. In 2010, only approximately 30 countries had mandates in place. By 2017, this number tripled. However, majority of the mandates require relatively low biofuel blending shares and countries' limited efforts to enforce them on the part of governments result in slower progression of renewables in the transport energy consumption.

At the same time, opportunities for the deployment of renewables in heating and cooling sector are vast as this is the largest end-use sector, accounting for more than half of total worldwide final energy consumption. However, renewable energy policy adoption visibly lags behind fort this sector. In 2010, only 13 countries had in place a renewable heat mandate with the majority located in Europe. By 2017, the number of countries with some form of a renewable heat mandate increased only to 22. European countries continued to lead policy adoption in this sector driven by obligations under the EU Renewable Energy Directive with mandatory 2020 renewable targets (IEA 2018).

Over time, renewable energy policies evolved, morphed and diversified in their structure, in particular policies targeting renewable electricity.

3.1.1 Targets

Target setting (in absolute or share levels) is often country's first step in strategy and policymaking for renewables deployment. Targets provide a clear objective where a country, region or city aims to get in terms of renewables deployment and may or may not consider how renewables will interplay with other energy technologies in the future energy mix and over a clearly defined period. They are often accompanied by a roadmap or an action plan in which the country enlists the measures it intends to use in order to achieve set objectives, translating targets into concrete steps and actions. This step is usually followed up by a secondary legislation, which adopts rules and renewable energy measures such as fiscal or financial policies, measures setting tariffs for renewable power generation, premiums or programmes supporting self-consumption.

In order to set ambitious yet achievable targets, target adoption is usually pre-empted by a series of studies and consultations such as resource availability study, technical potential and cost competitiveness assessments, grid integration and evolution of energy demand. These give a clear set of information on which targets can be set in an informed way by a country (IRENA 2015).

Renewable energy targets have four major characteristics:

1. **strength and obligation status**;
2. **structure and scope**;
3. **time scale**;
4. **context**.

The **first** element defines whether a target is of voluntary or of a binding nature. For the target to be legally binding it must be adopted into the national law. The target can be embodied in the overall energy act or adopted by a specific renewable energy legislation. Targets can also be set in the form of renewable obligation (RO), Fuel Mandate or renewable portfolio standard (RPS) that are adopted as secondary level laws and are renewable energy mechanisms per se.

Example of a legally binding target is the European Union's 2020 renewable target adopted in 2009 by Renewable Energy Directive 2009/28/EC. This target is binding for the overall of the European Union but it is also accompanied by legally binding country-level targets, which are expressed in a share of renewable energy consumed in country's total energy consumption in 2020. These targets are further split into sector-level objectives that are non-binding but are of an indicative nature on how countries intend to achieve their overall targets. Nonfulfillment of the overall targets is to be penalised with financial fines placed on member countries by the European Union.

Voluntary targets are not incorporated into national body of law but noted in various policies and strategies, and they do not entail financial burdens or fines.

Renewable energy targets can take a wide range of forms and **structures**. The first tier target is an overall target, applicable to country's entire energy mix (as before mentioned EU target). The second tier targets refer to a specific sector such as electricity, transport or heating and cooling sectors. The third granularity level of the target specifies if the target is technology-neutral or split into specific technology targets (separate target for solar PV capacity to be reached, wind, etc.). While a technology-neutral target allows markets to decide on the most cost-effective renewable technology option to be used in order to meet the target, it can also lead to overrepresentation of one technology on the market with the most mature technology dominating the deployment and preventing other technologies from maturing. There is a benefit in assuring deployment of a variety of technologies. Technology specific targets guarantee that technologies that are not yet fully mature will have their opportunity to develop within the country market.

While setting the target, a country should decide the metrics in which the target is expressed in and therefore chooses how the progression towards the target is to be measured and calculated. One of the most commonly used metrics by countries are either total primary energy supply (TPES) or total final energy consumption (TFEC). Targets can be also expressed as a share of energy demand or a specific amount of energy, power, heat or fuel delivered or consumed measured in a corresponding metric unit.

The **third** target element is its time scope. Typically targets are set for 10 or 15 years periods, however, many countries choose shorter horizons for their objectives. For example, China operates on the basis of carefully designed 5-Year Plans. Targets with horizons of 20 years and over are also practiced and are often accompanied by supporting mid-term targets.

Often, countries set renewable energy targets in a larger **context** of their overall energy and climate strategies. The EU 2020 and 2030 renewable targets are accompanied by the CO_2 reduction and energy efficiency targets. Countries often embed renewable targets in their overall energy mix objectives that are increasingly supplemented by climate targets.

Target setting is an important process that should be done carefully and supported with various studies. Regardless of what approach is chosen for a target setting, tracking progression towards the set objectives is a fundamentally important part of the process. This requires for the policymakers and regulators to develop a clear and coherent monitoring process.

However, the stand-alone target is an orphaned ambition if not supplemented with an appropriate blend of policies and measures triggering renewables deployment and putting countries on the path to meet the set targets.

The above-explained target setting process refers mostly to setting tangible objectives in the renewables sector. However, it can be replicated and adjusted accordingly to setting objectives in the area of energy efficiency or any other part of the energy system.

3.1.2   Fiscal and Financial Policies

Most low-carbon technologies face high capital costs, and therefore the conditions for finding such capital and the financing conditions can play an essential role in the deployment of these technologies. Fiscal and financial policies generally aim at reducing upfront capital costs and providing cheaper and more affordable financing conditions to investors. These can take a wide range of forms such as capital grants and rebates, soft loans, tax discounts, tax waivers or other tax benefits. These measures are easy to adopt and are often the first type of renewable support in place. They can also be complemented by other policies and measures directly supporting renewable production. These measures are rarely stand-alone policies available in a country, although stand-alone tax benefits and waivers are adopted in developing countries where no other measures are yet implemented.

Fiscal and financial measures are easy to manage from an administrative side as they can be amended, adjusted or removed within regular budgetary work of the government. These measures can address any particular renewable energy technology and are applicable to any sector. Measures supporting renewables are often embedded into energy efficiency policies and support systems. These types of policies can be used in renewables as well as in the energy efficiency efforts across all sectors and users. Their characteristics and principles remain similar but the beneficiary changes accordingly to an envisaged effect the measure supposed to achieve.

3.2   Renewable Electricity Policies

Much of the renewables policies over the last two decades have been aiming at increasing installed capacity in the electricity sector. Renewable capacity more than tripled from 850 GW in 2000 to 2 500 GW by the end of 2018, with more than 90% of this growth coming from hydro, wind and solar PV in almost equal shares. Hydropower was—and still is—the largest renewable power technology, both in terms of capacity and of electricity generation. For long, it has also been leading the annual additions of installed capacity. Nearly twenty years ago, total installed onshore wind capacity was less than 20 GW, offshore wind was an experimental technology with only demonstration projects in waters, and solar PV total capacity was standing at less than 1 GW.

Over the last two decades, wind and solar PV technologies experienced efficiency gains and large cost cuts. By 2018, solar PV additions grew to some 100 GW level reaching total installed capacity of around 500 GW and wind annual additions grew by 50 GW reaching 565 GW installed capacity. Hydropower was no longer a leading technology in terms of capacity additions, expanding by 20 GW in 2018; half of these new projects came online in China, as majority of viable sites for large hydropower projects in the OECD countries have been already used (IEA 2019e).

In 2000, half of the global annual renewable capacity additions were commissioned in Europe. In 2018, renewables growth was spread across all regions with China accounting for 43% of the global growth and Europe delivering third-highest level of new projects after APAC region. Renewables for power generation are turning from an expensive possibility for few to a mainstream across the globe (IEA 2019e).

There is a strong link between pace of additions and renewable energy policy adoption. Knowing the evolution of policy types and history of policy adoption across regions and sectors is important to understand the success and failures that have been a part of renewables take off and allow to draw lessons for the future (IEA 2018).

3.2.1 Evolution of Feed-in Tariffs and Feed-in Premiums

Feed-in tariffs (FITs) are administratively set price tariffs for electricity generated by renewable energy technologies. Countries started using FITs in the late 1970s as an incentive tool, gaining policy expertise over time (NREL 2010). With that and growing maturity of renewable technologies paired with dropping costs, the FIT systems started being more complicated with tariffs tailored to technology, installation size, mounting system type and often location of the installation. Remuneration levels are set by countries' regulatory office, separated from the market forces. Tariffs are assigned in long-term power purchase agreements (PPAs) signed between the generator and the responsible energy regulatory body or the electricity offtaker.

FIT contracts durations are usually of 12–20 years, reaching up to a maximum of 30 years in rare cases. Initially, the duration of FITs was of the same length for all technologies and project sizes. However, similar to remuneration levels, countries started gradually to adjust contracts' lengths to technology life cycles and how fast developers would get the return on their investments. Ultimately, on average, FIT contracts' durations become shorter and shorter oscillating currently around 10–15 years.

Initially used to support the deployment of all types and size renewable power stations, developed economies started limiting the use of FITs for renewables deployment in favour of other price-finding mechanisms. Conversely, in several developing countries FITs continue to be used, mostly for support of small installations for residential and commercial usage. They also continue to be an important mechanism for the deployments of all types and size of renewable installations in these countries, where access to financing instruments is more limited and investment is overall riskier.

One of the main limitations of FITs is the difficulty to accurately identify the correct level of remuneration for the amount of power generated, and to adjust it rapidly and in sync with the decrease of the technology costs. If the FIT is not sufficiently high, it will not trigger investors' interest, while if it is too high it will provide an excessive return on the investment, increasing the burden to electricity consumers or taxpayers. This second case can be exacerbated by the fact that, unless complemented with additional policies, there is no control on the volume of new capacity.

Feed-in premiums (FIP) are also set administratively, similar to feed-in-tariffs, but hold a greater exposure to electricity markets, as they are paid on top of the electricity price set by the market and collected by the generator. The overall remuneration of the project is therefore the sum of the electricity price received from the market and the FIP. The advantage of this mechanism is that generators are encouraged to react to the variation of electricity prices, while on the other side it increases their market exposure and therefore the risks associated to the project. FIPs can be set to remain constant over time regardless of the variation of the electricity price or sliding, with adjustments and eventual minimum and maximum levels. Similar to the FITs case, FIPs are contracted for long periods under PPAs.

Feed-in policies played a pivotal role in renewables deployment that was successfully adopted in Germany and later duplicated by many countries. In 2010, just over 50 countries had either a feed-in tariff or premium tariff systems in place, of which nearly half were European countries, followed by adoption in APAC region. By 2017, FITs or FIPs were in place in over 80 countries with Europe continuing to lead the adoption followed by APAC, Eurasia and other regions. The International Energy Agency reports that feed-in tariff and premium policies were the main policy bringing online around 80% of all commissioned utility-scale renewable projects over 2012–2017 period (IEA 2018). However, the policy paradigm shifted as renewables became more mature, and costs have experienced large cuts with countries turning towards auctions for price-finding mechanism for large-scale projects. Going forward, with China limiting its feed-in tariff policy, the IEA estimates that the feed-in mechanisms will drive around 40% of the global large-scale deployment over the next five years (2019–24) (IEA 2019e).

Administratively set tariffs are set to continue to play an important role for hydropower, concentrated solar panels (CSP), marine and geothermal technologies. Renewable auctions can drive the deployment, serving as a new go-to policy mechanism for large renewables, in particular for solar PV and wind projects. Nevertheless, countries start to increasingly use auction systems for awarding contracts to mid-size, or commercial size projects.

3.2.2   Renewable Auctions

Renewable energy auction is a selection process designed to procure new renewable electricity capacity (or generation volumes) competitively, in which a long-term PPA is granted to a qualified bidder based on a submitted financial offer, and in certain cases, additional criteria (e.g. the bidder's financial health, bank guarantees received, and previous experience in developing and operating renewable energy plants).

Auctions are, in fact, a price-discovery tool that takes advantage of competitive forces, shifting the burden away from the administrator who is responsible for setting the suitable framework within which private sector bids for a price and project.

Auctions are an excellent mechanism for a renewable capacity volume-control; however, their design must be carefully tailored to each country's context, as any other policy mechanism, in order to be effective. Auction mechanism must be designed so to attract sufficiently large pool of competition, which ultimately leads to low price discovery. Additionally, auctions must be accompanied by rules preventing unsuitable developers from participating or winning the auction and failing to deliver the allotted production capacity.

Auctions can be either open to all renewable technologies, wherein all projects compete with one another, or limited to one specific technology, for example to solar PV. PPAs can then be structured to incentivise production in desired locations more than in others, or at certain times that are more valuable to the system.

In auctions, bidders are invited to compete for a portion of the capacity up for auction (the minimum capacity size is usually specified in the auction rules), while in tender processes bidders must bid on the entirety of the sought-after capacity.

Winning developers are granted long-term PPAs, usually of up to 20 or 25 years depending on the country, project and technology. In effect, auctions are a price-finding mechanism for granting contracts similar to the way FITs and FIPs do. Renewable energy auctions are most suitable for procuring utility-scale projects. Large companies are most likely prone to participate, be eligible for participation and win the PPA contracts.

Set up of the auction system is difficult to cope with successfully by smaller companies to participate in, so there is a risk that competition will, in effect, be limited to several large players that will dominate the results and cut the smaller entities out of the market.

Selected project developer turn to financial institutions to secure loans. Once conditions of the loan are secured, the developer return to the relevant institution or the offtaker to sign the PPA. Time limits and penalties for late delivery of contracted projects are embedded in the PPA conditions.

In 2010, less than 20 countries had a renewable auction mechanism in place with Europe leading the adoption (seven countries), followed shortly by five countries from Latin America region. By 2017, number of countries with auctions increased to 88 and overtook number of countries with feed-in mechanisms in place. Europe is no longer a region dominating auctions activity as countries of Latin America, Eurasia, Africa and APAC are very active in adoption and running consecutive auction rounds as well. Within the next five years, auctions are to drive the large-scale renewable deployment bringing around two-thirds of global capacity growth. Nearly all large-scale renewable capacity to come online in Latin America is to come online winning bids in national auctions. In Europe, MENA and North America regions auctions are to drive around 70% of all renewable additions while in China and other Asia regions auctions are to drive just over half of the new deployment (IEA 2018).

Less developed regions, such as sub-Saharan Africa and Eurasia are to support the large renewables through a blend of feed-in tariffs and auctions policies as hydropower continue to be an important technology for new projects that are usually signed under administratively set tariffs, and non-hydropower renewables are still in early deployment levels and require strong policy framework (IEA 2019e).

Subsequently, annually awarded capacity volumes were increasing on an annual basis. In 2010, just under 3 GW of capacity was awarded, with majority of contracts going to onshore wind projects in Brazil. By 2017, auctions around the world allotted contracts to projects corresponding to 28 GW capacity to be commissioned across Latin America, APAC, Europe and North America regions. Nearly half of winning bids came from solar PV projects, followed by wind. In the first three quarters of 2019, a record of 40 GW renewable capacity was awarded through auctions tipped by China running renewable capacity auctions for utility and commercial size solar PV projects and completing country's first offshore wind auction round.

All together, as of the third quarter of 2019, auctions globally allocated long-term PPAs to around 175 GW of renewable capacity with the most recent wins to be commissioned by 2022.

Introducing factor of a strong competition pushed developers to drive cost cuts along the supply chain, in particular for solar PV and onshore wind technologies as they represent over 95% of the awarded capacity, with hydropower, bioenergy, geothermal and CSP accounting for the remaining portion (IEA 2019f). Capacity-weighted average auction prices for utility-scale solar PV projects declined from almost USD 160/MWh for those projects commissioned in 2014 to around USD 70/MWh for projects commissioned in 2018. Prices for projects selected in the 2019 auctions noted further decline with contracts awarded at USD 40/MWh. These projects are estimated to come online in 2022 given the usual project lead times for this technology and clauses written in the auction guidelines (IEA 2019e).

Average auction prices dropped for onshore wind fell from USD 65/MWh for projects commissioned in 2014 to around USD 56/MWh for projects coming online in 2018. Given the 2019 auction results, some projects to come online in 2022 will sell their generated electricity just below USD 40/MWh (IEA 2019e).

Going forward, auctions are expected to increasingly drive larger shares of annual additions as a tool that is able to secure attractive prices for new capacities, drive innovation and more policy know-how from various country setups are emerging. Increasingly, developing countries in Eurasia and sub-Saharan Africa are introducing auctions, often with a support from international financial agencies such as the World Bank, EBRD and other regionally relevant institutions that are able to assist with policymaking and often provide funds for soft financing for the initial auction rounds.

However, in countries where renewable energy policies will be increasingly phased out leaving renewables to play on the market independently, over the counter, or corporate PPAs can be an attractive tool for securing necessary financing outside of renewable energy auction schemes for the developers and for private companies as a way for decarbonising their energy needs.

3.2.3   Corporate Power Purchase Agreements (Corporate PPAs)

Corporate PPA is one of the tools increasingly used by private companies to decarbonise their energy demand, meet sustainability objectives while satisfying investors and the general public's pressure. As of mid-2019, three of the most significant platforms gathering private corporations that announced their sustainable commitments (TCFD, the RE100 and the science-based targets platform) reached 1400 members, of which majority joined post-2016. This level of engagement demonstrates a keen interest in the private sector in decarbonisation efforts and options to do so. In 2018, estimated 14 GW of renewable capacity was signed by the private sector under corporate PPAs, bringing the global total to 33 GW. The first three quarters of 2019 resulted in deals equivalent to 9 GW renewable capacity (BNEF 2019a).

A power purchase agreement is essentially a contract between two parties where one party sells both electricity and renewable energy certificates that might be a secondary product to the electricity generated to another party. In a corporate PPA, the seller is most often a renewable energy project developer or project owner. The buyer, or the offtaker, is a corporate or industrial entity with significant energy needs. The corporate PPAs can take two forms: physical or financial, often referred to as 'virtual'.

In a physical corporate PPA, the seller develops, owns and operates the renewable energy project and is responsible for delivering contracted power to the seller up to the delivery point. The offtaker buys the electricity directly from the seller. The offtaker takes ownership of the electricity from the delivery point as well as any corresponding renewable energy certificates and is responsible for moving bought electricity from the generator to its load.

The virtual corporate PPA (VPPA) is a financial contract rather than a contract for power purchase. In this type of contract, the offtaker buys the project's output and associated renewable energy certificates at a fixed price. The generator sells electricity on the market and passes the revenue collected to the offtaker. On the other hand, the offtaker agrees to pay the seller a fixed price, agreed upon in the signed PPA, for the renewable generation that developer sold to the grid. The fixed rate signed in the contract is the guaranteed price that the developer is to receive for generated power, irrespective of the ever-changing market price. When the market price exceeds the fixed contracted price, the developer passes the positive difference to the offtaker. When the market price is below the contracted price, the offtaker must pay the developer the difference (RE-Source 2019). This type of contract is often referred to as a contract for difference (CfD). The VPPA guarantees the seller a fixed price for generated and sold electricity, helping them in obtaining affordable financing to cover the initial investment costs. As a result, the VPPA can help commissioning of new renewable projects that would otherwise not be constructed. The VPPA resulting in a new capacity built can be particularly attractive to commercial offtakers that aim to decarbonise their demand and have electricity supply widely dispersed (Penndorf 2018).

The corporate PPAs are complex contracts bearing advantages and risks to both parties. Advantages to the seller are numerous, starting from guaranteed offtaker for electricity generated at a fixed price agreed for a long time durations. Increased creditworthiness can help to secure affordable financing, making the project economically viable resulting in new business for the developer. Benefits for the offtaker are long-term price stability, the company's budget stability, decarbonisation of energy demand and possible influence over renewable energy project specs and site.

Several risks are common for both the developer and the offtakes. Corporate PPAs have higher administrative costs. Possible regulatory changes rendering further contract execution impossible or at an increased cost are an essential risk factor to be accounted for by both parties and accommodated for upfront. In addition, in the case of long-term market distortion, the agreement has to be flexible enough to permit post-signature adjustments.

Corporate PPAs require the existence of certain regulatory and market conditions. These types of contracts are possible in countries with organised liberalised electricity markets where independent developers are allowed to build, own, operate and sell electricity to the grid or another party at its liberty with either regional transmission operator or an independent system operator in place. Additionally, signing corporate PPAs makes the most sense in countries where renewable energy policy support is minimal, evolves towards full market integration, or is to be increasingly phased out. In these conditions, developers are required to seek other streams of revenues and financing outside of administratively set prices or auction systems. For the offtakers, signing corporate PPAs requires a profound understanding of the electricity market, market trends and understanding of price evolutions over a short to mid-term (BNEF 2019b).

Thus far, the corporate PPAs predominantly developed in the United States of America, where the developers could rely mostly only on renewable support in the form of a tax exemption (the Production Tax Credit). However, as renewable energy measures in Europe are becoming limited or being phased out, corporate PPAs are becoming an increasingly attractive option for both developers and corporate entities. Corporate PPAs in Europe are more common in Nordic countries with several projects also signed in the United Kingdom, France and Denmark. Recently, developers and investors started to show interest in opportunities in Poland with two contracts signed so far by Mercedes-Benz and Grupa Azoty, amounting to 50 MW of renewable capacity together (Grupa Azoty 2019; WindEurope 2018). However, it is anticipated that going forward, more corporate PPAs will be signed in Poland as prices for baseload electricity are increasing rapidly and with estimations of further growth over the short-term period. This growth is mostly driven by increasing prices of the EU emissions trading system (ETS) auctions to which Poland is very susceptible as country's power generation is based on lignite coal (BNEF 2019c).

3.2.4   Policies for Self-consumption and Distributed Generation

The story of renewables for power generation is not only a story of large-scale installation previously driven by fixed feed-in tariffs and nowadays by auction systems or corporate PPAs going forward. In 2018, around 42% (or 210 GW) of the total installed solar PV capacity was coming from smaller (less than 1 MW capacity size) projects for commercial, residential and off-grid installations. For a long time, distributed PV projects were not cost-competitive with large-utility scale power plants. However, with the costs of PV dropping significantly over the last decade and expanding policy support in Europe and increasingly so in other regions across the globe, jump-started an important growth of distributed solar PV to the status we observe today. Currently, majority of the distributed solar PV capacity is located in Europe. Government policies, incentives and regulations have been central to this deployment in Europe and globally.

As indicated above, large portion of countries have renewable energy targets, which often include solar PV capacity objectives. However, several important markets have a specific distributed solar PV targets such as China (60 GW by 2020) or India (40 GW of rooftop PV by 2022). Targets alone often do not determine the deployment. The policies adopted along the target, at least until today, determined the deployment speed and capacity growth volume.

Similarly as in case of large-scale projects, distributed PV installations can benefit from policies decreasing investment costs. To repeat, these are grants, rates, various tax benefits and tax exemptions. According to the IEA classification, there are three main policy models targeting consumption and sale of electricity from the distributed solar PV. These are:

1. **buy-all, sell-all model**;
2. **net metreing system** and
3. **real-time self-consumption models**.

In the **first** model, all solar PV generation is contracted to be sold directly to the utility. Usually, the electricity is sold at a fixed tariff rate for the duration of the contract similar as FIT system. However, in some countries, auctions are used as a price-finding and contract awarding mechanism for commercial installations (e.g.: France). Often, a feed-in tariff mechanism is used as well.

In this model, the installation owners still source all their electricity from the grid. Their activity as generator is separated from their needs as customers. In order for this model to be functional, the private entity is obligated to have two metres installed. One to measure electricity generated and send to the grid, and one to measure electricity sourced from the network.

In the **net metreing model**, the PV owners are enabled to consume electricity generated by their installation reducing consumption from the network. The excess of the electricity is sent to the grid in return for an energy credit. Accumulated credit can be used to decrease future electricity bill accrued by electricity consumption at another time from the network. The duration of the validity of the credit is determined by the scheme in which the prosumer is participating. Usually, the validity period ranges from 6 months to one year. However, it is possible for this period to be longer. Recently, in Poland, the validity of the credit was extended from one to two years. The duration of the validity period strongly influences the economic attractiveness of the net metreing scheme for the PV investor as the output of the installation varies largely through a day and seasons. Installation of one bidirectional metre is required for this model to be possible on the owner side.

In the **real-time self-consumption model**, the PV owner is allowed to generate, self-consume and sell excess of the electricity produced to the network. The difference from the net metreing system is that the accounting of electricity procurement from the network and sending off the excess happens within short time intervals (hourly or less than hourly time spans). Each unity of the electricity sent to the network is paid for with a price level determined by the utility with which the contract is signed. Often, the price is based on the wholesale or retail electricity price. Similarly as in case of the net metreing model, installation of one bidirectional model is required.

Self-consumption policies continue to evolve and countries are working on tailoring them to their needs and particular conditions. In order to cap costs of the self-consumption policies, often additional limits and eligibility conditions are added. These can be put on a size of the installation eligible to participate in the scheme or on the amount of electricity sent to the network.

Other, less common self-consumption and electricity sell models exist where countries try to put a value of PV generation based on avoided large-scale generation capacity expansions, fuel expenditures or on benefits brought to the system or a society such as grid integration costs, CO_2 reduction or job creation.

The economic attractiveness of the participation in the above models is determined by several main factors. In the all abovementioned options, duration of the contract signed with the utility is pivotal. Additionally, in the buy-all, sell-all model, the LCOE is impacted by the contracted tariff level. In the remaining models length of the energy accounting period, remuneration type and price of excess generation and evolution of electricity retail tariff determine the attractiveness of the distributed PV project.

3.3   Renewable Heat and Transport Policies

The **heating and cooling** sector is complex and fragmented, and generally less well understood than the electricity sector. Its complexity makes effective policymaking challenging. Different thermal demand patterns in buildings (depending on climate, buildings efficiency, technology, occupancy purpose and others), and a multitude of technologies and fuel options availability on the supply side for water and space heating contribute to the complexity of the sector. Manufacturers range from large, multi-national corporations to small, local installers using different solutions. Additionally, different solutions are implemented across countries but also on regional and city-levels, which means that different institutions are responsible for policies and regulations-making. Due to this complexity, renewable energy policy adoption is more difficult and visibly lags behind in comparison to policy saturation in the renewable electricity sector.

In 2010, 13 countries, mostly in Europe and Latin America, had renewable heat mandates in place. By 2017 mandates spread to 22 countries across all regions, with Europe still leading, driven by the EU 2020 targets. Outside of mandates, almost 35 countries had some form of a capital grant in place facilitating purchase of the renewable heating equipment (IEA 2018). While the number of countries with availability of soft loans and tax incentive are difficult to track, these policies remain pivotal for the sector's decarbonisation due to their direct impact on the investment costs.

Many policy tools used for support of renewables in electricity sector are adjustable and applicable to support renewables in the heating and cooling sector. These are in particular different forms of targets and obligations as well as a range of measures affecting upfront investments costs. Heating and cooling sector is an area where policies for renewable energy and energy efficiency converge as many financial and fiscal forms of support for renewable heat technologies are embedded in the energy efficiency policies, programmes and standards.

Policies to support the use of renewables in heat can take a variety of forms, from mandates to heat generation incentives, from additional taxation to the ban of use of certain forms of energy. Mandates are the most commonly used type of policy, often requiring targeted types of buildings to satisfy a part of the heat demand needs through renewable energy (such as bioenergy or solar water heaters). The rate of implementation remains, though, more effective for newbuilds than for refurbishments, given the slow pace of the latter, due to the long lifetime of buildings.

District heating producers can benefit from incentives based on the amount of heat generated. This type of policy is very similar to feed-in-tariffs used in the power sector, providing a certainty of cashflow over a predetermined span of time. This form of policy support is still not widely spread; an example of a successful renewable heat generation-based policy is the Renewable Heat Incentive available in the UK since 2011 and currently scheduled for closure in the early 2021.

Countries can also adopt indirect forms of support for renewable heat. These are in particular carbon or additional tax obligations put on owners that generate heat using fossil fuels (wood and various coal products). These indirect measures provide important price signals, however design and implementation challenges remain, especially in contexts where energy-intensive industries are subject to strong international competition and may ask for exemptions.

Countries also have in their disposal introduction of bans on fossil fuel heating options. Bans can be very effective provided other suitable heat alternatives exist and are accessible to investors of various heat demand patterns and quotas. The effectiveness of bans heavily relies on monitoring and enforcement of such measures.

Renewables in **transport** take a form of biofuels, primarily used in road transport but increasingly applied in rail, shipping and aviation. The most common form of the support for biofuels in transportation is well spread biofuel blending mandates, often accompanied by fiscal incentives (IEA 2011). In 2018, around 80 countries had some form of a blending mandate. However, majority of these mandates required low blending levels (less than 10%) (IEA 2018) with few exceptions of Brazil (ethanol 18% blending mandate in 2019 cut from 27%), Paraguay (ethanol 25% mandate) and Indonesia (biodiesel 15%.mandate) (BiofuelsDigest 2019). Countries often provide tax waivers or tax cuts on biofuels production, distribution and consumption. Governments start to adopt sustainability mandates instead of blending mandates. The sustainability criteria are set based on avoided greenhouse gas emissions in comparison to conventional fuels.

4   Energy Efficiency

4.1   Energy Efficiency Policy Classification

Improvements in the energy efficiency can be the most cost-effective strategy to reduce emissions associated with energy consumptions in all sectors globally. However, various barriers such as financial, institutional, technical or lack of awareness often slow down the energy efficiency uptake. The energy efficiency policies and measures aim to overcome these barriers and as in the case of renewables support, blend of policies can be used in parallel in order to achieve energy efficiency gains in various sectors. Policies and tools can be implemented on regional, national, state or city-levels targeting appliances, equipment, improvements in buildings and vehicles in the end-use sector.

The IEA in its recent Energy Efficiency Market Report (IEA 2019a) groups the energy efficiency policies in three large categories:

(1) **mandatory policies**
(2) **energy efficiency,obligations** and
(3) **fiscal or financial policies**.

Outside of this main categorisation governments adopt other types of policies and programmes such as information provision, installers training and capacity programmes. Increasingly, private sector and various corporations adopt voluntary objectives and programmes to decrease the energy demand and improve efficiency.

4.1.1   Mandatory Energy Efficiency Policies Setting Performance Requirements and Standards

The first policy category, the mandatory policies and regulations set minimum energy efficiency performance requirements. Forms of such requirements are used extensively since the 1970s.

This category includes **mandatory minimum energy performance standards (MEPS)** for appliances and equipment, mandatory building codes, fuel economy standards and targets for industry. The MEPS for appliances is widely adopted and is considered an effective tool to improve the energy efficiency improvements in the end-use sector and reduce CO_2 emissions by banning the worst-performing appliances from the market, forcing manufacturers to innovate and improve their product. At the same time, they force customers to purchase more efficient appliances. The energy-efficient labels help the customers to make an informed choice but also provide educational value (Sonnenschein and van Buskirk 2019). In 2004, around 50 countries had these types of standards in place. By 2013, just over 80 countries used MEPS for energy efficiency improvements (IEA 2015).

The voluntary or mandatory **building energy efficiency codes (BEECs)** pertain specifically to the buildings sector and can be put on existing or newly build structures in the form of a standard. As buildings account for around 36% of global final energy use and nearly 40% of energy-related CO_2 emissions in 2017, the energy-saving opportunities, cuts in the CO_2 emissions and in costs are substantial. Number of building codes implemented grown from 54 countries in 2010 to nearly 70 countries in 2018 (IEA and UNEP 2018). Energy gains in the buildings sector can be achieved through reducing energy wastage, usage of energy-efficient appliances, lightening, space heating and cooling as well as through using materials and designing buildings to minimise the energy use.

The BEECs are often categorised as either prescriptive or performance based depending on the choice of an approach chosen for a compliance with the standard. The prescriptive compliance approach is a regulation requiring usage of specific materials in the construction of the building. These standards require minimum thermal performance level of each building envelope component such as walls, roofs, windows and doors as well as minimum energy efficiency requirements for heating, ventilation, water heating and lighting systems. The insulation of the building envelope or the insulation of pipes and ducts are also specified in the code. The performance-based approach sets annual levels of building energy consumption covering space cooling and heating, lightening as well as water heating.

The BEECs can be adopted on the voluntary or mandatory basis. However, even if the standard is adopted in the form of a mandatory measure compliance and enforcement of the codes is a key challenge. Achieving full compliance with set codes ab extend in time. Building strong compliance infrastructure around construction and building sector facilitates enforcement of the adopted codes and standards. Compliance with adopted codes should be measured during four stages of building project. That is to check compliance at the design stage before issuing the permit, during the construction phase, before issuing occupancy permit and once the building is occupied to measure actual energy performance. These checks help fixing potential issues and deviations from standards while it is still possible to do so (IEA 2013).

Energy efficiency standards also exist for transport sector. These are **transport fuel economy standards** and can be applied to passenger as well as to light, large trucks or to heavy duty vehicles. In these standards the governments or responsible relevant country regulatory body sets a minimum fuel standard for each given vehicle model produced or sold within its jurisdiction that must be respected. These standards pertain to fuel consumed by a vehicle in respect to travelled distance. Often expressed in miles per USgallon (mpg). Goal of such standards is to push manufacturers to produce more fuel-efficient machines while continuously reducing CO_2 emissions generated per usages of the vehicle. Standards also help to push out less efficient and more polluting cars or other types of automobiles from the market (US DOE 2019).

4.1.2 Energy Efficiency Obligation Policies

The second policy category, the energy efficiency obligation programmes are strongly established in the United States of America where are mostly known as the **energy efficiency resource standards (EERS).** These types of standards establish specific, long-term targets for energy savings to be met through energy efficiency measures. These standards play a role of a target, with a clear specification by whom the target is to be met. Obligation on meeting the standards are put on utility or non-utility administrators that are subsequently met through customer energy efficiency programmes. The standards can apply to electric or natural gas utilities adopted via legislation or secondary regulation (ACEEE 2019).

The standards do not indicate how the efficiency improvements are to be gained but indicate a minimum amount of savings to be reached. The obliged party has a flexibility of choosing how the requirement will be met giving the utility freedom of choice of the programme that would be the most effective for them. The objectives can be expressed in various units ranging from a percentage form (on electricity delivered) or in megawatt, gigawatt, or kilowatt-hours (ACEEE 2019). The EER standards are similar in their set up to the renewable energy portfolio standard (RPS). Often the EERS are of mandatory nature, they can be also established as a voluntary mechanism (C2ES 2019).

4.1.3 Fiscal and Financial Policies Targeting Energy Efficiency

Financial incentives. These include policies put in place to encourage the take-up of energy-efficient technologies and behaviour through financial or fiscal rewards, including grants and subsidies, tax relief, equity finance, loans and debt finance, guarantees, on-bill finance and other incentives (IEA 2019a). All this incentive types mirror fiscal and financial support mechanisms used for renewable energy solutions. Their main goal is to decrease level of capital investment bore by the investor.

5 Nuclear Energy

The deployment of nuclear energy power plants saw its peak during the 1970s and the first half of the 1980s, on the back of national programmes, in particular in the United States, in Canada, in several European countries, in Russia and Japan.

In the period 1965–1990, global nuclear-installed capacity increased 64-fold and, by 1990, more than 90% of the global capacity was installed in these five regions. Over the following years the growth reduced drastically, with global capacity increasing by some 15% in 20 years (IAEA 2019), as a result of several factors, including the fears that followed the Chernobyl disaster, the saturation in some markets and the significant investment costs needs.

The most recent years, in particular following the Fukushima accident, saw the emergence of three major trends. On one side, some countries (e.g. Germany, Belgium and Switzerland) decided the gradual phase-out of nuclear energy from the power mix. On the other side, several emerging economies have decided to continue to pursue the deployment of nuclear power; China has been leading this trend over the last decade (and is expected to continue over the coming years), but several other countries have plans to expand or start new nuclear energy programmes. Thirdly, several countries with existing and ageing nuclear power fleet are facing the question mark if closing, extending the lifetime or replacing the existing assets (IEA 2019c).

Nuclear policies are therefore very varied according to the intentions and the direction decided in each country. Some countries see nuclear power as an important component for the decarbonisation of their power mix—and of the energy transition at large—while others see it as a no-way forward. In several cases, major decisions on the timing of phasing out, on the allowed maximum lifetime of the existing power plants and on the eventuality and the extent of new builds are spurring important national debates and have not yet been set.

Significant differences emerge across countries depending on if liberalised or regulated electricity power markets are in place, but it has become increasingly clear that the construction of new nuclear power plants based solely on market forces is not a viable option, in particular due to financial and regulatory risks. Several countries have therefore decided to put in place support mechanisms (such as the Contract for Difference in the United Kingdom, or the inclusion of nuclear within the Zero Emission Credit in some States of the USA), while many others support directly the construction of new nuclear power plants.

The investment costs can be very different across countries, with the highest costs seen in first-of-a-kind plants in mature economies (e.g. Europe and the United States) and the lowest costs in China, thanks in particular to a continuous stream of new builds and low construction, labour and material costs. Despite these differences, one element is common to almost all new nuclear power plants: the very high upfront investment cost. This is due to the high unit investment costs (usually in the order of 2500–6500 USD/kW) and the very large capacities of the main reactors built (usually in the order of 1000–1600 MW).

Policies aimed at reducing the related financing costs can therefore play a key role for the deployment of new plants, as well as policies that can optimise and reduce the building process and therefore reduce the long construction times (another key element for new nuclear plants). To solve or reduce some of these financing aspects, but also to open new market opportunities for smaller and more flexible use of nuclear power, several constructors are now exploring the possibility of building so-called small modular reactors (SMRs). These plants, usually of the size of 30–300 MW, are now in the demonstration phase and are attracting significant attention and support from several governments.

6 Carbon Capture, Utilisation and Storage (CCUS)

Carbon capture, utilisation and storage (CCUS) is a technology that can provide a significant contribution to achieve a low-carbon—and in the long-run a zero-carbon—energy world. This technology has been applied in some industrial sectors for over two decades, and currently has almost 20 large-scale projects and several dozens of small-scale ones in operation. Looking ahead, it can play a significant role in the decarbonisation strategies of several sectors, ranging from enhanced oil recovery (EOR) to fossil-fuelled power generation (mainly coal, but also gas), from the production processes in heavy industry to the creation of so-called carbon sinks (e.g. bioenergy CCS or BECCS).

The importance of CCUS technology in the industry sector stems from the fact that the CO_2 emissions in this sector, and in particular in the cement, iron and steel, and chemical subsectors, are among the most challenging to abate, as a significant portion of these emissions result from chemical or physical reactions and there is a limit scope for fuel switching away from fossil fuels in processes that require high-temperature heat. The scope for policy action is therefore relevant, requiring concerted actions between governments, industrial and financial actors (IEA 2019g).

The power sector has a more varied availability of low-carbon generation technologies, ranging from dispatchable renewables (such as hydropower and bioenergy) to non-dispatchable renewables (mainly wind and solar PV) and nuclear power. Nonetheless, two-thirds of current global electricity generation is fossil fuel based. Given the long lifetime of power assets, and the very young age of the coal fleet (in particular in China); it faces the mutual challenge of 'emissions lock-in' and of possible stranded assets. Retrofitting with CCUS technology represents therefore a very important option, in particular for some countries, to avoid stranded assets and to keep flexible capacity in the power systems (IEA 2019h).

While the CCUS technology is already a competitive decarbonisation option in some process (such as the production of ammonia), it still needs to achieve cost reductions and the scale needed in the long-term decarbonisation scenarios. As it is the case for most other low-carbon technologies, CCUS is characterised by high upfront investment costs. The technology had significant momentum globally in the second half of the first decade of 2000s, but then lost some grounds and slowed down in terms of demonstration projects in subsequent years, as actual public funding support did not match previous announcements (IEA 2016).

Adequate funding and support measures for R&D and for demonstration projects are therefore key for a quick upscale of the technology. For the early phases, the policy options range from regulatory to financial, including grants, tax credits or low-carbon product incentives. In the longer run, CO_2 pricing can provide an important long-term investment signal. Particular attention should be provided to the industrial sectors that face global competition, while for other sectors such as cement or power generation, a fair level playing field should be established at national or regional scale.

Additional areas for policy intervention are also represented by the development of CO_2 transport and storage networks, the removal of regulatory and social barriers and obstacles to the deployment of the technology and removing uncertainties regarding the availability of storage. Increasing attention is also being given to the development of CCUS hubs, that can support new investment opportunities through economies of scale, and can be developed together with storage considerations (UK Government 2018).

7   Conclusions

Energy policies have been and continue to have a pivotal role in the energy transition triggering and fast-tracking renewables deployment, incentivising uptake of energy efficiency, supporting changes in the energy system and paving the role of CCUS and nuclear power in the energy mix of tomorrow.

Policies targeting renewables expanded fast over the last two decades, initially adopted by a small pool of countries with well-developed economies. Currently, nearly all countries across the globe have some form of a renewable energy policy in place effectively triggering high deployment levels and making solar PV and wind front-running green power technologies as cost continue to decrease and various market-entry barriers are tackled.

Over time, policies evolved growing in their complexity. Through increasing policymaking know-how, governments gained experience and skill in tailoring mechanisms to their needs and energy system specifications at decreasing costs. Policies with administratively set prices for renewable power generation triggered the renewables deployment. However, nowadays countries move away from these types of mechanisms moving towards policy frameworks that allow price setting through market competition.

Going forward, overall system integration in which all generators are able to secure sufficient level of revenues is to be in focus. Private sector is expected to play an important role in furthering decarbonisation of the energy system through greening their energy demand with help of corporate power purchase agreements becoming an active player in the energy domain.

Policies for energy transition go beyond renewables for power generation but support mechanisms are adaptable and can be applied to heating, cooling and transport sectors as well as in efforts aiming to decrease energy consumption through efficiency measures. Energy efficiency is a pivotal domain in the successful energy transition and often strongly interlinked with renewables and requires strong governments attention.

Role of nuclear energy in the energy transition depends on countries' energy strategies as we currently observe examples of states that decide to reduce (France) or eliminate (Germany) this source of energy, while others take their initial steps towards first reactors. According to World Nuclear Association, as of 2020 around 30 countries are either considering, planning or launching their nuclear power programmes. UAE, Belarus, Bangladesh and Turkey are currently working on the construction of their reactors (World Nuclear Association 2020). The CCUS provides important opportunities in achieving deep decarbonisation level with several proven projects in operation. However, in case of both technologies, strategy making, target setting and policymaking are indispensable to tackle high upfront costs.

Going forward, energy policies will continue to be at the heart of energy transition continuously evolving and adapting to countries needs and changing market realities.

References

ACEEE (American Council for Energy-Efficient Economy) (2019) Energy Efficiency Resource Standard (EERS). https://aceee.org/topics/energy-efficiency-resource-standard-eers

BiofuelsDigest (2019) Biofuels mandates around the World 2019. https://www.biofuelsdigest.com/bdigest/2019/01/01/biofuels-mandates-around-the-world-2019/

BNEF (Bloomberg New Energy Finance) (2019a) Corporate PPA Deal Tracker, BNEF, London

BNEF (2019b) A guide to corporate clean energy procurement, BNEF, London

BNEF (2019c) 2 h 2019 Corporate energy market outlook, BNEF, London

C2ES (Center for Climate and Emergy Solutions) (2019) Energy efficiency standards and targets. https://www.c2es.org/document/energy-efficiency-standards-and-targets/

UK Government (2018) Policy paper: the UK carbon capture, usage and storage (CCUS) deployment pathway: an action plan, London. https://www.gov.uk/government/publications/the-uk-carbon-capture-usage-and-storage-ccus-deployment-pathway-an-action-plan

Grupa Azoty (2019) PGE Energia Odnawialna zbuduje z Grupą Azoty Siarkopol farmę fotowoltaiczną. www.grupaazoty.com/pl/wydarzenia/pge-energia-odnawialna-zbuduje-z-grupa-azoty-siarkopol-farme-fotowoltaiczna.html

IAEA (International Atomic Energy Agency) (2019) PRIS database, IAEA, Vienna. https://pris.iaea.org/

IEA (2011) Technology roadmap: biofuels for transport, IEA, Paris

IEA (2013) Modernising building energy codes, IEA, Paris

IEA (2015) Market report: energy efficiency 2015, IEA, Paris

IEA (2016) 20 years of carbon capture and storage, Paris

IEA (2018) Renewables 2018—analysis and forecast to 2023, IEA, Paris

IEA (2019b) How2Guide to solar technologies, IEA, Paris

IEA (2019c) Nuclear power in a clean energy system, IEA, Paris

IEA (2019d) Policies database, online database, IEA, Paris. www.iea.org/policies

IEA (2019e) Renewables 2019—analysis and forecast to 2024, IEA, Paris

IEA (2019f) Renewables auctions database, internal database, IEA, Paris

IEA (2019g) Transforming industry through CCUS, IEA, Paris

IEA (2019h) World energy outlook 2019, IEA, Paris

IEA (International Energy Agency) (2019a) Energy efficiency 2019, IEA, Paris

IEA and UNEP (United Nations Environment Programme) (2018) 2018 Global status report, IEA and UNEP, Paris

IRENA (International Renewable Energy Agency) (2015) Renewable energy target setting, IRENA, Abu Dhabi

IRENA, IEA and REN21 (2018) Renewable energy policies in a time of transition. https://www.irena.org/-/media/Files/IRENA/Agency/Publication/2018/Apr/IRENA_IEA_REN21_Policies_2018.pdf

NREL (National Renewable Energy Laboratory) (2010) A policymaker's guide to feed-in tariff policy design, NREL, California

Penndorf S (2018) Renewable energy power purchase agreements, 3Degrees, London. https://3degreesinc.com/latest/ppas-power-purchase-agreements/

RE-Source (2019) Introduction to corporate sourcing of renewable electricity in Europe, Re-Source. http://resource-platform.eu/toolkit/

Sonnenschein J, van Buskirk R (2019) Minimum energy performance standards for the 1.5 °C target: an effective complement to carbon pricing. J Energy Effic 12:387–402. https://doi.org/10.1007/s12053-018-9669

US DOE (U.S Department of Energy) (2019) Vehicle fuel economy and greenhouse gas (GHG) emissions standards. https://afdc.energy.gov/laws/385

WindEurope (2018) First polish PPA is also Europe's first automotive renewables deal. https://windeurope.org/newsroom/press-releases/first-polish-ppa-is-also-europes-first-automotive-renewables-deal/

World Nuclear Association (2020) Emerging Nuclear Energy Countries. https://www.world-nuclear.org/information-library/country-profiles/others/emerging-nuclear-energy-countries.aspx

The Role of Policy Design and Market Forces to Achieve an Effective Energy Transition: A Comparative Analysis Between the UK and Chinese Models

Marco Dell'Aquila, Daniel Atzori, and Ofelia Raluca Stroe

The interplay between policy design and market forces has been crucial in driving the global energy transition from fossil fuels to renewable energy sources. However, models adopted by countries are based on different historical approaches and have therefore evolved differently over time, while also influencing each other.

In this chapter, the UK and China have been chosen as two countries which have achieved successes in decarbonising their respective economies through the deployment of significant levels of renewable energy resources; however, each has adopted a differing energy transition strategy.

The UK electricity market was one of the first globally to be privatised and deregulated in the late 1980s and early 1990s, following similar earlier initiatives in the natural resources and telecommunications sector. This experience and the subsequent market adjustments which were undertaken by the UK Government and the regulator laid the groundwork for similar processes which have taken place within the European Union and further afield. The entry of new market players initially in conventional gas-fired power generation during the 1990s and subsequently in renewables starting in the 1990s—but more substantially over the past 20 years—have brought non-utility private sector expertise into the deployment of new technologies. It was this track record which enabled the UK to design incentives for renewable energy—starting with tariffs and moving to auctions—with a strong likelihood that they would bring about successes in terms of deployment of new renewable capacity. It is for this reason that the UK has been selected to represent a noteworthy example of how regulation can harness market forces to play a decisive role in the shift towards renewable energy.

M. Dell'Aquila (✉)
The Johns Hopkins University - SAIS Europe, Bologna, Italy
e-mail: mdellaquila@jhu.edu

D. Atzori
Cornwall Insight, Norwich, UK

O. R. Stroe
British Academy of Management, London, UK

© The Author(s) 2020   227
M. Hafner and S. Tagliapietra (eds.), *The Geopolitics of the Global Energy Transition*,
Lecture Notes in Energy 73, https://doi.org/10.1007/978-3-030-39066-2_10

China, on the other hand, began its journey towards decarbonisation later, its motivation was driven partly by the availability of natural renewable resources—such as wind and sun—which in turn would reduce the country's need to import primary energy and secondly, abate emissions which were becoming particularly significant in urban areas risking a public backlash. China epitomises a very large and topographically diverse country in which the role of the state has been instrumental not only in policymaking aspects of the energy transition, but crucially also in funding, building, owning and operating renewable energy power plants.

1 The UK: A Case Study of Market-Led Energy Transition

As an island endowed with some of the best renewable resources in Europe, in particular wind, tidal and wave power resources, the UK has been chosen as a case study to show how a major economy is capable of achieving ambitious renewable energy targets whilst ensuring economic growth. However, Britain's transition away from coal, like many of its European neighbours, has been neither a smooth nor a linear process.

Importantly, in June 2019, the UK became the first G7 economy to commit to reducing greenhouse gas emissions (GHG) to net-zero by 2050, as compared to the previous less ambitious target of an 80% decrease from 1990s levels.

The UK is also one of a handful of markets around the world which is currently transitioning from traditional direct government support of renewable energy sources, using feed-in-tariffs for utility-scale renewable energy plants, to the private sector support of renewables. This is predominantly through the use of Power Purchase Agreements ('PPA'), entered into directly with corporations keen to purchase some or all of their power requirements from renewable energy sources.

This section argues that one of the key reasons for Britain's success in promoting a renewable agenda and transitioning away from direct government support has been policy flexibility combined with a commitment to allowing the market to play a crucial role in the energy transition. In this respect, the British model differs from those of other European countries, which were more heavily reliant on direct state support.

The UK has been a pioneer not only in the incentivisation of renewables through the Non-Fossil Fuel Obligation (NFFO) subsidies introduced in 1990s, but also the liberalisation of its electricity market, thanks to wide-ranging reforms implemented in the late 1980s and early 1990s. The role of successive UK Governments since the 1990s has been decisive in paving the way for the energy transition, latterly as capital costs have declined sufficiently enabling the private sector to take a leading role, thanks to the development of 'more competitive forms of price discovery such as auctions or tenders' (DECC 2011).

While these policies can be deemed to have been successful, they mainly supported commercially viable technologies, such as wind and solar, at the expense of promoting commercialisation of technologies such as tidal. Beyond that, they failed

Table 1 UK energy sector, key milestones

1983	The Energy Act encourages the private generation and supply of electricity
1986	Privatisation of British Gas
1988	'Privatising electricity' White Paper
1989	The Electricity Act calls for the privatisation of the UK electricity supply industry
1990	Privatisation of the Regional Electricity Companies (distribution and supply) and two generators National Power and PowerGen
1990	Non-Fossil Fuel Obligation (NFFO)
1990–1998	Liberalisation of electricity supply
1993	'The prospects of coal' White Paper
1997	British Gas splits into Centrica (gas trading and retail) and Transco (pipelines)
2000	The Utilities Act introduces the Renewables Obligation (RO)
2001	Climate Change Levy
2008	Climate Change Act
2013	Energy Act—Electricity Market Reform
2018	Launch of 'The Road to Zero' Strategy
2019	UK commits to net-zero emissions by 2050
2020	COP 26 to be hosted by the UK, jointly with Italy

to provide any incentive for research and development into earlier stage technologies such as wave, notwithstanding the higher priced banding available in recent Renewables Obligation (RO) tenders. Indeed, the government resorted to entering into a long-term PPA to support the construction of the £20 billion Hinkley Point C 3.2GW nuclear power plant at a price of £92.50/kWh but was unwilling to undertake the same for the smaller £1.3 billion 320 MW Swansea Bay Tidal Lagoon Project at a price of £89.90/kWh (UK Government 2011) (Table 1).

2 Early Processes of Decarbonisation

Decarbonisation of the UK energy industry started in the 1970s, well before any global initiative to halt climate change. The primary motivation was economic rather than environmental, and more specifically linked to the depletion of the UK's North Sea oil and gas reserves. A series of policies were launched to reduce the UK's reliance on fossil fuels in the electricity sector, including stronger support for new nuclear power plants as well as other clean-energy sources, the result of which was a 29% decrease in petroleum consumption between 1970 and 2012 (UK Parliament 2018).

As a result, the interplay between a set of economic and policy factors has significantly contributed to accelerating the energy transition. Hence, in order to understand the roots of Britain's decarbonisation processes, it is necessary to briefly analyse the historical context from which this transition emerged.

Despite having its own domestic sources of coal, the need for the UK to become less dependent on fossil fuels first became apparent as a consequence of the 1973 and 1979 oil shocks. Following the first oil shock, the UK Government formed a Department for Energy in January 1974 (Pearson and Watson 2012). The economic consequences of the two oil shocks were far reaching in the UK and translated into spiralling inflation rates which reached 24% in 1974 (The Guardian 2011). Wage demand began to rise as earnings were eroded and the Trades Unions representing some 280,000 miners played a pivotal role in wage negotiations with the government led by the Conservative Prime Minister Edward Heath (Gouiffes 2009). The UK economy was forced into a three-day working week as miners went on strike which led to frequent blackouts, as coal-fired power plants were shut down due to a lack of coal. Public opinion and the resulting public policy began to favour the search for alternative sources of energy such as renewables. However, these oil shocks also drove heavy investment into the UK North Sea to reduce the need for oil imports and this translated into the UK having access to large amounts of natural gas, which fuelled the boom in gas-fired power plants in the 1990s. At the same time the UK continued to build new nuclear power plants, an industry which had thrived in the UK, having been home to the world's first civil nuclear power station built at Calder Hall, in Cumbria, in 1953, which only closed in 2003. Both nuclear and gas increasingly drove down the overall share of coal, dampening the drive for non-fossil sources of energy. It would be another decade until this trend was to re-emerge (Fig. 1).

As in other countries, a sudden increase in global hydrocarbon prices led policy-makers to rethink the need for an energy policy focusing on the country's security of supply. As Elliott (2019) convincingly argues, the development of UK policies in this regard was also strongly influenced by a set of utopian ideas developed mainly in the USA and the UK by the Alternative Technology movement of the late 1960s and early 1970s (Elliott 2019). However, although the 1973 oil crisis introduced renewables into the policy debate, capital costs were still too high and the immediate consequence was to strengthen the UK's development of nuclear and gas-fired power generation sources.

As will later be demonstrated, the debate between nuclear and renewable energy as the most effective way to decarbonise the economy has raged in the UK over the last three decades and is yet to reach a conclusion. The UK has undoubted achieved remarkable successes in its building of nuclear capacity, but the rapid reduction in new nuclear builds over the past 20 years, as well as the well-publicised delays and cost overruns in the construction of new plants, has shifted the policy onus in favour of renewables (EDF Group 2019). In Britain, investment in nuclear has historically been framed primarily as necessary to ensure national energy security, and only more recently has the rationale been extended to cover the industry's contribution to the decarbonisation of the economy. However, Fig. 2 shows that nuclear as a share of the UK electricity sector actually grew from the 1970s and peaked at almost 30% in the late 1990s, before declining to below 20% today. This share is likely to decline

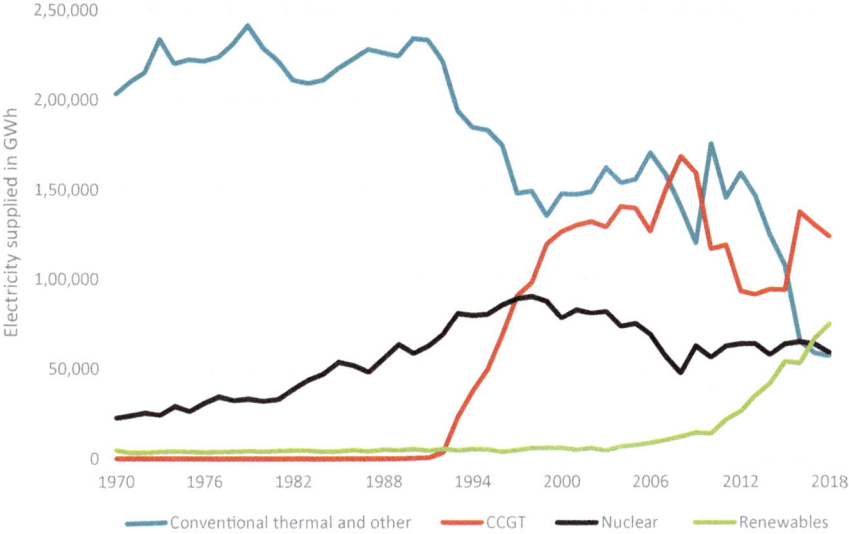

Fig. 1 Electricity supplied in the UK between 1970 and 2018, by source type (GWh). *Source* UK Government (2019) Digest of UK Energy Statistics (DUKES). Department for Business, Energy and Industrial Strategy. Available from https://www.gov.uk/government/statistics/digest-of-uk-energy-statistics-dukes-2019

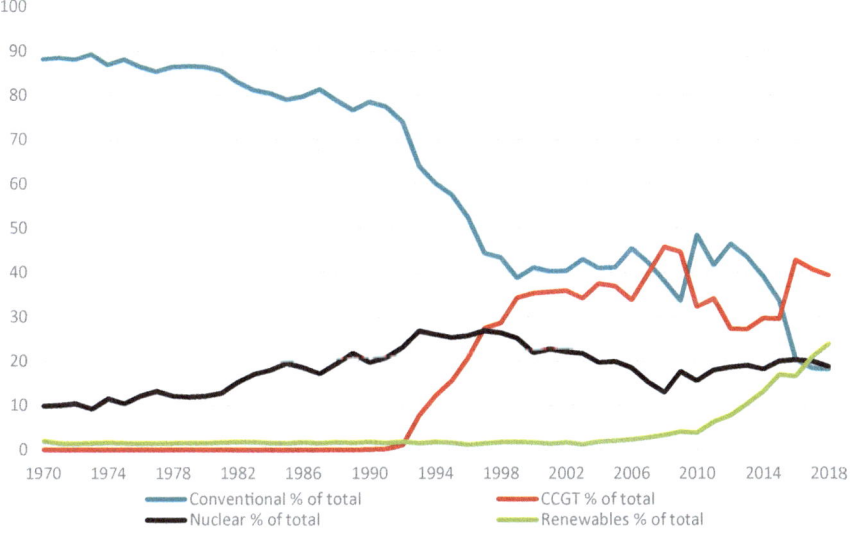

Fig. 2 Share of conventional, nuclear, CCGT and renewables in total electricity supplied between 1970 and 2018 in the UK (%). *Source* Dukes energy statistics, 2019

further despite the construction of the 3.2 GW Hinkley C plant, as many of the older plants are retired and not replaced.

In 1979, David Howell, Secretary of State for Energy in the government led by Prime Minister Margaret Thatcher, spearheaded a new nuclear programme, aimed at increasing the country's nuclear capacity in the face of growing geopolitical uncertainty that threatened energy imports (Pearson and Watson 2012). However, the government's overall support for nuclear in the context of energy security, even in the face of growing public hostility following the 1979 Three Mile Island and 1986 Chernobyl accidents, was not the only factor in promoting the first phase of decarbonisation of Britain's economy.

A key element of this early decarbonisation was the shift from public to private ownership of the energy sector, which took place throughout the 1980s and into the 1990s, starting with the privatisation of British Gas in 1986, followed by the break-up of the Central Electricity Generating Board (CEGB) and the sale of the Regional Electricity Companies (RECs) in 1990. Nuclear was retained in public hands until it was partly privatised through a flotation of 28% of the shares in 2006 (Horrocks and Lean 2011). It was subsequently fully privatised when it was purchased in its entirety by EDF 2 years later (World Nuclear Association 2019), although it could be argued that with EDF being fully owned by the French Government at that stage, it was not a pure privatisation. Throughout the 1990s, electricity companies, which had recently been privatised, undertook a shift in the building of new power plants from coal-fired to natural gas generation, as a result of the increased availability and competitive price of North Sea gas.

The so-called 'dash for gas' led to a replacement of old coal plants, which had efficiencies of between 20% and 30%, with newer and more efficient Combined Cycle Gas Turbines (CCGT), with efficiencies of between 40% and 60%, and which ultimately led to the closure of a significant number of deep mines owned by British Coal.[1] Overall between 2005 and 2016, the average UK power fleet improved efficiencies by 5.9% (European Environment Agency 2018). From 1913 when there were 3,024 deep coal mines in the UK employing 1,107,000 people, the numbers had plummeted to 20 mines employing some 12,000 people in 1999 and ultimately down to 5 mines employing 1,000 people in 2019 (UK Government 2019).

In the landmark 1993 white paper 'The prospects for coal—Conclusions of the government's Coal Review', the UK Government endorsed the principle that:

> Competitive markets provide the best means of ensuring that the nation has access to secure, diverse and sustainable supplies of energy in the forms that people and business want, and at competitive prices (Department of Trade and Industry 1993).

This argument, inspired by Thatcherism, would also influence later energy policies of the Conservative government of John Major (1990–1997) and also New Labour under Tony Blair. Indeed, the 1993 White Paper also stated:

> The coal industry must take its place within a competitive energy market. It must compete with other fuels and other suppliers to meet the needs of its customers at commercial prices. Its dominant market is in electricity generation (UNFCCC 2019).

---

[1] See: https://netl.doe.gov/sites/default/files/gas-turbine-handbook/1-1.pdf.

Another interrelated factor that contributed to the decreasing importance of coal in Britain's energy mix was the bitter confrontation between the government of Margaret Thatcher and Arthur Scargill, the then President of the National Union of Mineworkers (NUM), which culminated in the 1984 and 1985 miners' strike, ultimately leading to the shutdown of many mines in the UK.

The share of coal in the energy mix declined throughout the 1980s and the 1990s, as a consequence of the decline of the country's domestic production along with growing imports of natural gas making CCGTs more economically viable. This trend was also one of the main factors in the decentralisation of the electricity system, made possible by the emergence of Independent Power Producers (IPPs) in the aftermath of the privatisation and deregulation of the electricity industry.

While the need to transition to a low-carbon economy was increasingly reiterated at international summits throughout the 1990s, the country had already achieved significantly reduced levels of coal in the energy mix thanks to an increasing reliance on nuclear and gas. Overall, during this period, a substantial decarbonisation of Britain's economy was mainly driven by a need to ensure energy security through the support for nuclear, and by economic factors that made gas-fired generation increasingly cost-competitive with coal.

3 Reducing Greenhouse Gas (GHG) Emissions

As mentioned, in this first phase of decarbonisation of the UK energy mix, the main driver was not so much represented by environmental concerns, although this certainly played a part, but by geopolitical and market factors. However, in the 1990s, the need to reduce Greenhouse Gas emissions (GHG) was perceived as increasingly important in global public opinion and by governments.

Such a growing awareness was reflected in Britain's policymaking. The UK Government's white paper on the environment, entitled 'This common inheritance', was published in 1990, setting out a strategy to stabilise CO_2 emissions. Although the document was criticised by environmentalists for not going far enough, it was nevertheless hailed as a landmark, especially for stating the principle that polluters should pay.

Meanwhile, the UN Framework Convention on Climate Change (UNFCCC), held in Rio in 1992 to prevent human interference with the climate, increased the pressure on governments to stabilise greenhouse gas emissions.

1988	Establishment of the Intergovernmental Panel on Climate Change (IPCC)
1992	Earth Summit takes place in Rio de Janeiro
1995	COP1 takes place in Berlin
1997	Adoption of the Kyoto Protocol

(continued)

(continued)

2001	COP7, Marrakesh Accords
2005	EU Emission Trading
2005	Entry into force of the Kyoto Protocol
2015	COP21, Paris Agreement

Source United Nations Framework Convention on Climate Change

The origins of the UK promotion of renewable energy can be found in the Electricity Act of 1989. In it, the government established a mechanism whereby producers of non-fossil fuel based energy would participate in tenders submitting a minimum price for a 15-year contract which would underwrite their project. The scheme, known as the Non-Fossil Fuel Obligation (NFFO)—together with Northern Ireland NI NFFO and Scottish Renewable Obligation (SRO)—paid the generator the then wholesale electricity or pool price plus a technology premium linked to the specific technology being utilised (International Energy Agency 2013).

Buyers of the renewable electricity were the recently privatised Regional Electricity Companies (RECs) who were obliged to source a fixed percentage of their power from non-fossil sources. The RECs contracted collectively through the Non-Fossil Purchasing Agency Ltd (NFPA) with renewables generators, while the above market costs were reimbursed by a levy passed onto consumers (Department of Trade and Industry 1999).

Generators had up to 5 years to commission their plants, but no penalties were imposed for either delays or failure to build a power plant. The scheme was rolled out in five orders, in 1990, 1991, 1994, 1997, with a total of 3,639 MW contracted, but only 1,198 MW of capacity was actually built (The National Archives 2006). The NFFO mechanism supported nuclear, as well as renewables.

4 A Market-Led Decarbonisation

Britain pioneered the privatisation of energy assets and has led the way in developing market-based tools for the promotion of renewable energy. The advent in 1997 of the New Labour Government led by Tony Blair ensured a continuity with the market-oriented approach to the energy transition implemented by the previous Conservative governments. By the 2000s, the UK's decarbonisation was no longer only driven by economic factors, as was the case of the previous decade with the 'dash for gas', but by environmental concerns crystallised in international commitments, as demonstrated by the initiatives implemented by the governments of Tony Blair, who was in power from 1997 to 2007.

In 2000, the Utilities Act replaced NFFO with Renewables Obligation ('RO'), a market-based incentive based on green certificates to promote the development of large-scale renewable generation in the UK. Arguably, RO encapsulated Britain's attempt to promote the development of renewable energy through market-based mechanism, rather than more top-down approaches predominant in other European countries, and even more so later in China.

The RO, which came into effect in England and Wales, and Scotland, in 2002 and in Northern Ireland in 2005, required UK electricity suppliers to purchase an increasing share of their electricity in the form of Renewable Obligation Certificates (ROCs) either from their own renewable sources, or by purchasing from qualifying power producers. The RO was paid for over a 20-year period and was banded by technology, with landfill gas generators earning the lowest band of 0.25 of a ROC, increasing to double the price of a ROCs for wave, tidal, dedicated biomass and Combined Heat and Power producers. Overall renewable power producers would benefit from three key income streams. This consisted of the wholesale electricity price and the ROC payment, divided into the fixed buyout element as described above and a variable element paid by non-compliant suppliers and shared amongst renewable generators. Finally there was a levy exemption certificate paid by all industrial and commercial consumers to renewable producers (Climate Change Levy).

In 2017, the RO closed to new generation capacity. Overall, the market-based mechanism of RO ended up favouring specific technologies, such as wind and biomass, developed by large groups through utility-scale projects. Innovative technologies that were less market ready were discarded, such as the proposed tidal project over Severn Barrage in 2010. Other measures implemented in this period include, in 2001, the Climate Change Levy and the establishment of the Carbon Trust. A landmark step was represented by the 2003 White Paper, which set the future government priorities:

> In reducing carbon dioxide emissions, our priority is to strengthen the contribution of energy efficiency and renewables. They will have to achieve far more in the next 20 years than previously. We believe such ambitious progress is achievable, but uncertain (UNFCCC 2019).

While endorsing renewables, the government's support for nuclear had waned, as the issue of nuclear waste disposal began to be addressed:

> There are also important issues of nuclear waste to be resolved, including legacy waste and continued waste arising from other sources. We do not make specific proposals for building new nuclear power stations.[17]

Hence, in 2003 the UK seemed to be on the verge of a major shift to a whole new energy policy, focusing virtually exclusively on the development of renewable energy, even at the expense of nuclear, which had represented the backbone of the country's energy security in the previous decades. However, such an attitude was short-lived. At the beginning of his third term, Prime Minister Tony Blair again backed nuclear power, to ensure the attainment of Britain's decarbonisation goals whilst at the same time contributing substantially to its energy security. The renewed support for nuclear was contained in the 2007 White Paper, which set the new energy priorities of the government. It expressed concerns about the country being too dependent on a limited number of technologies and positioned nuclear as an important element in the diversification of energy sources, with the additional benefit of reducing carbon emissions (UK Government 2007).

This position was reiterated by Tony Blair's successor Gordon Brown in his foreword in 'Meeting the energy challenge. A white paper on nuclear power', published by the then Department for Business Enterprise and Regulatory Reform (BERR) in January 2008, stating:

> Nuclear power is a tried and tested technology. It has provided the UK with secure supplies of safe, low-carbon electricity for half a century. New nuclear power stations will be better designed and more efficient than those they will replace. More than ever before, nuclear power has a key role to play as part of the UK's energy mix. I am confident that nuclear power can and will make a real contribution to meeting our commitments to limit damaging climate change (UK Government 2008).

As Pearson and Watson (2012) point out, one of the main reasons behind the shift may have been the fact that, by 2004, the UK had again become a net energy importer as a result of dwindling cheaply recoverable gas supplies from the North Sea, after having been a net exporter for several years previously.

The new commitment to nuclear was part of a broader political will to address the decarbonisation of the economy. In this respect, in 2008, the government of Gordon Brown launched the Climate Change Act, which set forth legally binding targets to reduce GHG emissions by 80% from 1990 levels by 2050.

However, during the coalition government formed by the Conservative and the Liberal Democrat Parties and led by David Cameron between 2010 and 2015, Britain's path towards decarbonisation suffered some setbacks. In the Conservative Party Manifesto 2015, it was stated:

> Onshore wind now makes a meaningful contribution to our energy mix and has been part of the necessary increase in renewable capacity. Onshore windfarms often fail to win public support, however, and are unable by themselves to provide the firm capacity that a stable energy system requires. As a result, we will end any new public subsidy for them and change the law so that local people have the final say on windfarm applications (The Conservatives 2015).

This halted new onshore wind projects despite an abundance of wind resources. Concurrently, the cancellation of incentives for utility-scale solar saw activity plummet in that sector too.

In 2013 the UK Government passed the Energy Act, with the aim of reforming the energy sector to enable it to attract £100 billion of infrastructure required to bring it up to date (OFGEM 2019). These Electricity Market Reforms, which importantly introduced a Capacity Market to ensure long-term security of supply, also impacted renewables by phasing out the RO support and replacing it with a Contract for Difference (CfD):

> A Contract for Difference (CFD) is a private law contract between a low carbon electricity generator and the Low Carbon Contracts Company (LCCC), a government-owned company. A generator party to a CFD is paid the difference between the 'strike price' – a price for electricity reflecting the cost of investing in a particular low carbon technology – and the 'reference price'– a measure of the average market price for electricity in the GB market. It gives greater certainty and stability of revenues to electricity generators by reducing their exposure to volatile wholesale prices, whilst protecting consumers from paying for higher support costs when electricity prices are high (UK Government 2015).

A key element of these reforms was that the government was able to decide on when and the level of capacity it would make available for each CfD auction, which in turn is divided into two pots, one for established and one for emerging technologies. To date, three auctions have taken place with a total of almost 10GW allocated; the first round (AR1) ran from October 2014 to March 2015, the second (AR2) from March to September 2017 and the third was launched in May 2019. The only sector to have seen significant growth has been offshore wind, which reached a total of 19GW. AR1 awarded 1.2GW, comprising two offshore wind farms (the 714 MW EA1 and the 448 MW Neart na Gaoithe) (BEIS 2019), 11 projects were awarded in AR2 totalling 3.3GW, while in the third six projects totalling 5.5GW of installed capacity received CfDs. In fact, prices in the third-round saw offshore wind achieve £39.65/kWh, as opposed to £92.50/kWh which the government negotiated bilaterally with EDF for the construction of the new Hinkley C nuclear reactor (The National Audit Office 2017).

5 Towards Net-Zero

In June 2019, Theresa May, the then UK Prime Minister, passed legislation to cut emissions to zero by 2050. The announcement took place almost a month before the formal dissolution of her cabinet, on 24 July 2019, and it was likely made in order for her to secure a legacy beyond Brexit.

It is highly significant that her successor, Boris Johnson, setting out the priorities of his new government in his first speech at the House of Commons on 25 July 2019, supported this policy stating:

> Our kingdom in 2050—thanks, by the way, to the initiative of the previous Prime Minister—will no longer make any contribution whatsoever to the destruction of our precious planet, brought about by carbon emissions, because we will have led the world in delivering that net-zero target. We will be the home of electric vehicles—cars and even planes—powered by British-made battery technology, which is being developed right here, right now (House of Commons 2019).

Despite the optimistic tone, however, the UK's path towards decarbonisation is set to face a number of challenges. Currently, besides repowering existing wind farms, it is extremely difficult to develop onshore wind projects. Yet according to a report prepared by Vivid Economics for RenewableUK (2019) (Vivid Economics 2019), the deployment of 35GW of onshore wind—the UK's cheapest renewable technology—by 2035, could lead to a 7% decrease in electricity costs, together with a set of wide-ranging set of socioeconomic benefits.

6   Subsidy-Free Shift

Whilst auctions can be seen as a backward step in the UK's history of allowing the private sector to drive decarbonisation, in that it places a greater onus on government to decide when to allocate new capacity and how much, the primary drivers of new projects that will contribute towards the Net-Zero by 2050 target are likely to come from the private sector. Indeed, the UK is a pioneer in the development of projects that do not rely on government subsidies, or subsidy-free projects. The 10 MW Clayhill solar project with 6 MW of co-located storage situated in Milton Keynes and inaugurated in September 2017, exemplifies this new trend as it required no subsidy and instead signed a long-term Power Purchase Agreement (PPA) with EDF.

Such subsidy-free projects rely on a burgeoning market for corporates for whom the need to become greener and more sustainable is leading them to enter into PPAs directly with developers of renewable energy projects, effectively cutting out the middleman—i.e. the UK Government. As most projects must compete with low power prices, they attract a mix of predominantly debt to make them competitive, and lenders to these projects require the terms of the power purchase to be as robust as those previously in place when the projects were effectively underwritten by the government. As a result, corporate PPAs tend to favour large companies with deep balance sheets, capable of entering into long-term contracts, enabling the power developers to seek long-term debt funding. However, the financial and corporate sectors are leading the way in creating 'synthetic' PPAs, in which the obligation to purchase electricity is shared between several buyers at differing conditions and is wrapped or underwritten by utilities, banks and increasingly oil companies who are keen to enter into the renewable energy market. Whether all this potential results in many new projects coming online is still open to debate, as the whole corporate PPA sector is still in its infancy, with 7 projects totalling 804 MW having been built and operating as at Q4 2019.[2]

7   Electrification of Transport

Although the UK had been able to achieve significant success in the decarbonisation of its electricity production, it is increasingly being acknowledged that the main challenge is the need to achieve higher rates of decarbonisation of heat and transport, where progress had been much slower. For the electrification of transport, the launch in July 2018 of 'The Road to Zero' represented a major turning point. The document set the objective of 'all new cars and vans to be effectively zero emission by 2040' and of 'almost every car and van to be zero emission' by 2050 (UK Government 2018).

Consistent with the market-driven approach that, as we have seen, is a key feature of the UK model to energy transition, the document also clearly states that '[w]e

---

[2]www.inspiratiacom, dataLive database, 2019.

expect this transition to be industry and consumer led, supported in the coming years by the measures set out in this strategy' (UK Government 2018).

'The Road to Zero' was presented by the UK Government as part of its industrial strategy. Hence, its goal was not only to reduce pollution and decarbonise the economy, but also to ensure that the UK was going to be a leading manufacturer of zero emission vehicles. The importance of this document cannot be overstated, given the sheer scale of its ambition. However, the strategy presented several shortcomings.

According to a letter from Lord Deben, Chairman of the Committee on Climate Change (CCC) published on 11 October 2018, 'existing and newly agreed policies for road transport […] are insufficient to ensure the reductions in emissions necessary to meet the 5th Carbon Budget in the most cost-effective way' (UK Government 2018).

A report published by the House of Commons on 16 October 2018 conducted a much more abrasive critique of the 'Road to Zero', accusing the government of a 'lack of clarity on the meaning of the 2040 targets' which was 'unacceptable' (House of Commons 2018). The Parliament's report criticised the fact that the 'Road to Zero' did not clearly define the 'conventional' petrol and diesel cars to be phased out, as it did not specify whether conventional hybrids will be banned or not. With conventional hybrids potentially still available on sale after 2040, the goal of achieving a zero-emissions fleet target by then seemed impossible to reach. Such an ambiguity was seen as a major constraint for both car manufacturers and charging infrastructure providers. Hence, the House of Commons' report was asking for more clearly defined and more stringent targets, to allow the industry to make the appropriate investment decisions.

Despite its shortcomings, Britain was, however, effective in establishing a partnership between the public and private sector to develop charging infrastructure for Electric Vehicles (EVs). As announced by the Chancellor of the Exchequer in the 2017 Budget, a fund was later established to increase the roll-out of charging infrastructure in Britain. The Charging Infrastructure Investment Fund (CIIF) was a £400 million investment fund to be set up with £200 million raised by the private sector, matched by £200 million from the government. In February 2019, the private equity firm Zouk Capital was named CIIF's preferred bidder, entering in exclusive negotiations with the government to act as the manager of the fund.

8   China: A Case Study of Government-Led Energy Transition

A world leader in installed capacity of hydropower, solar PV and wind, China has been selected as a case study, as it represents one of the most iconic examples of a successful government-led energy transition. Moreover, the country's policy shifts have a considerable impact on the global energy environment, due to the size and growth of its economy and its high levels of electricity production and consumption.

This can be coupled with China's ambitious cross-border investments in energy, including those within the framework of the Belt and Road Initiative (BRI).

This section outlines the reasons behind China's success in promoting an ambitious renewable energy agenda. Four different drivers will be assessed, including the status of energy dependency, decarbonisation policies and initiatives, the swiftly developing manufacturing sector and the shift to subsidy-free renewables.

Firstly, energy dependency in China underwent significant changes in the 1990s, throughout the Gulf War and beyond. While annual oil imports accounted for less than 5% of total imports at the time of the Gulf War, the country gradually became a major global importer of oil by the 2000s reaching a level of 10.35 million b/d in December 2018 (Dannreuther Roland 2003; Li 2015; S&P Global 2019). This increasing dependency is one of the drivers behind the country's relatively swift transition to renewables, as China aims for energy self-sufficiency. According to a document published by China's National Development and Reform Commission (NDRC)—mentioning the targets set by NDRC—the country's self-sufficiency rate should be above 80% by 2020. In parallel, by 2030, clean energy is set to meet most of the demand, forecasted at 5.6 Btce (3.91 Btoe) between 2035 and 2040, with non-fossil fuel sources accounting for over 50% of the power production (IEA 2018; NDRC 2016).

Secondly, as a consequence of its economic growth, carbon emissions per capita have almost quadrupled in China since 1990 (Wang et al. 2015). As an emerging economy, the country strives to enhance quality of life and to achieve progress through further industrialisation and urbanisation, necessitating that sustainable development be at the forefront of its political agenda. Economic growth has driven up energy consumption significantly and thus, with demand on the increase, the Chinese Government has kick-started a series of decarbonisation initiatives, focusing on electricity generation using clean sources, and more specifically renewables.

Thirdly, the swiftly growing renewable energy manufacturing sector emphasises China's interest in boosting economic growth whilst creating new jobs and increasing exports. Therefore, China considers the manufacturing of renewable energy equipment a strategic industrial area, entitling it to preferential land policies, taxation and loans. China began to promote the use of locally manufactured products for renewables as early as 1999, when the 'Notice on Relative Problems of Further Supporting the Development of New and Renewable Energy' stipulated that projects using such equipment would be granted a preferential investment profit rate of 5% (Fan et al. 2018; NDRC 1999). The investment profit rate is the ratio of investment to rate of returns on capital.

The manufacturing of wind turbines and photovoltaic cells has been designated as a critical industrial sector in China, as a result of which the country has become a global manufacturing leader in both sectors.

Lastly, renewable energy incentives through feed-in tariff policies have played an important role in the rapid roll-out of such projects throughout China. Some of the most noteworthy policies include the 'Improving Policies on Feed-in Tariff of Wind Power' issued in 2009 by NDRC, which announced the launching of tariffs for wind power and dividing the country into four regions with tariffs ranging from 0.51to

0.61 RMB/kWh. Power costs above coal-fired generation were split between the central government and the operators of the provincial grid (Grau et al. 2012; Zeng et al. 2013; Lewis 2011). Another similar document was released in 2011, addressing solar photovoltaic (PV) power projects and awarding a fixed feed-in tariff of RMB 1.15/kWh for projects reaching the completion stage by the end of 2011 and RMB 1.00/kWh for those approved by July 2011(Fan et al. 2018).

This section analyses China's challenging transition process to subsidy-free renewables. The country witnessed substantial decreases in project capital costs due to technology-related manufacturing improvements and increasing competition in the renewables market. Between 2000 and 2010, China's total renewable energy supply grew at an average annual rate of 12%, approaching mass production (Zeng et al. 2013). On 31 May 2018, NDRC, the National Energy Board and the Ministry of Finance announced that all subsidies for utility-scale solar projects would be halted in favour of competitive bidding (NDRC 2018).

Below are some of the key milestones the country achieved in the energy sector since the 1950s, including the implementation of the country's iconic five-year plan, the adoption of the first minimum energy performance standards and the launch of the Belt and Road Initiative (Table 2).

On 28 February 2005 the decarbonisation of the Chinese electricity generation sector began, triggering debates amongst academics as to how realistic these targets would be, with a considerable number deeming non-fossil fuel based electricity targets of 20% by 2030 as feasible (Zhang et al. 2018).

An in-depth analysis of drivers of the government-led energy transition in China, including additional impetus necessary to preserve success on the long term, is detailed in the section below.

Table 2 China energy sector, key milestones

1953–1957	The first Chinese Five-Year Plan
1978	Deng Xiaoping commenced its ambitious programme of reforms, with an emphasis on attracting foreign trade and investment
1981	The State Energy Commission was established
1982–1983	The Ministry of Petroleum was divided in China National Offshore Oil Corporation, Sinopec Corporation and China National Petroleum Corporation
1984	China joined the International Atomic Energy Agency (IAEA)
1988	Formation of the Ministry of Energy
1989	China adopted the first Minimum Energy Performance Standards (MEPS)
1997	Establishment of the State Power Corporation
1997	The Energy Conservation Law entered into force
2003	Abolishment of the State Power Corporation
2003	Launch of the State Electricity Regulatory Commission (SERC)

(continued)

Table 2 (continued)

2005	Adoption of the Renewable Energy Law of the PRC on 28 February
2007	The National Climate Change Program was issued in June 2007
2008	China launched US$850 billion economic stimulus package in November 2008, with 35% allocated to low-carbon development
2008	Publication of the first white paper on energy, the country emphasising the development of renewable energy as a top priority
2010	The revised Energy Conservation Law entered into effect on 1 April 2010
2011	China overtook US becoming the largest power system world-wide, with an installed capacity of over 1TW
2013	The Belt and Road Initiative was launched by China's president Xi Jinping
2015	After connecting roughly 3 million people in remote areas, China reached 100% electrification rate
2018	China increased its renewables target in the electricity consumption mix from 20% to 35% by 2030
2021	No subsidies will be granted to offshore wind projects from 2021 onwards

9 Paving the Path to Energy Self-sufficiency

As an importer of oil since 1996, China's energy self-sufficiency levels plummeted, while electricity consumption was and continues to be on the rise[52], as shown in the figure below. To tackle this issue, the country's government implemented a series of strategies, meant to lower dependency on imported oil. Renewable energy sources have become a significant part of the solution, providing new avenues through which the country could develop its national power industry and build a reputation for the use and deployment of sustainable alternatives (Fig. 3).

When assessing China's success in paving a path to energy self-sufficiency, three main factors should be taken into consideration. The first focuses on the ability of renewable power to tackle substantial growth in energy demand, whilst diminishing the need for fossil fuel imports. The second relates to the historical progress China has made with regard to achieving renewable energy technology (RET) self-sufficiency. The third assesses policies and initiatives adopted by the government to enable energy producers, developers, advisers and financiers to learn from the experience of more mature markets in Europe.

China experienced rapid economic growth in a number of industries between 2005 and 2011, including thermal power generation, which grew by roughly 90%, steel production grew by 135% and automotive vehicles by 223%, with proportions of coal, gas and oil used in manufacturing overtaking the entire output from nuclear and renewables (Liu et al. 2013).

Annual electricity generation and consumption in China continued to grow dramatically between 2000 and 2019 (World Bank 2018; IMF 2019; UN Statistics 2019). This growth drove the Chinese Government to seek to accelerate the development and deployment of renewable energy sources, as this was widely perceived as the

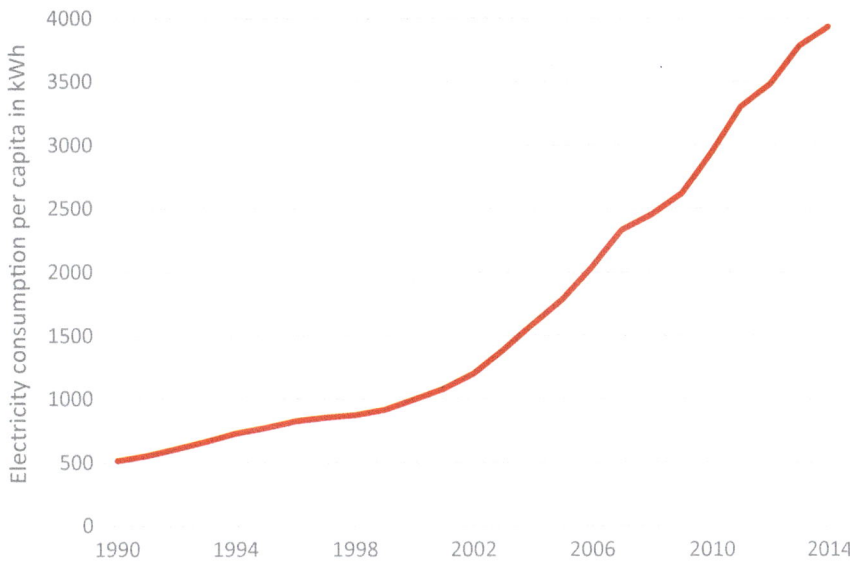

Fig. 3 China's electricity consumption per capita in kWh, between 1990 and 2014. *Source* World Bank (2014) Electric power consumption (kWh per capita). Available from: https://data.worldbank. org/indicator/EG.USE.ELEC.KH.PC

engine to swift and secure economic development (Liao and Wang 2019). A number of initiatives were implemented to boost renewables production and decrease dependency on oil, gas and coal imports.

Having identified several vulnerabilities within the energy supply sector, the June 2002 Law on Promoting Clean Production was enacted to increase security by diversifying the energy mix (SCNPC 2002). This initiative was not the first to be implemented, as the government had already offered a construction tax credit to the renewables generation sector since 2001 (Zhang et al. 2016).

In September 2007, the Medium and Long-term Program for Renewable Energy Development was published by the Chinese Government, announcing a target of 10% for energy consumption from renewables by 2010 and 15% by 2020 (Gao et al. 2011; NDRC 2007a).

In December 2007, the government's white paper on China's Energy Conditions and Policies also emphasised the importance of energy diversification for the country's security, which, according to the document, could only be achieved through boosting electricity production from renewables. Chinese President Hu Jintao also mentions this in the report of the 17th People's Congress (UN Statistics 2019; Chinese Government Official Platform 2007; IOSC-PRC 2007).

In an attempt to regulate energy consumption, China's Ministry of Finance introduced a tax targeting large vehicles with inefficient energy consumption in 2008 (MOF 2009). In 2010, the government amended the original Renewable Energy

Law, published in February 2005, emphasizing the urgency to roll-out renewable energy (Wang and Chang 2014; Chinese Government Official Platform 2009).

Between 2011 and 2015, throughout its 12th Five-Year Plan, China ramped up installation of offshore wind plants, announcing it aims to achieve 5GW of capacity by the end of 2015 (NEA 2019).

In the latest Five-Year Plan (2016–2020) the Chinese Government plans to increase the share of non-fossil fuels in primary energy demand from 14.3% in 2018 to approximately 20% by 2030 and, in addition, is also aiming to achieve 15% share of clean-energy in total primary consumption by 2020 (Lee 2019; NEA 2016).

Commentators, including S&P Global, agree that a slowdown in the growth of fossil fuels and more specifically oil in total consumption is likely to be achieved over the next 5 to 8 years, as a result of increasingly favourable government support for renewables and its rapid adoption, but also due to an increase in electric vehicles in the country (S&P 2019).

These strategies appear to be ever more important, now that China and the United States are in a mounting trade conflict, affecting energy commodities flow (Kempe 2019). This current political uncertainty is accelerating the deployment of renewable energy projects.

China's rapid deployment of renewables has seen installed capacity grow from 3 MW in 1994 to over 11GW of solar and 5GW of wind capacity in the first half of 2019[66]. Most of the technology used in these projects was either produced locally or nationally.

The Belt and Road Initiative (BRI) is one of China's best recognised international plans to increase its economic and political influence globally and, amongst other projects, it involves a multi-billion dollar investment program in the energy sector. While to date the main beneficiaries have been large-scale fossil fuels schemes, according to Greenpeace (2019), Chinese equity backed over 12GW of wind and solar projects under this initiative.

Initiatives such as the BRI are ushering a new era of Chinese investment globally, including in Asia, Europe, Africa and Latin America, potentially enabling it to learn from the experiences of early renewables adopters—especially in Europe–further driving its accelerated transition to renewables and thus reducing its dependence on energy imports. China has already showed substantial interest in exchanging knowledge on innovative solutions and legislative improvements in the clean-energy sector. In July 2018, China signed an agreement with the EU to cooperate on meeting power demand with alternative energy (EC 2018).

10   A Leader in RET Manufacturing

As a result of the significant scale that it has been able to achieve, China has evolved its renewables manufacturing base from being a supplier of domestic projects to an export-focused industry. The clean-energy sector expanded much faster than nuclear power and fossil fuels between 2012 and 2017, as the use of renewables gradually became more efficient, affordable and accessible (Standaert 2019; Tan and Mathews 2014).

To better assess the country's manufacturing sector, this section focuses on wind power, solar PV and hydropower in China.

The country's first grid-connected wind farm was constructed as far back as 1986, using equipment from Denmark (Shi 1997). By the end of 2020, China is set to employ 800,000 people in the sector, according to the Chinese Wind Energy Association (CWEA 2017).

Despite being a relative latecomer to wind generation, the Chinese Government has enacted a number of policies to facilitate the expansion of wind power manufacturing. Primary amongst these were the 2008 Interim Measure of Management of Special Funds for Wind Power Industrialization, the 2009 Notification of Improving Price Policy of Grid-connected Wind Electricity and the Accelerating Smooth Development of the Wind Equipment Industry published in 2010 (Zhao et al. 2012; Dai et al. 2014).

In addition, the Ministry of Science and Technology announced in 2011 that wind turbines are amongst the national key technologies and proceeded to fund the research and development (R&D) departments of both universities and emerging manufacturers (MOST 2012).

The country has also built its wind expertise through mergers and acquisitions, as well as international technology transfer (NDRC 1999; Standaert 2019). Some of the more noteworthy Chinese purchases abroad were the 2016 acquisitions comprised of Australia's Pacific Hydro's wind and hydropower assets by China State Power Investment Corporation for roughly US$2.1 billion and the UK's Beatrice and Inch Cape Offshore Wind farms by the State Development and Investment Corporation for US$260 million (Nicholas 2018).

The size of Chinese manufactured turbines has grown from 600 kW in 1997 to Dongfang Electric's 10 MW offshore model revealed in August 2019 (Shi 2007). Other manufacturers of wind turbines in China include Goldwind, Envision, Mingyang, United Power, Shanghai Electric and CSIC Haizhuang.[75]

The solar PV manufacturing sector's boom was been spurred by substantial subsidies, tax rebates, research grants and cheap land, benefits similar to those offered by the Chinese Government to both wind power and hydropower industries. According to multiple studies published in these sectors, the state played a central role in encouraging provincial governments to support local industries by investing in their renewables-associated products and technologies (Beeson 2009; Gang 2015a).

Between 1986 and 2005, advancement in solar PV was still relatively slow due to insufficient expertise in the PV industry and a lack of raw materials and advanced production equipment. However, China experienced a surge in productivity in this sector between 2011 and 2012 (Dai et al. 2014; Gang 2015b; Yang and Pan 2010).

As highlighted in the 2007 Medium to Long-term Renewable Energy Development Plan, industrial development of renewables including solar PV represented an essential national strategy (NDRC 2007b). The document specifically mentioned the need to promote the use of solar materials manufactured locally and to invest in associated R&D activities.

Whilst the country became one of the largest manufacturers and exporters of solar PV technologies globally between 2004 and 2008, the dissemination process within the country was slow (Zhang et al. 2015; Cao and Groba 2013). China's export-oriented strategy led to significant improvements in the quality of solar PV products manufactured domestically and gradually sped up the process of domestic renewables adoption.

Between September 2012 and August 2013, a series of policies, including a resource-based Feed-in-Tariff (FiT) scheme was introduced by the Chinese Government, to provide stronger support to the industry (NDRC 2016; IEA 2013). In 2015, China became the world's largest producer of solar panels, led by companies such as Jinko Solar.

The development of China's hydropower sector began in 1912, with a 0.48 MW power station in Yunnan, in 1912 (Li 2012; Du et al. 2008). The hydropower industry today– which has grown to a total capacity of approximately 352GW or over a quarter of the global installed capacity–occupies a pivotal position in Chinese manufacturing and represents one of the country's most valuable energy sources, as well as enabling it to develop a competitive export offering. The country's Medium and Long-term plan for all renewable energy, adopted in 2007, has a goal to achieve 800GW of installed capacity by 2020 (Beeson 2009), underlining the strong political support for hydropower within all renewable energy sources.

11   A Government-Led Decarbonisation

The decarbonisation process in China has been driven by a mix of climate concerns as well as the need to diversify away from fossil fuels, especially imports. Climate concerns touch upon social issues and how rapid industrialisation which expanded energy intensity has made pollution a major concern for the population (Zhang et al. 2010). In June 2007, China published its national climate change program aimed at speeding up decarbonisation, with wind, solar, hydropower and biomass sources acting as the key drivers to mitigate climate-related concerns (Beeson 2009).

Since 2006 the country's government outlined the importance of sustainable development, which was deemed to be an essential part of its national strategy. Amongst the targets announced was achieving a balance between emission reduction and preserving economic growth (Dai 2015). In 2009, the Chinese prime minister Wen Jiabao reiterated his intentions when he announced the country's aim to reduce CO_2 emissions by between 40% and 45% by 2020, in comparison to 2005 levels (Watts 2009).

Moreover, since 2007, China has seen growing population unrest caused by health, environmental and food safety concerns, including successful protests in Xiamen opposing the manufacturing of paraxylene, 2007 Shanghai demonstrations opposing the route of the Maglev rail project, 2008 and 2012 Sichuan protests and 2013 Kunming and Maoming demonstrations, all significantly contributing to China's shift towards sustainable development (Geall et al. 2014).

In 2018, the Key Work Plan for Industrial Energy Conservation Supervision was published by the Chinese Ministry of Industry and Information Technology. It addressed multiple energy-intensive industries such as the chemical, petrochemical and paper sectors (Chen and Li 2019), and stressed the importance of optimising transportation and improving sector efficiency across the country, while facilitating a shift to EVs.

Innovation and technology have played a significant role in the country's decarbonisation policies. In light of this, this section explores the electrification of transport as an essential driver of decarbonisation, as well as providing a boost to the clean-energy industry. EVs have become a key priority of China's transition to low-carbon mobility; the industry is dominated by major state-owned enterprises (SOEs) and, as well as more recently, joint ventures between international car automakers and SOEs.

In 2010, the Chinese Government announced an investment of RMB100 billion (US$14bn) in the EV sector and declared EVs a 'key strategic industry for the next 5 years' (Tyfield et al. 2014). Other favourable policies for e-mobility included the 2004 automobile industry development policy (amended in 2009), the 2007 comprehensive programme of work in energy saving and emission reduction and the 2009 eV initiative joint action plan, amongst many others (Chinese Government Official Platform 2013).

In 2010, the government released a notice on the expansion of energy saving and new energy vehicle demonstration to public services, which was one of the initiatives that has driven the emergence and growth of EV buses and taxi fleets (MOF 2010). Public authorities continued to implement favourable policies and initiatives to facilitate the roll-out of EVs nationally. This including the adoption of EVs in 2013 and the 2014 guidance on accelerating the deployment of EVs, as well as the implementation plan for government agencies and public institutions to buy new EVs (Du et al. 2008; Zhang et al. 2010; MIIT 2014).

China has become a global player in EVs and associated charging infrastructure. In the second half of 2018, the country had built 300,000 public EV chargers, accounting for more than 50% of the global EV charging market (Pyper 2019). In the same period, China announced the sale of approximately 1.3 million EVs, 62% more than in 2017, with roughly 2.6 million vehicles on the street by the beginning of 2019 (Hove and Sandalow 2019).

The surge in EV sales could be explained by the support schemes the government put in place, including allowing local governments to offer additional subsidies of up to 50% of the national subsidies already offered (MOF 2016; Manthly 2018). Non-financial incentives such as exemption from license plate lotteries and restrictions also had a major contribution to the swift increase in EVs popularity in the country.

12   From Tariffs to Zero-Subsidies

The path from tariff-based to subsidy-free renewable energy projects in China has not been a straight-forward transition, with the market sensitive to tariffs being phased out. It has been argued that the reason behind an apparent loss of momentum in China's renewables deployment has been the uncertainty surrounding government funding and regulatory support for green energy.

The feed-in tariff system in China has been based on competitive bidding, which takes place prior to the tendering process (Han et al. 2009) and obliges bidders to present the lowest possible prices. However, as many of the bids were unrealistically low, projects were not built and the country moved towards a fixed FiT, especially for technologies such as solar PV and wind (Martinot 2010; Chan 2009).

At present, China needs to tackle multiple issues, including low consumption problems, the pressure on the government to continue providing substantial subsidies and a very high percentage of fossil fuels in the overall energy mix.

To date, China has approved 224 subsidy-free wind and solar projects, which are estimated to add 21GW of installed capacity (Yu 2019). A joint statement by NDRC and NEA mentioned that roughly 60% of the capacity will be installed in six provinces, namely, Heilongjiang, Guangdong, Shanxi, Guangxi, Henan and Hebei (NDRC 2019).

The country's renewables industry and more specifically its wind and solar PV sectors are poised for long-term growth, with capital costs on a downward trajectory and likely to fall to levels equal or less than those of coal power stations.

13   Conclusion

This chapter showcases two examples of jurisdictions which have adopted differing strategies in achieving a decarbonisation of their respective economies.

The UK has been presented as a global leader in the deployment of renewable energy capacity primarily through onshore and offshore wind, as well as through solar. As one of the first countries in the world to privatise and deregulate its electricity sector, the UK has one of the longest track records in empowering the private sector to deliver a market-based approach which extended to decarbonisation.

Whilst other countries such as Germany have been extremely successful in the first phase of state-led deployment of renewables primarily through feed-in-tariffs, it is the UK's market-based approach which has stood the test of time. As technology costs continue to decline, governments are withdrawing direct support and instead favouring the direct participation of private sector buyers who are interested in purchasing renewable electricity through private PPAs. This recent development is therefore renewing the UK's position as a pioneer in the subsidy-free era.

The UK Governments' role in decarbonising the economy has by no means been a smooth process. Yet its reliance on market-led policies has ensured the engagement of the private sector throughout the various phases of its evolution. However, insufficient emphasis has been placed on newer technologies, which instead have been forced to rely on other forms of investment. Historically, the UK Government has preferred to remain technology agnostic. Whilst banding of ROCs and Pot 2 under the CfD regime has allowed some higher cost projects such as offshore wind and waste to energy projects to flourish, little of the UK's vast, untapped wave and tidal resource potential has been captured. In particular, for larger scale projects requiring longer time horizons like the Swansea Bay tidal lagoon project, a lack of support such as that provided to nuclear projects such as Hinkley Point C has effectively hampered the project's fruition, despite strong private sector interest.

One of the central hypotheses of China's decarbonisation strategy has been the rapid increase in energy dependency; decarbonisation can therefore be viewed partly as one of a series of energy policies and initiatives aimed at achieving greater energy self-sufficiency. The rapid deployment of renewables across the country has been at the centre of China's strategy, with the majority of involvement in these projects coming from domestic companies and more specifically major SOEs, including equipment manufacturers, investors, banks and utilities.

The expansion of China's renewables manufacturing sector, optimization of workforce and enhancement of equipment quality has led to the creation of an export-focused industry worth trillions of dollars– with solar PV module exports alone accounting for approximately £4 billion in Q1 2019 (CCCME 2019). Multinational conglomerates such as Jinko Power and Dongfang deploy 'made in China' renewables technologies, including solar panels and wind turbines globally. The growth in the RET manufacture sector is therefore an essential driver of the Chinese renewables market.

Moreover, global climate concerns and the pressure to decarbonise the energy and transport infrastructure sectors have contributed significantly to speeding up China's transition to renewables. The country boasts the world's largest fleets of electric vehicles and auxiliary charging infrastructure and it is on the right path to ensure an efficient transition to clean energy.

Notwithstanding their significant differences, the examples of the decarbonisation pathways undertaken by China and the UK analysed in this chapter, demonstrate how direct government involvement through SOEs together with direct incentives are equally effective as market-enabled policies with direct incentives. Both examples present viable strategies that can be adopted by governments either wishing to adopt top-down decarbonisation strategies, or those wanting to foster a greater role for the private sector as developers and owners and operators of renewable energy and other asset categories such as EV infrastructure.

Typically, emerging economies often need to attract foreign direct investment, as part of their strategies to open up their markets. For these countries and amongst the various policies that are in place, the government needs to foster a specific market-friendly approach to attract foreign companies. Furthermore, these policies need to remain in place for the duration of the specific renewable energy incentives. The temptation to change such policies to suit a government's budgetary constraints—as has happened in markets such as Spain and Italy over the past decade—will only have the effect of disincentivising private sector companies from making investments in emerging countries.

References

UK

Department for Business, Energy and Industrial Strategy (BEIS) (2019) Policy paper: Contracts for Difference. Updated 11 January 2019. Available from: https://www.gov.uk/government/publications/contracts-for-difference/contract-for-difference

DECC (2011) Planning our electric future: a White Paper for secure, affordable and low-carbon electricity. Department of Energy and Climate Change, London. https://assets.publishing.service.gov.uk/government/uploads/system/uploads/attachment_data/file/48129/2176-emr-white-paper.pdf. Accessed 3 Aug 2019

Department of Trade and Industry (1999) Promoting Renewable Energy: Experience with the NFFO: A Presentation to the OECD Experts Group, Richard Kettle. Available from: http://www.oecd.org/unitedkingdom/2046731.pdf

Department of Trade and Industry (1993) The prospects for coal conclusions of the Government's Coal Review. Available from https://assets.publishing.service.gov.uk/government/uploads/system/uploads/attachment_data/file/271969/2235.pdf

European Environment Agency (2018) Efficiency of the conventional thermal electricity and heat production in Europe. Available from: https://www.eea.europa.eu/data-and-maps/indicators/efficiency-of-conventional-thermal-electricity-generation-4/assessment-2

EDF Group (2019) Designing and building the nuclear plant of tomorrow. Global benchmark technology

Elliott D (2019) Renewable energy in the UK. Past, present and future. Palgrave Macmillan, London

Gouiffes P-F (2009) Margaret Thatcher and The Miners. 1972–1985: Thirteen years that changed Britain. Creative Commons

Horrocks S, Lean T (2011) An oral history of the electricity supply industry in the UK. In partnership with the British Library. National Life Stories. Available from: https://www.bl.uk/britishlibrary/~/media/subjects%20images/oral%20history/oral%20history%20and%20nls%20documents/nls_electricityindustryscopingstudy.pdf

House of Commons (2019) Priorities for Government. 25 July 2019. Hansard, vol 663. House of Lords, London. https://hansard.parliament.uk/commons/2019-07-25/debates/D0290128-96D8-4AF9-ACFD-21D5D9CF328E/PrioritiesForGovernment Accessed 3 Aug 2019

House of Commons (2018) Electric vehicles: driving the transition. Fourteenth report of session 2017–2019. House of Commons—Business, Energy and Industrial Strategy Committee, London. https://publications.parliament.uk/pa/cm201719/cmselect/cmbeis/383/383.pdf. Accessed 3 Aug 2019

International Energy Agency (IEA) (2013) Non-Fossil Fuel Obligation 1990. Available from: https://www.iea.org/policiesandmeasures/pams/unitedkingdom/name-21717-en.php

Ofgem (2019) Electricity Market Reform. Available from: https://www.ofgem.gov.uk/electricity/wholesale-market/market-efficiency-review-and-reform/electricity-market-reform-emr (Accessed on 12 Oct 2010)

Pearson P, Watson J (2012) UK Energy Policy (1980–2010) A history and lessons to be learnt. The Institution of Engineering and Technology, London. http://sro.sussex.ac.uk/id/eprint/38852/1/uk-energy-policy.pdf Accessed 3 Aug 2019

The Conservatives (2015) The Conservative Party Manifesto 2015: Strong leadership, a clear economic plan, a brighter, more secure future. Available from: http://ucrel.lancs.ac.uk/wmatrix/ukmanifestos2015/localpdf/Conservatives.pdf

The Guardian (2011) ackground: What caused the 1970s oil price shock? Available at: https://www.theguardian.com/environment/2011/mar/03/1970s-oil-price-shock

The National Archives, Policy, The Renewable Obligation, The Non-Fossil Obligation (2006) Available from: https://webarchive.nationalarchives.gov.uk/20060215070520, http://www.dti.gov.uk/renewables/renew_2.2.6.htm

The National Audit Office (2017) Hinkley Point C. Available from: https://www.nao.org.uk/wp-content/uploads/2017/06/Hinkley-Point-C.pdf

United Nations Framework Convention on Climate Change (UNFCCC) (2019) UNFCCC: 25 years of effort and achievement. Accessed 12 Oct 2019. Available from: https://unfccc.int/timeline/

UK Government (2019) Historical coal data: coal production, availability and consumption. Department for Business, Energy and Industrial Strategy. Available from: https://www.gov.uk/government/statistical-data-sets/historical-coal-data-coal-production-availability-and-consumption

UK Government (2018) Government launches road to zero strategy to lead the world in zero emission vehicle technology. Department of Transport. Available from: https://www.gov.uk/government/news/government-launches-road-to-zero-strategy-to-lead-the-world-in-zero-emission-vehicle-technology

UK Government (2015) Electricity market reform: Contracts for Difference. Department for Business, Energy and Industrial Strategy. Available from: https://www.gov.uk/government/collections/electricity-market-reform-contracts-for-difference

UK Government (2011) Hinkley Point C Project Wide Design and Access Statement. Available from: https://infrastructure.planninginspectorate.gov.uk/wp-content/ipc/uploads/projects/EN010001/EN010001-005318-8.1%20Hinkley%20Point%20C%20Project%20Wide%20Design%20and%20Access%20Statement%201.pdf

UK Government (2008) Meeting the Energy Challenge, A White Paper on Nuclear Power. Department of Business, Enterprise and Regulatory Reform. Available from: https://webarchive.nationalarchives.gov.uk/+/www.berr.gov.uk/files/file43006.pdf

UK Government (2007) Meeting the Energy Challenge, A White Paper on Energy. Available from: https://assets.publishing.service.gov.uk/government/uploads/system/uploads/attachment_data/file/243268/7124.pdf

UK Parliament (2018) Energy imports and exports, House of Commons Library, Briefing Paper No. 4046. Available at: http://researchbriefings.files.parliament.uk/documents/SN04046/SN04046.pdf

Vivid Economics (2019) Quantifying benefits of onshore wind to the UK. Vivid Economics for RenewableUK, London. http://www.vivideconomics.com/wp-content/uploads/2019/07/Quantifying_the_Benefits_ofO.pdf Accessed 3 August 2019

World Nuclear Association (2019) Nuclear Power in the United Kingdom. Available from: https://www.world-nuclear.org/information-library/country-profiles/countries-t-z/united-kingdom.aspx

China

Beeson M (2009) Developmental states in East Asia: A comparison of the Japanese and Chinese experiences. Asian Perspect 33(2):5–39

Cao J, Groba F (2013) Chinese Renewable Energy Technology Exports: The role of Policy, Innovation and Markets. DIW Berlin Discussion Paper No. 1263. Available from: https://papers.ssrn.com/sol3/papers.cfm?abstract_id=2205645

Chan Y (2009) China sets feed-in tariff for wind power plants. Available from: https://www.businessgreen.com/bg/news/1801182/china-sets-feed-tariff-wind-power-plants

Chen J, Li S (2019) An analysis of deep decarbonization trends in China's industrial sector. The Rocky Mountain Institute. Available from: https://rmi.org/an-analysis-of-deep-decarbonization-trends-in-chinas-industrial-sector/

China Chamber of Commerce for Import and Export of Machinery and Electronic Products (2019) Export Analysis Report of China's Photovoltaic Products in Q1 2019. Available from: http://www.cccme.org.cn/index.aspx

Chinese Government Official Platform (2013) 关于继续开展新能源汽车推广应用工作的通知/ Notice on the continuation and popularization and application of new energy vehicles. Available from: http://www.gov.cn/zwgk/2013-09/17/content_2490108.htm

Chinese Government Official Platform (2009) 全国人民代表大会常务委员会关于修改《中华人民共和国可再生能源法》的决定/ The revised renewable energy law of the People's Republic of China, to enter into effect on 1 April 2010. Available from: http://www.gov.cn/flfg/2009-12/26/content_1497462.htm

Chinese Government Official Platform (2007) 胡锦涛在党的十七次全国代表大会上作报告(摘要)/ Full text of Hu Jintao's report at 17th People's Congress. Available from: http://www.gov.cn/ldhd/2007-10/15/content_776431.htm

Chinese Wind Energy Association (CWEA) (2017) Chinese Wind Energy Association Industry Overview. In: He D, Du G, Lyu B. Available from: https://community.ieawind.org/HigherLogic/System/DownloadDocumentFile.ashx?DocumentFileKey=46f86d9b-7b90-ab73-f974-e23f97039fb1#targetText=During%20the%2013th%20Five%2DYear,had%20new%20installations%20in%20China. CNPC Economics and Technology Research Institute (2017) China Energy Outlook. Available from: https://eneken.ieej.or.jp/data/8192.pdf

Dai Y, Zhou Y, Xia D, Ding M, Xue Lan (2014) Innovation paths in the Chinese wind power industry. Available from: https://cedmcenter.org/wp-content/uploads/2014/06/Zhou-Yuan.pdf

Dannreuther R (2003) Asian security and China's energy needs. Int Relat Asia-Pac 3:197–219

Dai Y (2015) Who drives climate-relevant policy implementation in China? Institute of Development Studies, Brighton, UK

Du Z, Wu Y, Tong M et al (2008) Southwest water resources development status and problems. China Power 40(9):12–16

European Commission (2018) EU-China leaders' statement on climate change and clean energy. Available from: https://ec.europa.eu/clima/sites/clima/files/news/20180713_statement_en.pdf

Fan J, Wang J, Wei S, Zhang X (2018) The development of China's renewable energy policy and implications to Africa. In: IOP conference series: materials science and engineering, vol 394. Available from: https://iopscience.iop.org/article/10.1088/1757-899X/394/4/042034/pdf

Gang C (2015a) From mercantile strategy to domestic demand stimulation: changes in China's solar PV subsidies. Asia Pac Bus Rev 21(1):96–112

Gang C (2015b) China's solar PV manufacturing and subsidies from the perspective of state capitalism. Cph J Asian Stud 33(1):90–106. Available from: https://rauli.cbs.dk/index.php/cjas/article/viewFile/4813/5239

Gao X, Jin B, Li B, Yang K, Zhang H, Fan B (2011) Study on renewable energy development and policy in China. Energy Procedia 5:1284–1290

Geall S, Hilton I, Heroth T, Grune S, Chen Y (2014) China's environment: ambitions, challenges and opportunities for EU cooperation. Europe China Research and Advice Network

Grau T, Huo M, Neuhoff K (2012) Survey of photovoltaic industry and policy in Germany and China. Energy Policy 51:20–37

Han J, Mol A, Lu Y, Zhang L (2009) Onshore wind power development in China: challenges behind a successful story. Energy Policy 37(8):2941–2951. Available from: https://ideas.repec.org/a/eee/enepol/v37y2009i8p2941-2951.html

Hove A, Sandalow D (2019) Electric Vehicle Charging in China and the United States. Center on Global Energy Policy. Columbia Energy Policy. Available from: https://energypolicy.columbia.edu/sites/default/files/file-uploads/EV_ChargingChina-CGEP_Report_Final.pdf

Information Office of the State Council of the People's Republic of China (IOSC-PRC) (2007) China's Energy Conditions and Policies. Available from: http://en.ndrc.gov.cn/policyrelease/200712/P020071227502260511798.pdf

International Energy Agency (IEA) (2018) Energy Supply and Consumption Revolution Strategy (2016–2030). Addressing Climate Change: Policies and Measures Databases. Available from https://www.iea.org/policiesandmeasures/pams/china/

IEA (2013) IEA/IRENA Joint Policies and Measures database. Feed-in tariff support for solar PV in China

International Monetary Fund (IMF) (2019) IMF data: access to macroeconomic and financial data. Available from: https://data.imf.org/?sk=388DFA60-1D26-4ADE-B505-A05A558D9A42

Kempe F (2019) The US-China trade war has set in motion an unstoppable global economic transformation. CNBC. Available from: https://www.cnbc.com/2019/09/14/us-china-trade-wars-unstoppable-global-economic-transformation.html

Lee YT (2019). China's renewable energy policies. The Oxford Climate Review. Anthroposphere. Issue II. Available from: https://www.anthroposphere.co.uk/post/china-renewable-energy-policies

Lewis JI (2011) Building a national wind turbine industry: experiences from China, India and South Korea. Int J Technol Globalisation 5(3/4):281

Li J (2012) Research on prospect and problem for hydropower development of China. In: 2012 international conference on modern hydraulic engineering. Procedia Eng J 28:677–682

Li L (2015) China's energy security and energy risk management. J Int Affairs 69(1):Geopolit Energy 86–97

Liao M, Wang Y, (2019) China's energy consumption rebound effect analysis based on the perspective of technological progress. MDPI Sustain 11:1–15. Available from: file:///C:/Users/Ofelia/Downloads/sustainability-11-01461-v2.pdf

Liu Z, Guan D, Crawford-Brown D, Zhang Q, He K, Liu J (2013) A low-carbon road map for China. Nat Int J Sci. Nat 500:143–145. Available from https://www.nature.com/articles/500143a

Manthly N (2018) Change of mind: China to retain local EV subsidies. Electrive. Available from: https://www.electrive.com/2018/02/12/change-mind-china-retain-local-ev-subsidies/

Martinot E (2010) Renewable power for China: Past, present and future. Forthcoming in Frontiers of Energy and Power Engineering in China. Higher Education Press and Springer. Available from: http://www.martinot.info/Martinot_FEP4_prepub.pdf

Ministry of Finance China (MOF) (2016) 关于调整新能源汽车推广应用财政补贴政策的通知.Available from: http://jjs.mof.gov.cn/zhengwuxinxi/tongzhigonggao/201612/t20161229_2508628.html

Ministry of Finance China (MOF) (2010) 关于扩大公共服务领域节能与新能源汽车示范推广有关工作的通知/ Notice on the expansion of energy-saving and new energy vehicles demonstration to public services. Available from: http://www.mof.gov.cn/gp/xxgkml/jjjss/201006/t20100602_2499640.html

Ministry of Finance China (MOF) (2009) 发展新能源汽车符合我国国情/ Promoting new energy vehicles is in line with China's national conditions. Available from: http://www.mof.gov.cn/index.htm

Ministry of Industry and Information Technology (MIIT) (2014) 政府机关及公共机构购买新能源汽车实施方案/ The implementation plan for government agencies and public institutions to buy new energy vehicles. Available from: http://www.miit.gov.cn/n1146295/n1652858/n1652930/n3757018/c3763293/content.html

Ministry of Science and Technology (MOST) (2012) 关于印发风力发电科技发展"十二五"专项规划的通知/ The 12th Five-Year Plan for the Development of Wind Power Science and Technology. Available from: http://www.most.gov.cn/fggw/zfwj/zfwj2012/201204/t20120424_93884.htm

NDRC (2019) 国家发展改革委办公厅 国家能源局综合司关于公布2019年第一批风电、光伏发电平价上网项目的通知.594 号. Available from: http://www.ndrc.gov.cn/gzdt/201905/t20190522_936545.html

NDRC (2018) 国家发展改革委 财政部 国家能源局关于, 2018 年光伏发电有关事项的通知, 发改能源〔2018〕823号, 31 May 2018

NDRC (2016) 国家发展改革委关于调整光伏发电陆上风电标杆上网电价的通知. 2729号. Available from: http://www.ndrc.gov.cn/zcfb/zcfbtz/201612/t20161228_833049.html

NDRC (2007a) 国家气候变化方案. China's national climate change programme. Available from: http://en.ndrc.gov.cn/newsrelease/200706/P020070604561191006823.pdf

NDRC (2007b) Medium and long-term development plan for renewable energy in China. Available from: http://www.martinot.info/China_RE_Plan_to_2020_Sep-2007.pdf

NDRC (1999) 国家计委、科技部关于进一步支持,可再生能源发展有关问题的通知. 计基础44号 1999年1月12日. Available from: http://pkulaw.cn/%28S%28pvgj1x45wuwnwm5555df4e55%29%29/fulltext_form.aspx?Db=chl&Gid=877ef7598fa5c4e3bdfb

National Energy Administration (NEA) (2019) 国家能源局发布2018年全国电力工业统计数据. Available from: http://www.nea.gov.cn/2019-01/18/c_137754977.htm

National Energy Administration (NEA) (2016) 可再生能源发展"十三五"规划/ The 13th Renewable Energy Plan. Available from: http://www.nea.gov.cn/2016-12/19/c_135916140.htm

Nicholas S (2018) China is investing heavily in European wind. institute for energy economics and financial analysis. Available from: http://ieefa.org/wp-content/uploads/2018/08/China_Research_Brief_August-2018.pdf

Pyper J (2019) The Dynamics of China's Rapidly Expanding EV Charging Market. Available from: https://www.greentechmedia.com/squared/electric-avenue/china-rapidly-expanding-ev-charging-market

Shi P (2007) Various years of Chinese installed wind capacity statistics (in Chinese)

Shi P (1997) Chinese Installed Wind Capacity Statistics 1997. China Renewable Energy Association Deputy Director

S&P Global (2019) China's 2018 crude oil imports rise 10% to 9.28 mil b/d. 14 January 2019. Available from: https://www.spglobal.com/platts/en/market-insights/latest-news/oil/011419-chinas-2018-crude-oil-imports-rise-10-to-928-mil-b-d

Standaert M (2019) Why China's renewable energy transition is losing momentum. Yale School of Forestry and Environmental Studies. Available from: https://e360.yale.edu/features/why-chinas-renewable-energy-transition-is-losing-momentum

Standing Committee of the National People's Congress (SCNPC) (2002) 中华人民共和国清洁生产促进法(2012修正)/Cleaner Production Promotion Law of the People's Republic of China (2012 Amendment). Available from: http://en.pkulaw.cn/display.aspx?id=9255&lib=law&SearchKeyword=&SearchCKeyword=

Tan H, Mathews J (2014) Economics: manufacture renewables to build energy security. Int Wkly J Sci 513(7517):166–168. Available from: https://www.nature.com/news/economics-manufacture-renewables-to-build-energy-security-1.15847#/b1

Tyfield D, Zuev D, Li P, Urry J (2014) Low carbon innovation in Chinese urban mobility: prospects, politics and practices. Available from: https://steps-centre.org/wp-content/uploads/WP-71-E-Mobility.pdf

United Nations Statistics (UN stats) (2019) UN stats energy statistics website. Available from: https://unstats.un.org/home/

Wang C, Yang Y, Zhang J (2015) China's sectoral strategies in energy conservation and carbon mitigation. Clim Policy J 15(sup1):Climate Mitigation Policy in China. S60–S80. Available from https://www.tandfonline.com/doi/full/10.1080/14693062.2015.1050346

Wang N, Chang Y-C (2014) The development of policy instruments in supporting low-carbon governance in China. Renew Sustain Energy Rev 35:126–135

Watts J (2009) China sets first targets to curb world's largest carbon footprint. The Guardian UK. Available from: https://www.theguardian.com/environment/2009/nov/26/china-targets-cut-carbon-footprint

World Bank (2018) Electricity production from renewables, nuclear, oil, gas and coal sources. Available from https://data.worldbank.org

Yang M, Pan R (2010) Harvesting sunlight: solar photovoltaic industry in China. Background Brief. No. 562. Singapore, East Asian Institute

Yu Y (2019) China approves first 21GW of subsidy-free renewables. Available from: https://www.rechargenews.com/transition/1788759/china-approves-first-21gw-of-subsidy-free-renewables

Zeng M, Liu X, Li N, Xue S (2013) Overall review of renewable energy tariff policy in China: evolution, implementation, problems and countermeasures. vol 25, pp 260–271. Available from: https://www.sciencedirect.com/science/article/pii/S1364032113002785

Zhang J, Mauzerall D, Zhu T, Liang S, Ezzati M, Remais J (2010) Environmental health in China: challenges to achieving clean air and safe water. US National Library of Medicine. National Institutes of Health. vol 375(9720), pp 1110–1119. Available from: https://www.ncbi.nlm.nih.gov/pmc/articles/PMC4210128/

Zhang S, Zhao Y, Steigenberger M (2018) A star for China's Energy Transition. Published in collaboration with China National Renewable Energy Centre (CNREC) and Deutsche Gesellschaft fur Internationale Zusammenarbeit (GIZ). Available from: https://www.agora-energiewende.de/fileadmin2/Projekte/2017/China_Star/144_ChinaStar_WEB.pdf

Zhang S, Andrews-Speed P, Ji M (2015) The erratic path of low-carbon transition in China: evolution of solar PV policy. Available from: http://www.andrewsspeed.com/wp-content/uploads/2015/05/Solar-PV-paper.2014.pdf

Zhang X, Wang D, Liu Y, Yi H (2016) Wind power development in China: an assessment of provincial policies. J Sustain 8:1–12

Zhao X, Wang J, Liu X, Liu P (2012) China's wind, biomass and solar power generation: what the situation tells us? Renew Sustain Energy Rev 16:6173–6182

Financing the Sustainable Energy Transition

Alexander Van de Putte, Akshu Campbell-Holt, and George Littlejohn

1 Indroduction

In *The Perfect Storm: Navigating the Sustainable Energy Transition*, the authors argue that for the energy transition to be sustainable, the five capital stocks—natural, manufactured, human, social and financial—need to be grown (or maintained) and balanced simultaneously (Van de Putte et al. 2017).

What is considered sustainable energy though is not necessarily well understood. For the purpose of this chapter, we will argue that the end-game for the sustainable energy future is a global renewable energy internet comprised of five integrated layers: (1) largely distributed renewable energy generation technologies, such as wind, solar and hydro, (2) electricity transmission and distribution network infrastructure, (3) energy storage solutions and smart energy routers, (4) active network management software to manage and balance loads and (5) mobile and stationary sensors and smart terminals (Van de Putte and Nematova 2017).

To make the transition to a global renewable energy internet a reality, it is also necessary to make the current energy system more sustainable by employing Circular Economy (CE) ideas. According to the Ellen MacArthur Foundation (2013), a circular economy 'seeks to rebuild capital, whether this financial, manufactured, human, social or natural'.[1] The rebuilding of capital is achieved by reusing, reducing and recycling waste and the shift from a 'cradle-to-grave' to a 'cradle-to-cradle'

---

[1] https://www.ellenmacarthurfoundation.org/circular-economy (accessed 4 July 2016).

A. Van de Putte (✉) · A. Campbell-Holt
Astana International Financial Centre, Nur-Sultan, Kazakhstan
e-mail: alexander.van.de.putte@sustainable-foresight.com

G. Littlejohn
Chartered Institute for Securities and Investments, London, UK

A. Van de Putte
IE Business School, Madrid, Spain

M. Hafner and S. Tagliapietra (eds.), *The Geopolitics of the Global Energy Transition*,
Lecture Notes in Energy 73, https://doi.org/10.1007/978-3-030-39066-2_11

257

philosophy. The benefits of the circular economy cannot be ignored in the transition towards a sustainable global energy environment. McKinsey estimates that between 2016 and 2030, apart from the environmental, social and human capital benefits, the net economic benefits are estimated at €1.8 trillion,[2] or in excess of €120 billion per year or 0.8% of global GDP. For developing countries, especially natural resource-rich economies, this could be twice as much.

Most of the energy demand growth will come from developing countries. Developing countries are home to more than 6 billion people,[3] and especially natural resource-rich countries hold the key to sustainable development. This is because these countries tend to be at the early stages of climbing the 'energy ladder'. The energy ladder, first described by Shell in the early 1980s, shows the relationship between primary energy consumption per capita and Gross Domestic Product (GDP) per capita and follows an S-curve (Van de Putte 2010).

As capital markets and venture capital financing is well developed in OECD markets, and given that energy demand growth will primarily come from developing countries, this chapter largely focuses on financing the sustainable energy transition in the developing world. If financial resources would be properly channelled, developing countries could leapfrog the sustainable energy transition.

2 The Tables Are Turning

The ratification of the Paris Agreement by nearly 94% of the world demonstrates a strong global commitment from most countries to achieve the United Nation's Sustainable Development Goals (UNFCCC 2019). The challenge, however, is the enormous investment requirement which is estimated between $5 trillion and $7 trillion per year (UNCTAD 2014) and channelling investments were required.

Additionally, a growing body of studies demonstrates that the new generation of investors with changing mindsets provides increasing optimism for the inclusion of sustainability in investment decisions. This, together with the superior returns potential from sustainable investments compared to traditional investments, emphasises a strong match between requirements and investors—is it possible to leverage this match and what challenges lie ahead?

2.1 Countries Have Made Global Commitments

United Nations-led initiatives including the Paris Agreement, the Kyoto Protocol and the Sustainable Development Goals have seen widespread global commitment

---

[2]https://www.mckinsey.com/business-functions/sustainability/our-insights/europes-circular-economy- opportunity (accessed 4 July 2016).

[3]https://population.un.org/wpp (United Nations 2019a).

in recent years. While both the Paris Agreement and the Kyoto Protocol are targeted towards addressing climate change, the Kyoto Protocol established emissions reduction commitments for developed nations while the Paris Agreement applied to all countries—developing and developed. Out of 197 Parties to the United Nations Framework Convention on Climate Change (UNFCCC), 185 have signed up to the Paris Agreement and undertaken commitments and identified their Nationally Determined Contributions (NDCs) (UNFCCC n.d.).

There are several bodies that have recommended solutions to tackle the CO_2 emissions and climate change but now is the time to take collective action. One such solution is identified in the *World Energy Outlook 2018* as the Sustainable Development Scenario[4] that presents an integrated approach to maintain CO_2 levels at the same level as in 2017 and achieve internationally agreed objectives on climate change, air quality and universal access to modern energy. The *WEO 2018* highlights that a sustainable development scenario which is aligned with the Paris Agreement can be made possible but will be heavily determined by the actions undertaken by governments. The prerequisites for this will not only include the joint political will globally but also the mammoth funding required for the implementation of NDCs. Some of the key considerations in this scenario include (International Energy Agency 2018):

- renewable energy technologies to lead the way in providing universal access to energy and thereby increasing the share of renewables in the power mix from one-quarter in 2017 to two-thirds in 2040;
- implementation of economically viable options to improve efficiency in the energy sector thereby maintaining overall demand in 2040 at the same level as at 2017 and
- for the first time in the *WEO*, clean water is examined as a dimension including the energy required to provide universal access to clean water and sanitation.

2.2 Investors Are Changing, Mindsets Are Changing

A world that is faced with growing challenges is also in the midst of an intergenerational wealth transfer estimated at $30 trillion from baby boomers to their children, a majority being millennials, and is expected to take place over the next two or three decades (MSCI 2017). The millennial generation is seeking far more than mere financial factors—responsible investments and positive environmental, social and governance impacts alongside financial returns as evidenced by some of the recent revolutionary findings.

---

[4]According to the *World Energy Outlook (WEO)*, the scenarios (Current Policies Scenario, New Policies Scenario and Sustainable Development Scenario) do not aim to forecast the future but provide a way of exploring different possible futures, the levers that could bring them about, and the interactions that arise across a complex energy system. The base year for projections in all scenarios is 2016.

According to a Morgan Stanley study (MSCI 2017), the three common objectives among sustainably-minded investors are

1. integration of environmental, social and governance (ESG) factors because they believe that companies with strong ESG factors are better managed and that this may improve their investment results;
2. reflection of personal values in relation to ethical, social, religious or political beliefs and
3. selection of investments with a positive impact on environmental, social and political challenges and thereafter the ability to monitor those investments based on norms set through frameworks (e.g. Paris Agreement, UN Sustainable Development Goals).

The *Global Sustainable Investment Review 2018 (GSIR 2018)* (Global Sustainable Investment Alliance 2017) highlights the widespread global interest and engagement in sustainable investing which is an investment approach that considers environmental, social and governance (ESG) factors in making investment decisions. *GSIR 2018* covers information for five markets (Europe, USA, Canada, Japan, and Australia and New Zealand) collectively managing sustainable investing assets worth $30.7 trillion as of early 2018. This is a 34% increase since 2016 with Japan and USA leading the way. In Japan, sustainable assets grew at a staggering 308% since 2016 and the corresponding growth rate in the USA was 38%, with other regions continuing to rise but experiencing growth at a slower pace. On the other hand, at the start of 2018, the proportion of sustainable assets in relation to total assets grew in almost every region with Canada and Australia and New Zealand consisting of sustainable assets as the majority of their overall assets. The only region that had a slight decline in this proportion was Europe, however, with nearly half of the global sustainable and responsible investing assets domiciled in Europe, it continues to manage the highest proportion of sustainable assets (GSIA 2017).

GSIR 2018 also indicated that the top sustainable investment strategy globally continues to be 'negative or exclusionary screening'[5] with $19.8 trillion assets under management, followed by 'ESG integration'[6] and 'corporate engagement and shareholder action'[7] with $17.5 trillion and $9.8 trillion assets under management, respectively. Impressive growth rates in assets across almost all other strategies were reported with the exception of 'norms-based screening'[8] which declined by 24%

---

[5]GSIA defines the 'negative or exclusionary screening' strategy as the 'exclusion from a fund or portfolio of certain sectors, companies or practices based on specific ESG criteria'.

[6]GSIA defines the 'ESG integration' strategy as 'the systematic and explicit inclusion by investment managers of environmental, social and governance factors into financial analysis'.

[7]GSIA defines the 'corporate engagement and shareholder action' strategy as 'the use of shareholder power to influence to influence corporate behaviour, including through direct corporate engagement (i.e. communicating with senior management and/or boards of companies), filing or co-filing shareholder proposals, and proxy voting that is guided by comprehensive ESG guidelines'.

[8]GSIA defines the 'norms-based screening' as the 'screening of investments against minimum standards of business practice based on international norms, such as those issued by the OECD, ILO, UN and UNICEF'.

from $6.2 trillion to $4.7 trillion in assets. These trends support a growing change in mindsets among investors globally.

Despite the USA's intent to withdraw from the Paris Agreement, there are strong signs of increasing interest and awareness in sustainable investing among individual investors in the USA. A survey conducted with 1,000 US individual investors by Morgan Stanley's Institute for Sustainable Investing (2017) provides for a positive and pro-sustainable future as the results showed that

- awareness and interest in sustainable investing has grown steadily since 2015;
- among individual investors, 75% are interested in sustainable investing and the level of interest among millennials is even higher at 86%;
- almost three-quarters (71%) of individual investors believe companies with robust sustainability practices are better long-term investments and
- the USA saw a spike in sustainable, responsible and impact investing between 2014 and 2016 growing at a rate more than 33% amounting to $8.72 trillion.

These trends are on a rise and if applied across the world and across all stakeholders far beyond the investor community, it might rest with the voters' ability to choose an administration which is pro-sustainable—an interesting area to look out for especially in the 2020 elections in the USA.

ESG investments, Socially Responsible Investing (SRI) and impact investing, albeit differing from one another in their definitions and used interchangeably, are all components of sustainable investing and are strong contenders to become the new norm in making investment decisions. This also makes a strong argument for the likelihood of a future by 2030, if not earlier, where companies that deviate from ESG, SRI and impact investing principles would become the outliers and therefore not favoured by stakeholders—not only by the investor community but across the entire ecosystem including customers, suppliers, shareholders, consumers and the wider community. Studies and surveys demonstrate that the interest in ESG, SRI and impact investing is more pronounced among millennials and as they increasingly take up decision-making roles, positions of influence and control assets, they are bound to become a strong driving force for sustainability. Do these trends bring about a perfect match between investors and investment requirements for the sustainable development of economies? This is a question to be addressed later in this chapter.

2.3 ESG, SRI and Impact Investments Outperform Traditional Investments

As described above, the recent trends and developments in ESG, SRI and impact investing demonstrate that sustainability is increasingly becoming mainstream. While investors are being virtuous by increasingly seeking ESG, SRI and impact investments, they are also making money by doing what they believe to be the right thing. According to a growing number of studies, ESG, SRI and impact investing are providing better returns compared to traditional investments. 'Traditional investments' in this chapter indicate investments that do not consider pro-sustainable factors including, but not limited to, environmental, social, governance, ethical and norms, alongside financial factors.

Europe's largest asset manager by assets under management, Amundi, that has yet again received the top rating (A +)[3] in 2018 for its responsible investment approach,[9] has championed responsible investing by including it at the core of its identity. Between 2010 and 2017, Amundi conducted a study (Amundi 2018) which considered investment universes covered by MSCI indices (MSCI North America, MSCI EMU, MSCI Europe ex EMU, MSCI Japan and MSCI World) and applied three different strategies—active management, passive management and factor investing—and based its analysis on ESG criteria. The analysis found that while ESG investments produced negative excess return compared to non-ESG investments during the period from 2010 to 2013, 2014 was a turning point where ESG investments outperformed non-ESG investments. The results of the analysis showed that, between 2014 and 2017, being a 'responsible investor' and managing portfolios based on the ESG criteria would have resulted in annualised excess return of 3.3% in North America and a remarkable 6.6% in Europe. While Amundi promotes ESG-induced portfolios, it cautions while applying these factors to avoid reducing the investment universe beyond the point that could negatively impact diversification and performance.

In another study (Giese et al. 2019), a research team at MSCI looked at the attributes of ESG investments that led to positive financial effects and found that companies with robust ESG practices demonstrated the following characteristics:

- Higher profitability: When compared with low ESG-rated companies, higher ESG-rated companies had a competitive edge and generated better returns that usually resulted in higher profitability and dividend payments.
- Lower risk: An observation of companies within the MSCI World Index over a 10-year period demonstrated that high ESG-rated companies had a lower frequency of severe incidents such as drawdowns more than 95% or bankruptcy.
- Lower volatility: Because of better risk controls associated with high ESG-rated companies, they have fewer severe incidents of fraud, corruption, embezzlement and litigation cases.

In addition to Amundi, other global asset managers including Blackrock, BNP Paribas, Vanguard and Fidelity Investments have launched ESG funds. According to data provided by Morningstar, mutual funds based on ESG criteria surpassed the $1 trillion mark in 2018 with assets under management rising by 60% since 2012 from $655 billion to $1.05 trillion.

January 2019 witnessed another milestone for sustainable investment when S&P Dow Jones Indices launched the S&P 500 ESG Index. In the first half of 2019 alone, S&P Dow Jones Indices has launched six ESG indices (S&P 2019) with global

---

[9]https://www.amundi.com/int/ESG.

coverage.[10] Similar to the S&P 500 Index, in the S&P 500 ESG Index companies are ranked by their ESG scores and excluded if they do not meet the criteria as defined in the Index. Companies can be excluded from the S&P 500 ESG index for reasons including:

- failure to meet the required ESG scores;
- disqualifying United Nations Global Compact (UNGC) scores or
- business activities listed in the exclusion list such as tobacco production/sales or controversial weapons (e.g. cluster weapons, landmines and nuclear weapons), or an ownership stake of 25% or more in another company involved in these activities.

In April 2019, several notable companies were removed from the S&P 500 ESG Index, including Facebook, Wells Fargo, Oracle and IBM—Facebook being the largest with a weight of 2.5% in the index a day before it was excluded (Steadman 2019). As a result of the recent privacy protection concerns, such as a lack of transparency on collecting and sharing private user information, Facebook did not meet the requirement to remain in the S&P 500 ESG Index. Despite Facebook scoring high on the environment score, its aggregate was brought down by the social and governance factors which are weighted higher for tech companies.

In summary, a growing number of studies demonstrate that high ESG-rated companies are associated with positive characteristics that all stakeholders look for in any company or while making investment decisions. Together with the pronounced interest from investors who are seeking to make a positive impact through their investments, ESG, SRI and impact investing all have the likelihood of being the new norm in making investment decisions and in voting practices. Additionally, with an increasing adoption rate of ESG, SRI and impact investing among investors and leading global asset managers, sustainability will soon become the central topic among institutional investors, policymakers, regulators, government agencies and corporations alike. Importantly, investors can now do the right thing—socially responsible and sustainable investing—without settling for sub-optimal returns.

3   Channelling Investments to Meet Global Energy Demand Sustainably Remains a Challenge

With the world population projected to reach 9.2 billion by 2040 (UN 2019a)—a nearly 20% increase from the 2019 levels—an upsurge in the demand for energy consumption is inevitable and the demand is estimated to increase by more than a quarter by 2040 predominantly in developing countries led by India. Despite efforts towards combating climate change, 2017 recorded a rise in carbon emissions from

---

[10]The six indices are S&P 500 ESG Index, S&P Global 1200 ESG Index, S&P Europe 350 ESG Index, S&P Japan 500 ESG Index, S&P/ASX 200 ESG Index and S&P Developed Ex-North America & Korea LargeMidCap ESG Index.

the global energy industry at the fastest rate since 2011 and the upward trajectory is continuing. Global CO_2 emissions are forecast to rise by a near-record amount in 2019 (The Guardian 2019). These trends pose alarming global threats and they attest that efforts thus far are inadequate to achieve global commitments and are a reminder that it has never been more critical to transition to a cleaner and more efficient energy industry in the context of meeting global sustainability targets.

3.1 Current Overreliance on Fossil Fuels

The global energy demand is still being served mostly by fossil fuels, such as oil (30%), coal (27%) and gas (20%) (ENI 2018), which poses a growing concern in relation to carbon emissions and, consequently, climate change.

The Paris Agreement becomes binding only through ratification. Despite being signatories to the Paris Agreement, as of June 2019, 11 countries[11] were yet to formally ratify the Paris Agreement and these countries are collectively responsible for more than a tenth of global emissions which presents added concern and raises questions regarding their commitment to achieve global sustainable goals. Of the 11 countries, Russia, Turkey and Iran are notably some of the largest Global Greenhouse Gas (GHG) emitters (Apparicio and Sauer 2018). Although the USA has notified its intention to withdraw from the Paris Agreement, it can only officially withdraw after 3 years from the date of the Agreement which means that the USA must stay in the deal until at least 2019. Nonetheless, this demonstrates a lack of commitment from one of the world's biggest offenders in terms of GHG emissions and, according to the Climate Action Tracker, it is almost certain that the USA will miss its NDC target for 2025 due to the lack of implementation of new policies planned by the Obama administration. However, there are studies and surveys that display positive signs among individual investors with their keen interest in sustainable investments.

The Climate Action Tracker, which has been monitoring 32 countries that account for 80% of the world's greenhouse gas emissions, has reported (Erickson 2018) some alarming findings which establish that actions, thus far, are far from adequate in meeting the climate change goals. In fact, the world's major polluters have made minimal efforts, if any, to meet their commitments and that only seven countries (Morocco, The Gambia, Bhutan, Costa Rica, Ethiopia, India and the Philippines) have made efforts and commitments to class them under either of the two acceptable categories,

---

[11] As of 1 June 2019, the countries yet to formally ratify the agreement were Angola, Eritrea, Iran, Iraq, Kyrgyzstan, Lebanon, Libya, Russia, South Sudan, Turkey and Yemen.

'1.5 °C Paris Agreement Compatible' and '2 °C Compatible', with the other categories on the scale being 'Insufficient',[12] 'Highly Insufficient'[13] and 'Critically Insufficient'.[14]

In its latest working paper on global fossil fuel subsidies, the International Monetary Fund (Coady et al. 2019) estimates that the direct[15] and indirect[16] fossil fuel subsidies in 2017 amounted to $5.2 trillion, or 6.5% of global GDP, and that especially the indirect subsidies are likely to increase unless drastic measures are taken to move the world away from fossil fuels. But weaning economies from fossil fuels also requires huge financial investments.

3.2 Capital Markets in Developing Countries Are Underdeveloped

It is estimated (UNCTAD 2014, p ix) that the investment requirement to implement the Sustainable Development Goals (SDGs)[17] in developing countries alone is between $3.3 trillion and $4.5 trillion per year and faces an annual funding gap of $2.5 trillion. In addition to climate change mitigation and adaptation, these costs also account towards basic infrastructure, health, food security and education—elements of sustainable economic development.

Governments and public funds simply cannot meet these astronomical investment requirements making the role of the capital markets and private sector instrumental in bridging the gap. UNCTAD's Investment Policy Framework for Sustainable Development recognises stock exchanges and market infrastructure to be instrumental in meeting these investment needs. However, capital markets in most developing countries still remain underdeveloped. For instance, more than half of the Asian countries are not recognised in the MSCI indices (MSCI 2019a). The development of stock markets will be crucial in mobilising investments since stock exchanges

---

[12]Countries classified under 'Insufficient' are Australia, Brazil, EU, Kazakhstan, Mexico, New Zealand, Norway, Peru, Switzerland and UAE.

[13]Countries classified under 'Highly Insufficient' are Argentina, Canada, Chile, China, Indonesia, Japan, Singapore, South Africa and South Korea.

[14]Countries classified under 'Critically Insufficient' are Russia, Saudi Arabia, Turkey, USA and Ukraine.

[15]Direct subsidies are defined as the cash transfer from governments to industry and amounted to $296 billion in 2017 (IMF 2019). Direct subsidies are often referred to as pre-tax subsidies, because they reflect the difference between what consumers pay for fuel and the cost to produce the fuel.

[16]Indirect subsidies or post-tax subsidies reflect the difference between what consumers pay for fuel and the full environmental and societal costs of fossil fuels. Indirect subsidies may be perceived as less tangible, but they are both real and increasing.

[17]The United Nations' Sustainable Development Goals (SDGs) are a call for action by all countries to promote prosperity while protecting the planet. 'SDGs are the blueprint to achieve a better and more sustainable future for all. They address the global challenges we face, including those related to poverty, inequality, climate, environmental degradation, prosperity and peace and justice' (United Nations 2019b).

are positioned in the intersection of key stakeholders including the government, policymakers, investors and corporations.

3.3 Matching Financing with Projects Remains Problematic

In a recent working paper (Tyson 2018), the Overseas Development Institute (ODI) finds that the lack of bankable projects that meet investment criteria is a key barrier, not the lack of financing. This is found even in countries where international financial institutions play a significant role in providing infrastructural support including technical assistance in developing innovative and new policy approaches, providing new project-preparation facilities, and financing and co-financing of funds and de-risking for private investors. This is largely due to (1) the early stages of development of projects, (2) lengthy planning and construction phases, (2) requirements of non-standard financing and (3) the bespoke and complex nature of the projects making it excessively difficult for investors to understand. These challenges are further compounded especially in developing countries due to the lack of an adequate ecosystem to foster the growth of micro, small and medium-sized businesses, in particular, owing to limited early stage investors, such as venture capitalists, private equity and angel investors, who not only provide financing but also the strategic direction and platforms to grow.

4 Could There Be a Perfect Match?

The investment needed to finance a Global Renewable Energy Internet is estimated to be in the range of $100 trillion globally until 2050 or close to $3 trillion per year (Van de Putte and Nematova 2017; Liu 2015). This seems to be a very significant amount but is relatively small compared to the annual direct[18] and indirect[19] subsidies that are channelled into the fossil fuel industry globally.

There is also no shortage of capital or sources of capital to finance the sustainable energy transition. According to the Boston Consulting Group (BCG 2019), global wealth now exceeds $200 trillion and since the 2008 global financial and economic crisis, this global wealth has struggled to find bankable projects anywhere in the world. The Global Renewable Energy Internet could provide this opportunity for global investors to make a game-changing contribution to sustainability, while at the

---

[18]Direct subsidies are defined as the cash transfer from governments to industry and amounted to $296 billion in 2017 (IMF 2019). Direct subsidies are often referred to as pre-tax subsidies, because they reflect the difference between what consumers pay for fuel and the cost to produce the fuel.

[19]Indirect subsidies or post-tax subsidies reflect the difference between what consumers pay for fuel and the full environmental and societal costs of fossil fuels. Indirect subsidies may be perceived as less tangible, but they are both real and increasing.

Table 1 Financing landscape for sustainable projects

Commercial	Development institutions	Government
Banks − Debt/equity, Guarantee on leans, Funds, Asset Managers − Debt/equity, Funds, SPVs, Leases, Private placements, Private Equity, Venture Capital − Equity, Funds, SPVs, Private placements, Institutional Investors − Syndicated loans, private placements, debt	Multinational Bilateral National	− Grants − PPPs − Guarantees on loans − Sovereign debt
	Private	
	Project developers − Equity/leases	

same time providing superior returns in line with the findings of the 2019 Amundi study.

Sources of capital are diverse and growing and include banks, asset managers, private equity, institutional investors, development institutions and government financing (Table 1).

We argue that a perfect match exists between global capital and sustainable projects to rapidly scale the sustainable energy transition if at least three things come together: (1) capital market development in developing countries, (2) government regulation to create a level playing field and (3) a venture capital approach to stimulate sustainable investments and entrepreneurship (Fig. 1).

4.1 Developing Countries Need to Develop Capital Markets to Enhance Liquidity and Increase Capital Flows[20]

The Morgan Stanley Capital International (MSCI) Emerging Markets Index is an index used to measure equity market performance in global emerging markets and is the de facto index used by investors to channel investments to growth markets. The MSCI Emerging Markets Index grew from 10 countries in 1988 to 24 countries today and represents 13% of world market capitalisation. Most developing countries are currently not part of this index and need to explore ways how to upgrade from Frontier to Emerging Market status. To upgrade from Frontier to Emerging Market is relatively straightforward (Table 2), while the benefits are significant.

The benefits include as follows:

---

[20]This section is based on a presentation made by Prof. Alexander Van de Putte at the Belt, Road & Bridge: Creating New China-Europe Connections conference on 1 May 2019 in London and was published in the CISI *Review*, the journal of the Chartered Institute for Securities & Investments, as 'Growing the Digital Economy', by Alexander Van de Putte and George Littlejohn.

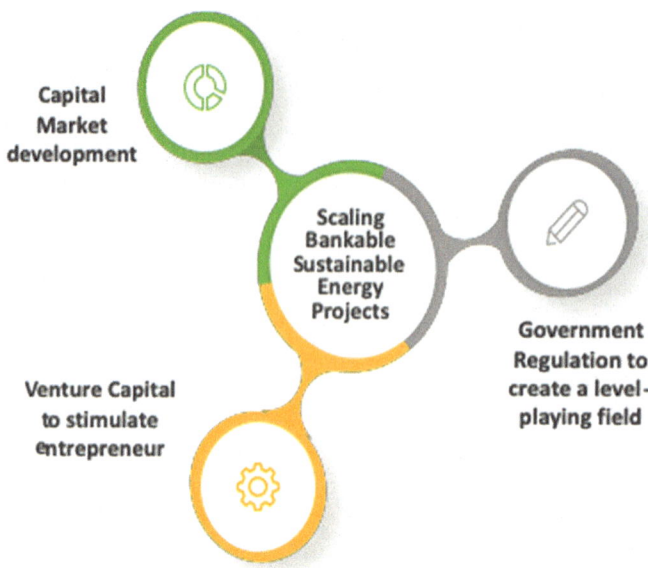

Fig. 1 Scaling the Sustainable Energy Transition

- Increased capital inflows: The MSCI Emerging Markets Index has over $2 trillion of assets benchmarked against it. Inclusion in the Index would not only increase the exposure of developing countries' stocks to international investors, but also lead to passive inflows from funds that follow the Index's progress.
- Enhanced liquidity: Liquidity in developing countries' stock markets is typically very low. Capital inflows resulting from inclusion in the Index will substantially boost liquidity in developing countries' stock markets and economies.
- Reduced cost of capital resulting from increased trading volumes: Being part of the Index should lead to a fall in the equity risk premium because of risk diversification. This will help draw in new investors, such as endowment funds and hedge funds. As stock prices swell as a result, even more investors are drawn to developing countries, effectively creating a virtuous circle.

Apart from equity, capital markets have an important role to play in the development of debt capital markets, in particular a green bond market. The issuance of green bonds has been growing rapidly to reach $162 billion in 2017 globally (Chartered Banker Institute 2019). Green bonds are issued to finance environmentally friendly projects, such as renewable energy projects, sustainable buildings and circular economy projects. According to the Climate Bonds Initiative (2018), most of the green bond proceeds were allocated to renewables energy projects (39%), followed by buildings (25%) and transportation (15%).

To facilitate the growth of green bonds, many stock exchanges and/or regulators around the world have released national green bond guidelines and listing rules

Table 2 MSCI Market Classification Framework and Requirements

Criteria	Frontier	Emerging	Developed
A. Economic Development A.1 Sustainability of economic development	No requirement	No requirement	Country GNI per capita 25% above the World Bank high income threshold[a] for 3 consecutive years
B. Economic Development B.1 Number of companies meeting the following Standard Index criteria • Company size (full market cap)[b] • Security size (float market cap)[b] • Security liquidity	3 $776 mm $776 mm 2.5% ATVR	3 $1,551 mm $776 mm 15% ATVR	5 $3,102 mm $1,551 mm 20% ATVR
C. Market Accessibility Criteria C.1 Openness to foreign ownership C.2 Ease of capital inflows/outflows C.3 Efficiency of operational framework C.4 Availability of investment instrument C.5 Stability of the institutional framework	At least some At least partial Modest High Modest	Significant Significant Good and tested High Modest	Very high Very high Very high Unrestricted Very high

Source MSCI, 2019 (https://www.msci.com/documents/1296102/1330218/MSCI_Global_Market_Framework_2019.pdf/57f021bc-a41b-f6a6-c482-8d4881b759bf (accessed 17 July 2019))
[a]High income threshold for 2018: GNI per capita of $12,056 (World Bank, Atlas method)
[b]Minimum in use for the May 2019 Semi-Annual Index Review, updated on a semi-annual basis

(Sustainable Banking Network 2018). Other markets, such as the EU, are also in the process of developing their green bond standards (EU 2019) or trying to encourage greater uptake of their green bonds as in the case with Japan (EF 2019).

Small and Medium-Sized Enterprises (SMEs) play a critical role in the economy. They tend to be nimbler and more dynamic compared to larger companies and therefore drive innovation and more sustainable business growth. SMEs, however, need access to growth financing to scale their business. In most developing countries, a capital market for SMEs does not exist. There are, however, important benefits for developing countries to develop an SME market:

- An SME market provides access to equity capital to scale the business beyond what would be possible through Venture Capital (VC) funding;
- An SME market provides exit options for VC firms which are thus more likely to provide risk capital at earlier stages of the SME development process;
- A listing often requires SMEs to bolster their corporate governance, including the recruitment of independent non-executive directors. Thus, a well-composed

board will help SMEs identify viable strategic growth options beyond what the
founders are typically able to identify;

- Over time, a lower the cost of capital which in turn will help SMEs to grow more
sustainably and
- Increased visibility to a diverse group of stakeholders, which could result in
additional revenue growth.

In order though for developing countries to develop the capital markets that
enable their sustainable energy transition, government policies that help create a
level playing field must also be implemented.

4.2 Government Regulations to Create a Level Playing Field

Based on the latest US Energy Information Administration (2019) report on levelised
cost of new electricity generation technologies entering service in 2023, electric
renewables are and will increasingly be competitive against fossil fuel alternatives
from a plant investment perspective. The levelised 'investment' cost in $/MWh of
selected new generation resources entering service in 2023 is as follows: hydroelectric (39.1), onshore wind (55.9), solar PV (60), conventional combined cycle (46.3),
coal with 30% carbon capture and storage (104.3) and advanced nuclear (77.5).[21]

When both the direct and indirect subsidies will be phased out, both wind and
solar integrated as part of smart grids and combined with advanced electric storage
solutions will be competitive in most geographic locations, even when the various
load factors are taken into account.

But more needs to be done by governments to stimulate the uptake of renewable
energy projects. Governments have a variety of policy tools available to de-risk
renewable energy projects. Now that the Paris Agreement and the UN Sustainable
Development Goals have been ratified by 185 states plus the EU[22] and 193 countries,
respectively,[23] governments around the world could provide an important stimulus
to sustainable economic development.

The first step is to gradually phase out the direct subsidies. The second step is to
internalise the externalities. Given that the externalities constitute by far the largest
share of the subsidies that benefit the fossil fuel industry, this would make the fossil
fuel industry instantly uncompetitive. A rapid internalisation of the externalities is
very unlikely though because many natural resource-rich countries depend on it for
their economic development and they need time to diversify their economies away
from fossil fuels. A gradual internalisation of the externalities combined with scaling
CE projects should attract most of the investment in the short to medium term. For

---

[21] Note that comparing different technologies using LCOE alone evaluates only the cost to build
and operate a plant and the value of the plant's output to the electric grid. This because different
generation technology have different load factors.

[22] https://www.unfccc.int/process/the-paris-agreement/status-of-ratification (accessed 6 July 2019).

[23] https://www.sustainabledevelopment.un.org/memberstates.html(accessed 6 July 2019).

natural resource-rich countries, a CE should be particularly appealing during the transition period because it would allow for making existing assets more sustainable and competitive. This, because there are many opportunities to reduce, reuse and recycle waste in the extraction industries value chain by leveraging skills, enabling infrastructure and SMEs. The circular economy in natural resource-rich countries will create skills, jobs, maintain and improve (or at least help maintain) natural capital, and create financial capital that is not dependent on the volatility of the demand for natural resources.

4.3 A Venture Capital Approach to Sustainable Investment Is Needed[24]

Foreign Direct Investment (FDI) into especially natural resource-rich countries still tends to target the extraction industries. Of the $4.1 trillion of global FDI flows in 2017, $2.3 trillion (56.1%) went into the primary sector[25] (UNCTAD 2018, p. 75). These global FDI flows therefore do not help in making the energy transition a reality. Instead, a Venture Capital (VC) approach could help accelerate the sustainable energy transition, and stimulate innovation and entrepreneurship.

The life cycle of green finance projects can be broken down into three discrete yet integrated stages: (1) Ideation and business case development, (2) Independent verification and project financing and (3) Implementation and post-implementation monitoring. Note that the green financing life cycle is a non-linear process with many feedback loops. The feedback loops exist to make the business case more robust or allow the entrepreneur or investor to revisit or abandon the project at any stage of its life cycle.

Stage 1: Ideation and business case development
During this stage (Fig. 2), entrepreneurs generate ideas for sustainable energy projects. Sustainable energy projects typically fall into two categories: (1) new renewable energy generation, storage and transmission technologies and solutions and (2) CE technologies to ensure that waste produced during one process is used as an input into another process. This eliminates, or at least dramatically reduces, waste while creating a financial return and higher skilled jobs.

For example, technology to convert the flue gases of coal-fired power plants into cement could potentially be considered as 'green financeable', subject to results from the independent review that shows that by doing so all the capital stocks are grown and balanced simultaneously.

Most projects will not meet both the financial and ESG viability criteria (Phase 3). A potential project may be a great idea from the perspective of the inventor, but not

---

[24]This section is derived from the Astana International Financial Centre (AIFC) Unified Strategy developed in 2018.

[25]UNCTAD defines the primary sectors as the extraction of crude petroleum and natural gas and the mining of metal ores.

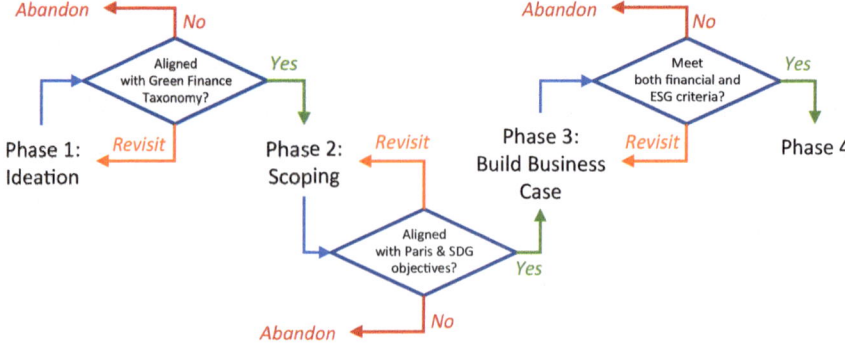

Fig. 2 Ideation and business case development stage. *Source AIFC Unified Strategy, 2018*

necessarily from the perspective of other stakeholders, including society at large and investors. That is the nature of entrepreneurship—it carries important risks. Stage 1 is, however, a critical stage, because entrepreneurs do not shy away from challenges. Instead, they learn from past experience and make improvements along the way, even starting afresh if need be.

Stage 2: Independent verification and project financing

After the project has demonstrated potential from a sustainability and financial return perspective, the project moves into the independent verification and project financing stage (Fig. 3). For green bonds, strict requirements need to be met to ensure that the bonds meet the local and/or internal ESG standards. For this, an external review and independent verification are required. The climate bonds initiative has issued guidelines for the issuance of climate bonds.[26] Independent verification is a key

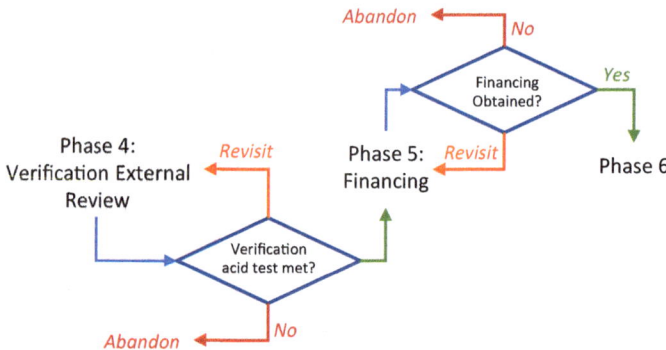

Fig. 3 Independent verification and project financing stage. *Source AIFC Unified Strategy, 2018*

---

[26]https://www.climatebonds.net/certification/get-certified-1 (accessed 27 July 2019).

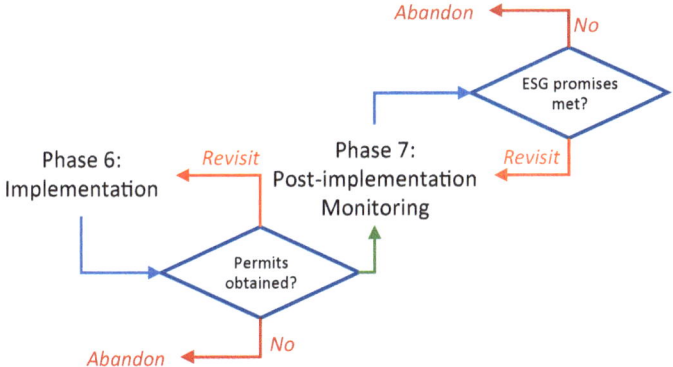

Fig. 4 Implementation and post-implementation monitoring stage. *Source AIFC Unified Strategy, 2018*

requirement before issuing a green bond in order to ensure that the investment meets the stated sustainable benefits, and thus to avoid 'greenwashing'.[27]

VC-backed green investments do not typically require an independent verification, after all the fund, and not bond market investors, bare the risk of the investment decision. However, VC funds have their own experts who conduct pre-investment due diligence, including ESG compliance of projects.

Only after the pre-investment due diligence process will an appropriate VC fund[28] consider investing in a green finance project. Usually, financing will come in stages so that the VC fund can better manage the investment risk.

Stage 3: Implementation and post-implementation monitoring
After financing has been obtained the hard work starts and the focus now switches to the implementation and post-implementation monitoring phase (Fig. 4). Monitoring is important to ensure that both compliance and performance criteria are met. Compliance criteria typically relate to ESG factors, while performance criteria typical relate to financial return factors, such as Internal Rate of Return (IRR), cost overruns and delays.

Note that apart from the aforementioned feedback loops, the flexibility in the life cycle of sustainable energy finance projects remains to also abandon the project. Abandonment may take many dimensions—from stopping a project to handing it over to a different type of investor. VC funds typically bring more than just equity finance to the sustainable energy project; they often lie in the trenches and assist in strategic and operational matters.

---

[27]Greenwashing can be defined as 'the practice of making an unsubstantiated or misleading claim about the environmental benefits of a product, service, technology or company practice'. https://whatis.techtarget.com/definition/greenwashing (accessed 24 July 2019).

[28]Various types of VC funds exist. Some will invest in early stage ventures, some during the growth stage and yet others help companies to prepare for a public offering.

The life cycle of green finance projects is designed to simultaneously achieve several objectives: (1) stimulate entrepreneurship; (2) obtain financing for ESG compliant projects; (3) reduce risk through independent review and verification, thus improving the bankability of green finance projects and (4) continuously monitor progress made to allow for taking corrective action as needed and develop a knowledge base for future green finance projects.

5 Conclusion

There is no shortage of capital, nor is there a shortage of bankable ideas to make the sustainable energy transition a reality. The challenge is that systematic change is needed for governments around the world to deliver on the commitments they have made following the ratification of the SDGs and the Paris Agreement. It is also important to create a level playing field and gradually eliminate the direct subsidies and internalise the indirect subsidies benefitting the fossil fuel industry. Finally, especially developing countries need to develop both capital markets and make venture capital financing available to entrepreneurs. Top-down forward-looking government policy combined with bottom-up entrepreneurs would dramatically accelerate the sustainable energy transition because capital would increasingly be channelled towards bankable sustainable projects.

Acknowledgement The authors would like to thank Dr. Kairat Kelimbetov, Governor of the AIFC for his insights and for allowing us to use some of the AIFC development strategy ideas and concepts.

References

Amundi (2018) How ESG Investing Has Impacted the Asset Pricing in the Equity Market (online). Available from http://research-center.amundi.com/ezjscore/call/ezjscamundibuzz::sfForwardFront::paramsList=service=ProxyGedApi&routeId=_dl_MjcxZGQxOGJkMGEyYmMwZTdiNGY1ODNiMjMyMzdiN2I. Accessed 18 July 2019

Apparicio S, Sauer N (2018) Which countries have not ratified the Paris climate agreement? Available from https://www.climatechangenews.com/2018/07/12/countries-yet-ratify-paris-agreement. Accessed 17 July 2019

Boston Consulting Group (2019) Global Wealth: Reigniting Radical Growth. Available from https://www.bcg.com/en-be/publications/2019/global-wealth-reigniting-radical-growth.aspx. Accessed???

Carrington D (2019) Worrying' rise in global CO2 forecast for 2019. The Guardian (online). Available from https://www.theguardian.com/environment/2019/jan/25/worrying-rise-in-global-co2-forecast-for-2019. Accessed 16 July 2019

Chartered Banker Institute (2019) Principles and Practice of Green Finance 2019

Climate Action Tracker (2017) *Trump's climate policies would see US climate action rating drop from "medium" to "inadequate"*. Available from https://climateactiontracker.org/publications/

trumps-climate-policies-would-see-us-climate-action-rating-drop-from-medium-to-inadequate. Accessed 17 July 2019

Climate Bonds Initiative (2018) Q2 2018 Green Bonds Market Summary. Available from https:// www.climatebonds.net. Accessed 21 July 2019

Ellen MacArthur Foundation (2013) Towards the Circular Economy: Economic and business rationale for an accelerated transition. Ellen MacArthur Foundation, Cowes

ENI (2018) World Energy Outlook: looking ahead to 2040. Available from https://www.eni.com/ en_IT/investors/global-energy-scenarios/world-energy-outlook.page#. Accessed 15 July 2019

Environmental Finance (EF) (2019) The green bond market looks east. Available from https:// www.environmental-finance.com/content/the-green-bond-hub/the-green-bond-market-looks-east.html. Accessed 18 July 2019

Erickson A (2018) Few countries are meeting the Paris climate goals. Here are the ones that are. The Washington Post. Available from https://www.washingtonpost.com/world/2018/10/11/few-countries-are-meeting-paris-climate-goals-here-are-ones-that-are/?utm_term=.e9b6ad570815. Accessed 17 July 2019

European Union (EU) (2019) Report of the Technical Expert Group (TEG) Subgroup on Green Bond Standard: Proposal for an EU Green Bond Standard. Available from https://ec.europa.eu/info/ sites/info/files/business_economy_euro/banking_and_finance/documents/190306-sustainable-finance-teg-interim-report-green-bond-standard_en_0.pdf. Accessed 18 July 2019

Giese G, Lee L-E, Melas D, Nagy Z, and Nishikawa L (2019) Foundations of ESG investing: how ESG affects equity valuation, risk, and performance. J Portfolio Manag 45(5):4–11. Available from http://info.msci.com/foundations-of-ESG-investing-part1. Accessed 19 July 2019

Global Sustainable Investment Alliance (GSIA) (2017) Global Sustainable Investment Review 2018. Available from http://www.gsi-alliance.org/wp-content/uploads/2019/03/GSIR_Review2018.3. 28.pdf. Accessed on 17 July 2019

International Energy Agency (IEA) (2018) *World Energy Outlook 2018*. Available from https:// www.iea.org/weo2018/scenarios. Accessed on 17 July 2019

Coady D, Parry I, Le N-P, Shang B (2019) Global fossil fuel subsidies remain large: an update based on country-level estimates. In: International Monetary Fund (IMF). Available from https://www.imf.org/en/Publications/WP/Issues/2019/05/02/Global-Fossil-Fuel-Subsidies-Remain-Large-An-Update-Based-on-Country-Level-Estimates-46509. Accessed on 1 Aug 2019

Liu Z (2015) Global energy interconnection. Academic Press, London

McKinsey and Company (2015) Europe's circular-economy opportunity. Available from https:// www.mckinsey.com/business-functions/sustainability/our-insights/europes-circular-economy-opportunity. Accessed 17 July 2019

Mooney A (2018) Rising investor interest pushes ESG funds past $1tn. *Financial Times*. Available from https://www.ft.com/content/f1e98ec7-083e-3b95-8c6b-ecc4810b988e. Accessed on 19 July 2019

Mooney C, Dennis B (2018) The world has just over a decade to get climate change under control, UN scientists say. *The Washington Post*. Available from https://www.washingtonpost.com/ energy-environment/2018/10/08/world-has-only-years-get-climate-change-under-control-un-scientists-say/?utm_term=.93c8bb06fd33. Accessed 17 July 2019

Morgan Stanley Capital International (MSCI) (2017) Swipe right to invest: Millennials and ESG, the perfect match? Available from https://www.msci.com/documents/10199/07e7a7d3-59c3-4d0b-b0b5-029e8fd3974b. Accessed 16 July 2019

Morgan Stanley Capital International (MSCI) (2019a) MSCI announces the results of the 2019 annual market classification review. Available from https://www.msci.com/market-classification. Accessed on 24 July 2019

Morgan Stanley Capital International (MSCI) (2019b) MSCI Market Classification Framework. Available from https://www.msci.com/market-classification. Accessed on 23 July 2019

Morgan Stanley Institute for Sustainable Investing (2017) Sustainable Signals, New Data from the Individual Investor Available from https://www.morganstanley.com/pub/content/dam/msdotcom/ideas/sustainable-signals/pdf/Sustainable_Signals_Whitepaper.pdf. Accessed on 17 July 2019

S&P (2019) S&P Dow Jones Indices. Available from https://us.spindices.com/indices/equity/sp-asx-200-esg-index-$. Accessed 19 July 2019

Steadman R (2019) Why Facebook Was Dropped from the S&P 500® ESG Index. Indexology Blog. Available from https://www.indexologyblog.com/2019/06/11/why-facebook-was-dropped-from-the-sp-500-esg-index. Accessed 19 July 2019

Sustainable Banking Network (2018) *Creating Green Bond Markets—Insights, Innovations, and Tools from Emerging Markets*. IFC, Washington

The Guardian (2019) 'Worrying' rise in global CO2 forecast for 2019. Available from https://www.theguardian.com/environment/2019/jan/25/worrying-rise-in-global-co2-forecast-for-2019. Accessed on 18 July 2019

Tyson JE (2018) Private infrastructure financing in developing countries: Five challenges, five solutions. Overseas Development Institute Working Paper 536. Available from https://www.odi.org/sites/odi.org.uk/files/resource-documents/12366.pdf. Accessed 24 July 2019

United Nations (2019a) World Population Prospects 2019. Available from https://population.un.org/wpp/DataQuery. Accessed 15 July 2019

United Nations (2019b) Sustainable Development Goals. Available from https://www.un.org/sustainabledevelopment. Accessed 16 July 2019

United Nations Conference on Trade and Development (UNCTAD) (2014) *The World Investment Report 2014. Investing in the SDGs: An Action Plan*. United Nations, NY and Geneva. Available at https://unctad.org/en/PublicationsLibrary/wir2014_en.pdf

United Nations Conference on Trade and Development (UNCTAD) (2018) *World Investment Report: Investment and New Industrial Policies*. United Nations, NY and Geneva. Available at https://unctad.org/en/PublicationsLibrary/wir2018_en.pdf

United Nations Framework Convention on Climate change (UNFCCC) (2019) Available from https://unfccc.int/process/the-paris-agreement/status-of-ratification. Accessed on 16 July 2019

US Energy Information Administration (2019) Levelized Cost and Levelized Avoided Cost of Generation Resources in the *Annual Energy Outlook 2019*. Available from https://www.eia.gov/outlooks/aeo/pdf/electricity_generation.pdf. Accessed 25 July 2019

Van de Putte A (2010) Trade and energy: a new clean energy deal. In: Lehmann F, Lehmann J-P (eds) Peace and prosperity through world trade. Cambridge University Press, Cambridge, pp 201–206

Van de Putte A, Kelimbetov K, Holder A (eds) (2017) *The perfect storm: navigating the sustainable energy transition*. Sustainable Foresight Institute, Brussels

Van de Putte A, Nematova S (2017) Electric renewables: the storage, intermittency and scalability challenge In: Van de Putte A, Kelimbetov K, Holder A (eds) *The perfect storm: navigating the sustainable energy transition*. Sustainable Foresight Institute, Brussels, pp 165–183

Zhou M (2019) ESG, SRI and Impact Investing: What's the Difference? Available from https://www.investopedia.com/financial-advisor/esg-sri-impact-investing-explaining-difference-clients. Accessed on 23 July 2019

Minerals and the Metals for the Energy Transition: Exploring the Conflict Implications for Mineral-Rich, Fragile States

Clare Church and Alec Crawford

1 Introduction

The transition to a low-carbon economy is accelerating, in part, due to two recent, landmark international agreements. To date, 185 parties have ratified the Paris Agreement since coming into force in November 2016, which aims to keep global temperature increase below 2 °C this century (UNFCCC 2018). In addition, the Sustainable Development Goals (SDGs), adopted by the UN General Assembly in 2015, lay out a global agenda for eliminating poverty, protecting the environment and ensuring that all people can enjoy equality, peace and prosperity. SDG 13, in particular, commits UN member states to take urgent action to combat climate change and its impacts, while SDG 7 calls for affordable and clean energy for all (United Nations 2018).

With the transition underway, many actors have stepped up their efforts to contribute to climate change mitigation through the adoption of green energy technologies. The Chinese government pledged to spend USD 360 billion on clean energy projects by 2020, creating 13 million new jobs in the process (Forsythe 2017). From 2015 to 2019, more than 95% of Costa Rica's electricity was generated from renewable energy sources (Rodriguez 2019). And it isn't just states that are driving this shift; consumer preferences and the private sector are also fuelling this change. Four of Europe's biggest primary insurers have restricted or limited insurance cover for coal, for example, and the car manufacturer Volvo announced in 2018 that by 2019 all of its new vehicles would at least be partially electrified (Bosshard 2018; Watts 2017). The pace of change has been so significant that the Bloomberg New Energy

This chapter is adapted with permission from Green Conflict Minerals: The fuels of conflict in the transition to a low-carbon economy publication, authored by Clare Church and Alec Crawford, International Institute for Sustainable Development (IISD).

C. Church (✉) · A. Crawford
International Institute for Sustainable Development (IISD), Vernier, Switzerland
e-mail: cchurch@iisd.ca

© The Author(s) 2020
M. Hafner and S. Tagliapietra (eds.), *The Geopolitics of the Global Energy Transition*,
Lecture Notes in Energy 73, https://doi.org/10.1007/978-3-030-39066-2_12

Outlook predicts that wind and solar will make up 50% of the world's electricity by 2050 (BloombergNEF 2019a, b).

The demand for green energy technologies—and corresponding demand for the materials and minerals needed to build, transport and instal these technologies—is predicted to grow dramatically in the years and decades ahead. In a recent report, the World Bank estimated that demand for the minerals required for solar panels—including copper, iron, lead, molybdenum, nickel and zinc—could increase by 300% through 2050 should the international community stay on track to meet its 2 °C goal (Arrobas et al. 2017). Similarly, demand for minerals like cobalt, lithium and rare earths[1] is expected to grow at unprecedented rates due to their strategic role in the production of wind turbines, electric vehicles (EVs) and energy storage.

This increased demand should be an economic boon to those countries that are home to the principal reserves of minerals like cobalt, lithium and bauxite; increased investments in their extraction should, in a well-governed sector, result in growing revenues to the state from taxes and royalties, improved infrastructure, more jobs and increased spending on local businesses, health and education. Unfortunately, not all strategic reserves of these minerals are found in countries applying international best practice to mining sector management. As such, while green energy technologies may contribute to the achievement of SDGs 7 and 13, failure to engage in responsible sourcing practices could increase conflict and fragility risks along the green energy supply chains of these key minerals and metals, stalling or reversing local development gains. This would jeopardize the achievement of another, foundational SDG: specifically, SDG 16, which prioritizes promoting peaceful and inclusive societies, providing justice for all, and building effective, accountable, inclusive institutions (United Nations 2018).

In countries struggling with political instability, where governance for the mining sector is weak, the extraction of these minerals can be linked to violence, conflict and human rights abuses. The mining of cobalt in the Democratic Republic of Congo, for example, has so often been connected to violence that the mineral has been dubbed as the "blood diamonds of this decade" by various news outlets (Wilson 2017; Safehaven.com 2017). The extraction of nickel, a mineral critical for both solar panels and energy storage, has been linked to murder, sexual violence and forced displacement in Guatemala (Kassam 2017). And while supply chain governance for certain minerals, including tin, tungsten, tantalum, gold and diamonds, is improving, such initiatives have not yet been expanded to include most of the minerals and metals central to green energy technologies.

The technologies require to facilitate the shift to a low-carbon economy, including wind turbines, solar panels and EVs, which all require significant mineral and metal inputs and, absent any dramatic technological advances, these inputs will come from the mining sector. How they are sourced will determine whether this transition

---

[1]The term "rare earths" refers to 17 elements often found in the same ore deposits, including cerium, dysprosium, erbium, europium, gadolinium, holmium, lanthanum, lutetium, neodymium, praseodymium, promethium, samarium, scandium, terbium, thulium, ytterbium and yttrium.

supports peaceful, sustainable development in countries where strategic reserves are found or reinforce weak governance and exacerbate local tensions and grievances.

This chapter seeks to understand the extent to which increased demand for the minerals critical to green energy technologies could affect fragility, conflict and violence in producing states, and explores what would be required of the international community to mitigate these local and national threats. It builds on extensive desk-based research, as well as a mapping analysis, case studies and findings from consultations with key stakeholders and experts.

2 Context and Background

2.1 Green Energy and the Demand for Minerals

The release and accumulation of greenhouse gases in the atmosphere is severely affecting the global climate. Higher temperatures, increasing variable rainfall, rising sea levels, more droughts and floods, coral bleaching and crop failure are some of the ways in which a changing climate will affect people and ecosystems. Scientists predict that temperatures will continue to rise in the coming decades and that the impacts will be felt across the globe, with varying severity and frequency depending on the region (NASA 2018). Current climate models project increases in mean temperatures, hot extremes in most inhabited regions, heavy precipitation in several regions and droughts and precipitation deficits (IPCC 2018).

In an effort to combat the impacts of climate change, 185 parties have ratified the 2016 Paris Agreement, which aims to keep global temperature increase below 2 °C above pre-industrial levels and to pursue initiatives that limit the temperature increase even further to 1.5 °C or lower (UNFCCC 2018). The key to this agreement will be a shift to a low-carbon economy, equipped with green energy technologies to decarbonize existing industries.

Within the past few years, global investments in the green energy sector have surged. Companies like Google and Amazon have made major commitments to renewable technologies, purchasing enough wind and solar power in 2017 to compensate for their energy needs (Donnelly 2017). In 2016, the deployment of new energy storage technologies—most notably batteries—grew by more than 50% (International Energy Agency 2017). And car companies—most prominently Tesla, Volvo and BMW—have taken significant strides to electrify their fleets.

This chapter will analyze four green energy technologies: solar panels, wind turbines, EVs and energy storage batteries. These technologies are already on the market, have made the biggest gains in the past decade, and are projected to increase in demand exponentially through 2050.

As the demand for these technologies grows, so too does the demand for a number of minerals required to develop and facilitate them. Solar panels, for example, were the fastest-growing source of renewable energy in 2016 (Levin Sources 2017a, b).

According to the World Bank, solar technologies could represent somewhere from 2 to 25% of total global energy production in a low-carbon economy through 2050 (Arrobas et al. 2017). While the minerals required for solar technologies vary depending on the type and make of the panel, key minerals including gallium, germanium, indium, iron, nickel, selenium, tellurium and tin.

Wind technologies are also becoming more widespread and price-competitive with traditional fossil-fuel-based energy. In Europe, wind accounted for 44% of all new power installations in 2018 (WindEurope 2019). The minerals required for wind technologies also vary, depending on whether or not they located off or onshore and whether they use geared or direct-drive technologies.[2]

EVs are expected to be as affordable as gas-powered cars by 2022, with the greatest demand coming from China (NetworkNewsWire 2017). Bloomberg New Energy Finance estimates that 57% of all passenger vehicle sales will be electric by 2040 (BloombergNEF 2019a, b). Lithium-ion batteries dominate the market for EV batteries, due to their excellent energy-to-weight ratio (Arrobas et al. 2017). Due to the increasing demand for EVs and energy storage batteries, the demand for and prices of minerals like lithium, cobalt and manganese—all used in lithium-ion batteries—are already rising. The price of manganese, for example, nearly doubled from 2015 to 2017 (USA News Group 2017). Other estimates suggest that, in order to meet upcoming lithium demand, at least one new lithium mine will need to begin operations each year through 2025 (Baystreet Staff 2017).

The predicted mineral requirements for each green energy technology are included in Table 1. These minerals were determined based on data from the World Bank, Levin Soures, the U.S. Geological Survey, Bloomberg New Energy Finance and the American Exploration & Mining Association. Although other minerals may be required, the minerals in the following graphic were verified and cross-referenced by multiple sources and represent strategic components of the technologies in question.

Table 1 Minerals required for green energy technologies

Green energy technology	Minerals required
Solar	Bauxite & Alumina, Cadmium, Copper, Gallium, Germanium, Indium, Iron, Lead, Nickel, Selenium, Silicon, Silver, Tellurium, Tin, Zinc
Wind	Bauxite & Alumina, Chromium, Cobalt, Copper, Iron, Lead, Manganese, Molybdenum, Rare Earths, Zinc
Electric vehicles and energy storage	Bauxite & Alumina, Cobalt, Copper, Graphite, Iron, Lead, Lithium, Manganese, Nickel, Rare Earths, Silicon, Titanium

Source Data primarily from the World Bank (2017), Levin Sources (2017a, b), USGS (2017), Bloomberg New Energy Finance (2018) and the American Exploration & Mining Association (2013)

---

[2]Geared technologies, for example, do not require as much lead or rare earths as direct-drive, but are generally less reliable and have a lower capacity to handle intense wind speeds (Arrobas et al. 2017).

It is important to note that, due to the rapid rate of technological advances and possible opportunities for metal substitutions, these minerals are subject to change and are dependent on market fluctuations.[3]

The variety of minerals and metals required, and the quantities of each that will be needed, place stakeholders from across the mining life cycle—including exploration, extraction and processing entities—in a strategic position to contribute to the shift to a low-carbon economy. Exploration is expected to surge in order to meet the blooming demand and projected mineral supply deficits. However, the rate of change in the transition to a low-carbon economy has so far been too rapid for the exploration industry to keep pace; while the price of metals can increase quickly, it takes anywhere from 10 to 15 years from the discovery of a new deposit to the presence of fully operating mines at the site (Allen 2017).

Mineral recycling could alleviate some of the pressure placed on extractive operations; however, to date most of the listed minerals have poor end-of-life collection and recycling rates. Even in a scenario where recycling rates improve, studies have shown that supplies of lithium and cobalt, in particular, will still deplete significantly by 2060 (Manberger and Stenqvist 2018). As a result, the development of new technologies and metal substitution pathways are likely to play a far greater role in addressing potential supply deficits.

2.2 A Note on Mining and Conflict

Mineral resources—their extraction and the responsible investment of the revenues generated—can be a key driver of sustainable development. Well-managed minerals and metals can be a source of significant revenue for developing countries, revenue that, when collected and distributed transparently, can support national investments in health, education, infrastructure and other sectors crucial to a country's growth and prosperity. Large-scale mines (LSM) can be a significant source of foreign investments, jobs, shared infrastructure and procurement for local goods and services, while artisanal and small-scale mining (ASM) can provide viable livelihoods in regions where opportunities may be limited. Ensuring that mining contributes to sustainable development will depend on the presence of strong laws and policies for the sector, as well as mechanisms and institutions in place for their implementation and enforcement.

---

[3]For example, there has been some recent speculation that vanadium redox flow batteries could become a substitute for stationary lithium-ion batteries, thereby placing a greater demand on vanadium (Church 2019).

The potential for conflict,[4] however, always exists in the mining sector—a function of the impacts of mining activities have on communities, economies and the environment (Andrews et al. 2017). These conflicts are rarely the result of a single actor, but rather the result of interactions between multiple actors including companies, governments (local, district, national), communities and civil society organizations (Andrews et al. 2017).

Conflict minerals are defined by the European Union as those minerals that "finance armed groups, fuel forced labour and other human rights abuses, and support corruption and money laundering" (European Commission 2017). Diamonds in Sierra Leone and Angola are a prominent example: gaining control of the country's rich alluvial diamond deposits was a key incentive for rebel groups to carry out violence during the country's civil war, and the stones were used as a funding source for their ongoing operations. Similarly, illegal tin, tungsten, tantalum and gold (3TG) mining continues to fuel violence in the DRC. Beyond Africa, armed groups exert control over 3TG operations in South American countries as well; notable examples include the ELN in Colombia and drug smugglers and illegal groups in Venezuela (Jamasmie 2017; Diaz-Struck and Poliszuk 2012). Prolonged resource conflicts contribute to further human rights abuses and facilitate corruption, as well as undermine state legitimacy and resource governance institutions.

Conflict minerals are, of course, not the sole source of tension in the sector. As mining activities expanded from 2000 to 2012 due to high commodity prices, driven by increasing global demand for raw materials, incidences of social conflicts around mining increased in parallel, driven and experienced by a diverse range of state, non-state and private actors (Dietz and Engels 2016). Between January 2006 and July 2013, 843 large-scale protest movements—relating to a range of societal issues— took place in 87 countries (Andrews et al. 2017). These protests were and continue to be most prevalent in Latin America, Africa and Asia, and have continued despite downturns in both commodity prices and mining activities. In 2017, for example, thousands of protesters in Jerada, Morocco, called for government intervention and regulation in the country's coal mining sector, known locally as the "mines of death," according to news reports (Leotaud 2017). In the same year in Boké, Guinea, one person died and 20 were injured by Guinean forces during protests against the impact of local bauxite mining operations.

This chapter will analyze the fragility, conflict and violence implications of both LSM and ASM operations. While the prevalence of conflict minerals is more commonly associated with ASM sites, the high rates of protests, civil unrest, environmental degradation, corruption and other financial crimes associated with LSM operations necessitate that both be examined to fully understand the range of conflict implications associated with an increased demand for minerals required for green energy technologies.

---

[4] It should be noted that conflict in and of itself is not necessarily a bad thing; disagreements among stakeholders can lead to dialogue, debate and constructive change. Violent conflict, conversely, is never an optimal solution to differing opinions and approaches.

3 Identifying Mineral-Rich Fragile States Critical to the Low-Carbon Transition

The minerals and metals identified as critical to the development and deployment of four key green energy technologies—solar, wind, EVs and energy storage—are presented in Table 1. These minerals include, but are not limited to aluminium, cadmium, chromium, cobalt, copper, gallium, germanium, graphite, indium, iron, lead, lithium, manganese, molybdenum, nickel, rare earths, selenium, silicon, silver, tellurium, tin and zinc. Recycling minerals—or secondary minerals—are not yet in sufficient supply to meet the predicted demand, and therefore the majority of these minerals and metals will continue to be sourced from mining sites.

Given the historical links between conflict and mining, it is essential to determine if increasing extraction of these minerals has the potential to aggravate grievances and tensions at current and future sites of extraction. Regions that are vulnerable to these dynamics were identified by overlaying fragility indicators with global reserves of identified minerals. Fragility was determined by using both the Fund for Peace's Fragile States Index as well as Transparency International's Corruption Perceptions Index. The Fragile States Index defines fragility using 12 indicators relating to internal cohesion, the economy, politics, cross-cutting factors including demographic pressures, refugees and internally displaced persons, and external intervention (The Fund for Peace 2018). Transparency International calculates perceptions of corruption in the public sector using 13 different data sources from 12 different institutions (Transparency International 2017).

Fragility and corruption measures are presented in Fig. 1; the darker the shading, the more fragile and corrupt the state is, according to 2017 and 2018 data sets. As

Fig. 1 Fragility and corruption indicators. *Source* Fund for Peace (2018), Transparency International (2017)

can be seen, fragility and corruption are not endemic to any one particular region; all countries struggle with these issues, to varying degrees. However, higher rates of fragility and corruption are found in the Sahel and Central Africa, the Middle East and North Africa, Southeastern Asia, Central America and some parts of South America.

Fragility indicators can next be overlaid with established mineral reserves[5]— defined as resources known to be economically feasible for extraction—to determine if any of the strategic reserves are found in countries already plagued by instability. The U.S. Geological Survey collects and disseminates data on global mineral production and reserves every year in its *Mineral Commodities Summaries*. While extensive, the report has notable data gaps resulting from reporting and collection inefficiencies, particulary in developing countries. For this reason, it is important to note that even if a country has been designated as having minimal to no mineral reserves, this does not necessarily indicate the country lacks the mineral in question, only that data is unavailable to determine if there are significant reserves.

Figure 2 overlays the location of reserves of 18 minerals that are strategic for green energy technologies with indicators of fragility and corruption, highlighting areas that may be vulnerable to conflict with the proliferation of green energy technologies. The size of the circle corresponds to the relative global quantity of the country's reserves in metric tonnes. Mineral data was unavailable for cadmium, gallium, germanium,

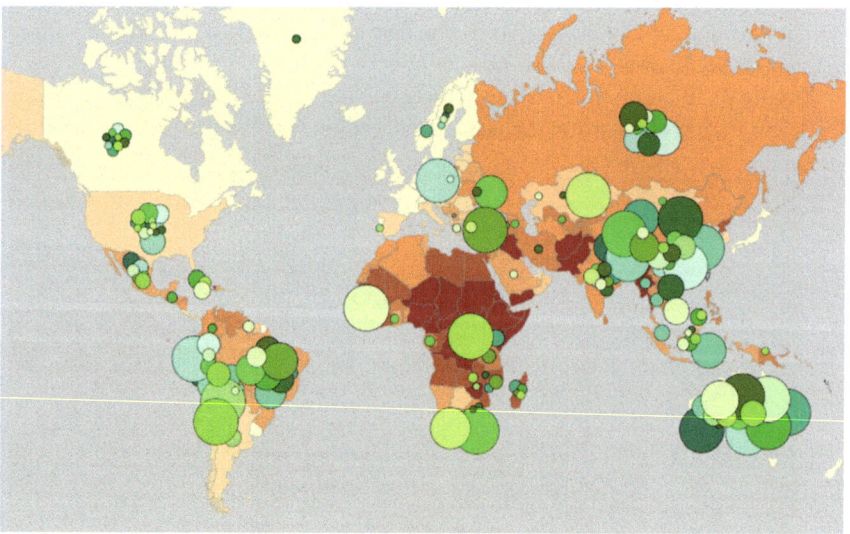

Fig. 2 Global reserves of minerals required for green energy technologies overlaid with fragility and corruption measures. *Source* Fund for Peace (2018), Transparency International (2017), U.S. Geological Survey (2018)

---

[5]While data on existing mineral deposits is readily available and may indicate large quantities of select minerals, these deposits have not yet been deemed economically viable and are therefore excluded from analysis.

indium and silicon. It is important to note as well that the circles signifying the reserves are geotagged to the country only, and not to a particular region within the country. Further, the colour of the circles has no significance, other than to differentiate between the different minerals.

While the map does not reflect current levels of production, the presence of strategic minerals in fragile states, coupled with the expected increase in demand for these minerals in the decades to come, point to the emergence of a number of potential hotspots for increased conflict or tension. Figure 2 demonstrates that there are significant reserves of strategic minerals in South America, sub-Saharan Africa, Southeast Asia and Australia. These regions, with the exception of Australia, have middle to high measures of fragility and corruption, highlighting areas that are potentially vulnerable to conflict with the proliferation of green energy technologies.

Table 2 Mineral reserves in states with high fragility and high corruption

Mineral	Global reserves located in a **fragile** or **very fragile state**[a] (%)	Global reserves located in states perceived to be **corrupt** or **very corrupt**[b] (%)
Bauxite and Alumina	44	68
Chromium	55	100
Cobalt	70	70
Copper	41	41
Graphite	73	100
Iron	42	60
Lead	49	49
Lithium	21	34
Manganese	66	86
Molybdenum	70	72
Nickel	42	59
Rare Earths	58	94
Selenium	76	76
Silver	52	52
Tellurium	67	67
Tin	69	84
Titanium	57	62
Zinc	52	59

Source Fund for Peace (2018), Transparency International (2017), U.S. Geological Survey (2018)
[a]Labelled as "elevated warning," "high warning," "alert," "high alert," or "very high alert" on the 2018 *Fragile States Index*: receiving a score of 70.00 or higher (113.4 is the highest score, held by South Sudan).
[b]Receiving a score of 43.00 or lower on the 2017 *Corruption Perceptions Index*. A score of 1 denotes a highly corrupt state; a score of 100 denotes a very clean state.

Table 2 further illustrates the possibility of conflicts emerging around these minerals. The table lists the percentage of known global reserves located in either fragile or corrupt states. Twenty-eight per cent of bauxite and alumina reserves, for example, are found in a very fragile state: Guinea. Fifty-six per cent of cobalt reserves are located in one very fragile and very corrupt state: the DRC. Notable, 100% of chromium and graphite reserves are found in states perceived to be either corrupt or very corrupt. In fact, substantial reserves of all 18 studied minerals are found in states perceived to be either corrupt or very corrupt in 2017 (Transparency International 2017).

In order to understand these conflict risks better, three case studies will be explored in further depth: cobalt in the DRC, rare earths in China and nickel in Guatemala. Each demonstrating a different geography, mineral and types of conflicts. These minerals were selected based on their importance to the development and deployment of green energy technologies and the rates of fragility and corruption where major reserves are found. Each case study examines the mineral's use in green energy technologies, as well as the conflict, fragility and violence implications of increased mineral extraction in one country, typically the country with the most reserves.

4 Case Studies

4.1 Cobalt in the DRC

Cobalt is used in the batteries of most modern electronics, including smartphones, digital music players and laptops. Critical to the low-carbon economy, cobalt is also instrumental for the development and facilitation of EVs and energy storage technologies. For EVs in particular, cobalt is found in three out of the four major lithium-ion batteries on the market: lithium cobalt oxides, nickel manganese cobalt and nickel cobalt aluminium (Levin Sources 2017a, b). Lithium-ion batteries are also used to store energy derived from solar, wind and other green technologies, thereby making the batteries—and the minerals required to power them—an integral part in the transition to a low-carbon economy.

Surging demand for cobalt has been a recurring theme in recent headlines. The World Bank estimated that if the international community is able to keep global temperature rise to 2 °C (instead of the project 6 °C) through the widespread adoption of green energy, demand for cobalt could see an increase of an exponential magnitude (Arrobas et al. 2017). Future demand and pricing are largely predicated on changes in the automotive industry, namely, the shift toward lithium-ion batteries. Some estimates suggest that the price of cobalt could increase to USD 100,000 per tonne by 2030 (compared to USD 60,000 per tonne in 2017) as a result of the transition to EVs and green energy storage technologies (NetworkNewsWire 2017). Much of this demand comes from Chinese manufacturing and consumer markets.

Global cobalt supply is subject to potential shortages. Some estimates suggest a 20% global cobalt supply gap by 2025, even with new mining operations in Canada coming online in the interim (Safehaven.com 2017). The British Geological Survey's 2015 Risk List gave cobalt a score of 8.1 out of 10, indicating a relatively high supply risk (British Geological Survey 2015). In addition to unprecedented demand for the mineral, this risk and supply gap is in part due to the way that cobalt is mined; cobalt is mined as a by-product of either nickel or copper and therefore can be dependent on price fluctuations and demand of the two. Cobalt can be mined through both ASM and LSM operations.

Cobalt supply is also designated as high risk because fragile countries host a large majority of the cobalt reserves and production. The DRC has the largest global reserves of cobalt, as demonstrated in Fig. 3, with an estimated 3,500,000 metric tonnes—50% of world reserves (U.S. Geological Survey 2018). Other, less significant reserves are found in Australia, Cuba, the Philippines, Zambia, Russia, Canada, Madagascar, Papua New Guinea, South Africa and the United States (U.S. Geological Survey 2018).

Despite a vast wealth of mineral resources and biodiversity, the DRC's recent history has been defined by fragility, corruption and violence. The centre of what was called Africa's World War, or the Second Congo War, from 1998 to 2003, legacies of human rights abuse, weak governance and exploitative practices still permeate the lives of many Congolese citizens. The country still scores high on global indicators of fragility and corruption, ranked the 5th most fragile country in the world and the 19th most corrupt (The Fund for Peace 2019; Transparency International 2019). And

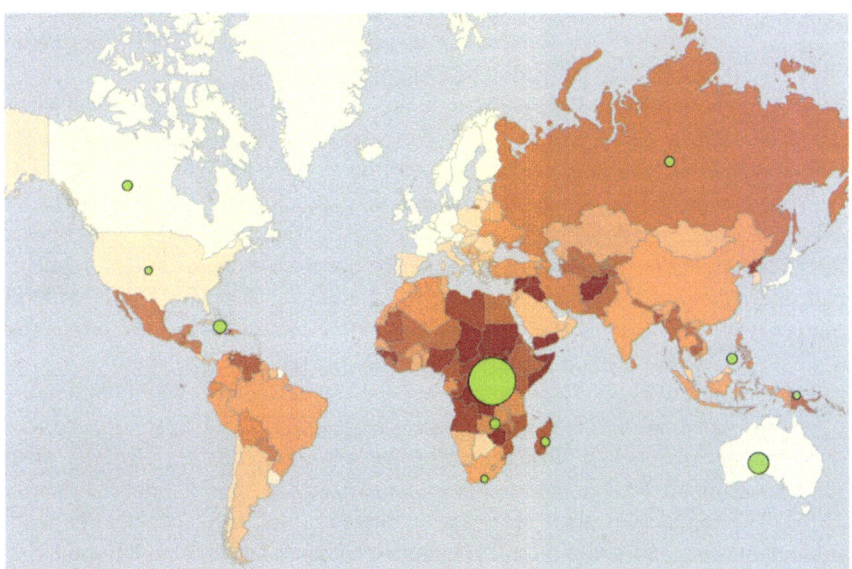

Fig. 3 Global cobalt reserves overlaid with fragility and corruption measures. *Source* Fund for Peace (2018), Transparency International (2017), U.S. Geological Survey (2018)

although the Ibrahim Index of African Governance demonstrates that, in recent years, the DRC has shown increasing improvement in developing a sustainable economy, it also highlights a continued deterioration of human development, political participation and human rights (IIAG 2016). Its positive peace ranking is low, indicating that despite marginal economic improvements, the state's stability is still prone to shocks (Institute for Economics & Peace 2017).

The DRC's mining industry is its largest source of export income (BBC News 2018a, b). In addition to cobalt, the country is a major global producer of copper, tantalum, tin and gold. Despite the potential the DRC's mineral wealth holds for its economy and development, the country's history of war, weak governance and grand corruption pose a risk to the responsible management of ongoing mining operations.

The DRC's resource governance scores have either been poor or on the cusp of failing for years, indicating that there are minimal established procedures and practices to govern the country's minerals (Natural Resources Governance Institute 2017). Throughout and after the Second Congo War, mineral resources, including 3TG, fuelled violence and human rights violations. Illegal armed groups fought for control of the mines, exploited miners and used profits from the minerals to fund their continued violence. These activities, in addition to other global examples, contributed to the emergency of thinking on "conflict minerals," referring to minerals that are extracted from conflict zones and fund continued violence.

In addition to conflict and violence, mineral resources in the DRC can be subject to corruption schemes and financial crimes. Corruption can erode the social contract between the state and its citizens, diverting funding away from the core services that need it most and impoverish communities, further exacerbating fragility.

The continued permeation of fragility, conflict and violence into the mining sector as well as ongoing records of corruption and weak governance pose considerable risks for the responsible extraction of Congolese cobalt for green energy technologies. Already, cobalt has been tied to some of the same exploitative and violent practices seen in the mining of 3TG; these mines have been connected to child labour, dangerous working conditions, extortion and human rights abuses.

Approximately 10% of the global supply of cobalt, and 20% of the DRC's total exports, comes from ASM operations (Amnesty International 2017). ASM sites are not inherently dangerous, but are prone to risk due to minimal oversight, regulation and safety measures (Reuters 2018). ASM operations can be associated with high rates of death and injuries, due to the lack of safety equipment and protective gear.

In 2016, Amnesty International visited ASM operations in the south of the DRC and interviewed workers at five mining sites. Researchers interviewed 17 children, all of whom were employed at the mining sites for less than USD 2 per day, and found that several children had been beaten by the mining companies' security guards for trespassing on the companies' concessions (Amnesty International & Afrewatch 2016). The researchers also found that workers did not have access to protective equipment, were exposed to harmful chemicals and that state officials extorted illegal payments from the artisanal miners (Amnesty International & Afrewatch 2016).

As a result of these risks, some companies have tried to avoid sourcing from the DRC altogether, looking instead to Australia, Canada or the Philippines. Given the

DRC's rich reserves, however, leading producers of cobalt will most likely continue to work in the African country. Efforts should be focused on mitigating conflict risks by addressing its root causes and stopping human rights abuses around the DRC's mining sites, which will ultimately improve certainty in the supply chain.

Despite the relatively widespread reporting of these ongoing conflict implications, only marginal improvements have been made to secure the responsible sourcing of cobalt. Cobalt is not officially classified as a conflict mineral in legislation like the U.S. Dodd–Frank Act or the European Union's Conflict Mineral regulation. Any regulations aimed at curbing the illegal flow of conflict minerals, therefore, may not explicitly apply to cobalt. Some groups—like the World Economic Forum's Global Battery Alliance and the Responsible Cobalt Initiative—have started to take measures to address these inefficiencies and gaps in international legislation and supply chain governance. However, given the ongoing human rights abuses and surging demand for cobalt, additional improvements in the responsible sourcing of cobalt are still sorely needed.

4.2 Rare Earths in China

The term "rare earths" refers to 17 different elements, often found together in the Earth's crust. Of the 17, three are of particular importance to the development of green energy technologies: dysprosium, neodymium and praseodymium. These minerals are necessary for the production of specialized magnets used in both EVs and energy storage technologies as well as wind turbines. The magnets are favoured for EVs because they are generally lighter, stronger and more efficient than induction motors that rely on copper coils (Desai 2018). Similarly, use of these magnets has significant advantages in the production of wind turbines, cited for their efficiency, weight, size and maintenance properties (Pavel et al. 2017). The World Bank notes that the use of these magnets in wind turbines is preferred, particularly for offshore turbines, due to their reliability and capacity to handle higher wind speeds (Arrobas et al. 2017). Some substitutions are available for rare earths; however, most of these are still in the research phase and in general have been found to be less effective.

The prices of wind turbines and EVs are increasingly competitive, making the deployment of both a rapid reality. The demand for rare earths to meet this reality, and for neodymium and praseodymium in particular, is expected to surge in the coming years with this transition. The global demand for neodymium in 2017 was approximately 31,700 tonnes, outstripping supply by 3,300 tonnes (Desai 2018). And without viable substitutions, demand for neodymium will need to increase by more than 250% through 2050 for the international community to meet its Paris Agreement goals (Arrobas et al. 2017).

Both the demand for and supply for rare earths are concentrated in China. China accounted for 80% of rare earth production in 2017 and is home to 36% of world reserves, as shown in Fig. 4 (U.S. Geological Survey 2018). Reserves are also located in Vietnam, Brazil, Russia, India, Australia, the United States, Canada, South Africa,

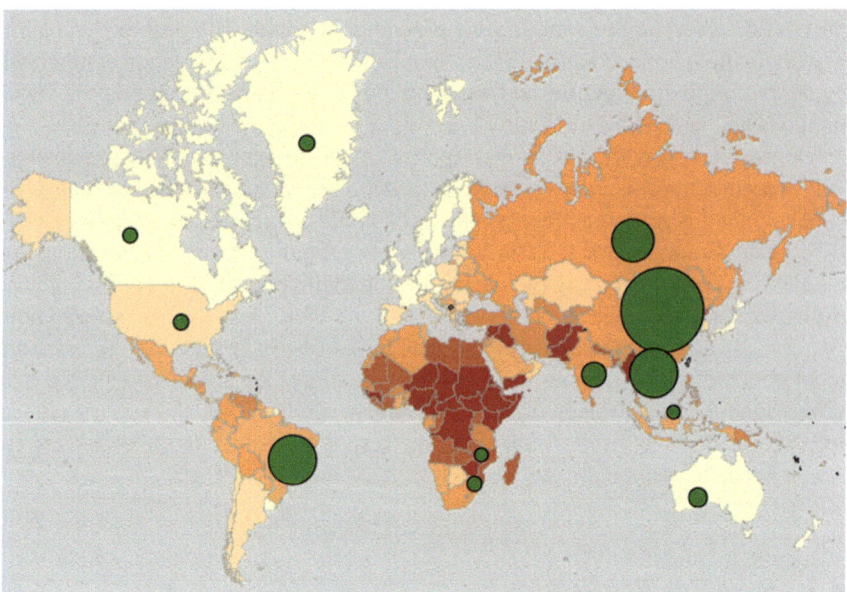

Fig. 4 Rare earths reserves overlaid with fragility and corruption measures. *Source* Fund for Peace (2018), Transparency International (2017), U.S. Geological Survey (2018)

Malawi and Malaysia, but these are yet to be developed to the same extent as the Chinese reserves (Australia hosts the only other major rare earths operations in the world). Increasing demand for and prices of rare earths are spurring the rapid development of rare earths projects around the world; however, most of the projects will not come online until the last 2020s. China's monopoly of the rare earths market, however, is set to continue for the coming decade at a minimum (Shaw 2017).

In addition to rare earths, China is rich in multiple mineral resources key to green energy technologies. Most notably, China has some of the largest global reserves of lead, selenium, tellurium, tin and zinc, which are critical to solar technology, as well as graphite, lithium and titanium, which are critical for EVs and energy storage technologies. The country also has reserves of bauxite and alumina, copper, iron, manganese, nickel and silver. This vast mineral wealth places China in a unique global position, in some cases allowing the state to exert a quasi-monopoly on several critical minerals. A number of producers and refiners for minerals required for green energy technologies, including cobalt, are also located in China. Virtually all lithium-ion battery production is done in China: the country is the largest global importer of cobalt, nickel and manganese, as well as lithium, despite having large reserves of its own (RCS Global 2017). As such, most discussions of green energy supply chains must include China.

China has mid-range scores of fragility and corruption (Fund for Peace 2019; Transparency International 2019). It has high positive peace scores, indicating high levels of resilience and the appropriate attitudes, institutions and structures needed to sustain a peaceful society (Institute for Economics & Peace 2017). However, the Natural Resource Governance Institute labels China's resource governance as "weak," indicating that the sector has a mix of strong and problematic areas of governance; indicators like value realization, revenue management and establishing an enabling environment for extraction have ample room to improve (Natural Resources Governance Institute 2017).

Mining and industrial development more generally in China have often come at the cost of the environment, growing alongside dangerous levels of air and water pollution. Since the 1990s, environmental activism, driven by localized grievances against pollution, has grown and manifested in the form of protests, petitions and, in some cases, violence (Ho and Yang 2018). In 2011, communities from the Qinghai province in China, for example, urged the government to take action against lead mining in the Ganhetan Industrial District, which was known to cause high levels of water pollution and endanger the lives of local residents (Environmental Justice Atlas 2018). Similar protests and pleas have also been recorded against controversial copper mining in Tibet, gold mining along the Gu Chu River, cadmium extraction in the Guangdong Province and cement production in the Madang Province (Environmental Justice Atlas 2018).

Rare earth mining can be both destructive and toxic to surrounding environments. Almost all rare earth ores contain the radioactive elements thorium and uranium (Huang et al. 2016). As a result, the extraction and processing of rare earths can be highly toxic and have a negative effect on soil, water and human health. In 1958, the Baotou Iron and Steel Company began producing rare earths near the city of Baotou in Inner Mongolia; by 1980 crops in the nearby villages had already started to fail due to pollution of soil and groundwater attributed to rare earth mining and processing (Bontron 2012). Today, the lands surrounding Baotou are stripped of topsoil while streambeds contain thousands of gallons of acid (Bradsher 2010a, b). Dalahai village, located close to a Baotou rare earths tailing pond, has been named a "death village" due to the high incidence of lung cancer, brain cancer, respiratory illnesses and cardiovascular diseases suffered by local residents (Huang et al. 2016). Ganzhou, the so-called "rare earths kingdom," has been described as a "site of devastation" by the *ChinaDialogue*, plagues as it is with crude open air mines, smelters, polluted water supplies and reduced crop yields (Hongqiao 2016). Coupled with the growth of environmental activism, rare earths mining in China could lead to increasing tensions at the local level.

The nature of deposits located in the southern province of Ganzhou makes rare earth extraction relatively easier than in Inner Mongolia. These deposits are also free of radioactive thorium. However, as a result of the ease of extraction and raising global prices for rare earths, a substantial number of illegal rare earth mines emerged in the area. These mines are cited to sell to organized crime syndicates and exploit workers, some of which are children (Bradsher 2010a, b; Schlanger 2017). Some estimates suggest that tens of thousands of tonnes of rare earths are illegally mined and sold on China's black market every year (Hongqiao 2016). In response, both the central and provincial governments have taken measures against illegal mining operations, including instituting new regulations against illegal exploration as well as dispatching police to outlaw the illegal mines (Yan 2012). The traceability of rare earths supply chains, however, is still relatively unexplored, and is not regulated to the same extent as other conflict minerals.

In addition to the risks of exacerbating local and global grievances surrounding pollution and public health, China's majority share of rare earth production has been used as political–economic leverage in past state-level disputes. In 2010, amid a territorial disagreement over disputed islands with Japan, China suspended its shipments of rare earths to its neighbour (Bradsher 2010a, b). Chinese officials later lifted the embargo and denied that the ban was in response to the dispute with Japan, but that they had instead reduced export quotas to mitigate pollution and environmental concerns (Bradsher 2010a, b). Nevertheless, the period raised a number of concerns with companies and countries around the world regarding the diminishing supply, rising prices and implications of China's dominant position in the rare earths market. More recently, concerns have been raised that China could use this position for leverage in the trade war between the U.S. and China (Reuters 2019).

4.3 Nickel in Guatemala

Cobalt, lithium and rare earths tend to dominate the discussions around the minerals required for the green energy transition. This is in part due to the fact that base metals like nickel are not exclusively produced for green energy technologies; nickel is used in more than 300,000 products worldwide, including those with consumer, industrial aerospace, marine and architectural applications (Nickel Institute n.d.). Of the 2.1 million metric tonnes of nickel content produced in 2017, approximately 65% was used to manufacture stainless steel, while only 6% was devoted to the production of coins, electronics and rechargeable batteries (U.S. Geological Survey 2018; Nickel Institute n.d.). However, its use in the production of steel is also expected to benefit from green energy technologies, and its use in rechargeable batteries is expected to comprise a growing share of nickel production annually. Nickel can only be mined through LSM operations.

Nickel is required for multiple green energy technologies. Currently, two types of lithium-ion batteries make up the majority of the EV market due to their efficiency, price and ease in manufacturing: nickel manganese cobalt and nickel cobalt aluminium batteries (Hunt 2018). A rate of 30 and 80% of nickel is required for nickel manganese cobalt and nickel cobalt aluminium batteries, respectively. As such, nickel will be crucially important to the green energy transition, regardless of which EV battery leads production in the coming decade (Hunt 2018). Nickel is also instrumental in solar technologies; according to the World Bank, the transition to solar could increase the demand for nickel by 300% through 2050 (Arrobas et al. 2017). This number increases dramatically when EVs and energy storage technologies are included, with a predicted increase in demand for nickel of up to 1,200% (Arrobas et al. 2017).

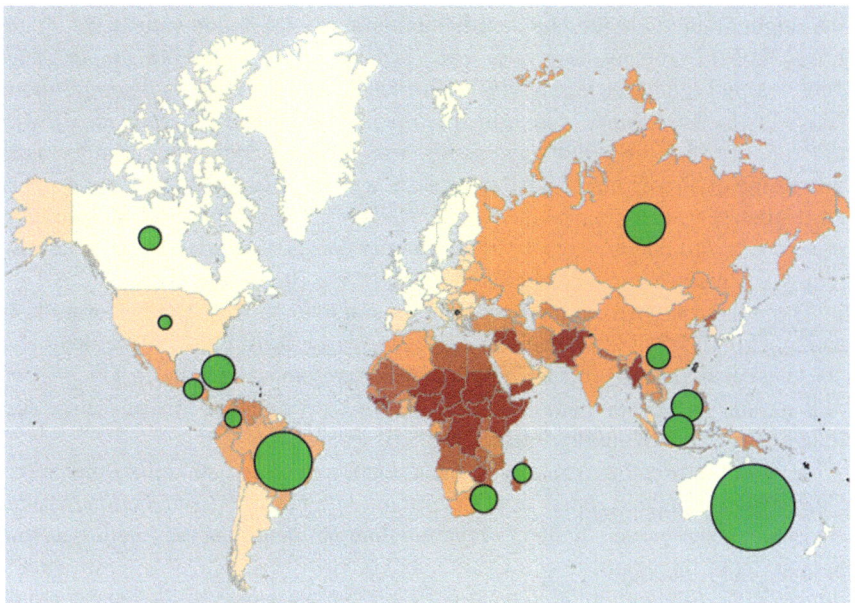

Fig. 5 Nickel reserves overlaid with fragility and corruption measures. *Source* Fund for Peace (2018), Transparency International (2017), U.S. Geological Survey (2018)

Nickel is mined in more than 40 countries, with significant quantities of reserves in 13, shown in Fig. 5 (U.S. Geological Survey 2018). However, of these reserves, 38% are found in states given an "elevated warning" label or worse on the Fragile States Index (Fund for Peace 2018), and 54% of reserves are located in states perceived to be corrupt or very corrupt (Transparency International 2017). Guatemala is currently ranked 10th in terms of world reserves, with an estimated 1,800,000 metric tonnes of nickel content (U.S. Geological Survey 2018).

Guatemala is one of the most resource-rich and populated countries in Central America. Despite this richness, the country ranks poorly on indicators of fragility, corruption and violence (Fund for Peace 2019; Transparency International 2019). Although the Indigenous Maya comprise almost 50% of the population, they are disproportionately affected by development challenges and the negative impacts of the country's mining industry (BBC News 2018a, b).

Mining activity currently makes up approximately 3% of Guatemala's overall GDP (MICLA n.d.). Along with nickel, Guatemala possesses rich deposits of gold, silver and copper, although not all have been economically viable to mine. On the 2017 Resource Governance Index, Guatemala's mining sector was given a "poor" ranking, having established minimal procedures and practices to govern resources (Natural Resources Governance Institute 2017).

Nickel mining is relatively new to Guatemala, having emerged during the country's armed internal conflict, which began in the early 1960s and lasted until 1996. At the height of this violence, the Canadian mining firm INCO (previously the International Nickel Company, now Vale) won a monopoly on nickel extraction in Central America, and controlled nearly 54% of the nickel market in North and South America (Price 2015). New mines were primarily opened in Guatemala's rural areas, where the majority of Indigenous Maya people reside. Indigenous opposition to mining during the civil war manifested as protests against illegal land use, environmental degradation, and connections to colonialism associated with LSM operations. In response, armed groups and private security forces were sometimes deployed to remove indigenous-led opposition to mining (Price 2015).

Throughout the armed internal conflict, more than 200,000 civilians were killed and more than 1.5 million displaced (Agence France-Presse 2015; PBS News Hour 2011). Guatemala's Historical Clarification Commission later found that the majority of these abuses were perpetrated by the state military and that 83% of victims were Indigenous Maya (Amnesty International 2014).

A recent report by Amnesty International found that mining companies in Guatemala throughout the civil war neglected to address existing community tensions and in many cases "failed to adhere to international standards on business and human rights" in regards to consultation and security operations (Amnesty International 2014). Although the civil war ended in 1996, the legacy of violence continues to affect Guatemalan society, oftentimes manifesting around the mining of nickel.

Fenix Mining project is one of the largest nickel mines in Central America (Hill 2014). Throughout the past decade, the project—located near the Lote Ocho Indigenous community in El Estor—has been linked to allegations of forced displacement, murder and sexual violence. Due to poor access to dispute resolution mechanisms and a weak judiciary, victims—the majority of which are Indigenous—have up to date been unable to successfully bring criminal cases against their perpetrators within Guatemala (Kassam 2017). Some victims, however, have been able to bring forward civil lawsuits in the home jurisdictions of the foreign mining companies that own the mine (and their subsidiaries) (Kassam 2017).

In the case of Caal v. HudBay, 11 Maya Q'eqchi' women are suing the mining company HudBay in its home jurisdiction of Ontario (Klippensteins et al. 2018). The women allege that in January 2007, private security personnel from the Fenix mining project forced villagers off their ancestral lands, burned their homes and sexually assaulted the women (Klippensteins et al. 2018). In addition to this lawsuit, community groups are suing HudBay and its subsidiaries for the 2009 murder of the community leader Adolfo Ich Chamán and the shooting of German Chub Choc. HudBay denies culpability, stating that the sexual assaults were carried before their involvement in the mine and that mining personnel were not involved in the 2009 evictions; HudBay bought the mine from the Guatemalan Nickel Company in 2008 (Kassam 2017; Price 2015; Hudbay Minerals Inc. 2017).

While these cases represent a landmark move to bridge the disconnect between foreign mining companies and the actions of its local subsidiaries, they also reflect how increased mineral demand in resource-rich states like Guatemala can result in an exacerbation of land ownership disputes between private companies, governments and community groups. A decade after the aforementioned incidents, nickel mining in Central America is still associated with the destruction of surface land resources, a primary source of livelihoods for many Indigenous communities (Fox 2014).

If the demand for nickel increases at its predicted rates, it is essential to ensure that mining companies and governments adhere to international human rights and business law, as well as actively respect the human and land rights of Indigenous persons. This consideration, as well as improved access to effective dispute resolutions, will be necessary to mitigate ongoing community tensions and grievances surrounding extraction, and to contribute to the responsible supply chain governance of nickel and other base metals.

5   Supply Chain Governance

Some states, international agencies and private sector entities have introduced legislation and guidance to curb the flow of conflict minerals and promote responsible and transparent mineral supply chains. This includes the OECD Due Diligence Guidance for Responsible Supply Chains of Minerals from Conflict-Affected and High-Risk Areas. In line with the UN Guiding Principles of Business and Human Rights, the OECD guidance provides a framework for companies operating in contexts of fragility to conduct due diligence on their supply chains by assessing potential risks, preventing and mitigating these identified risks, and adopting a risk management plan (OECD 2016). The guidance applies to all minerals, but has specific supplements for 3TG.

The European Union (EU) Conflict Minerals Regulation comes into force on January 1, 2021 and is designed to ensure that all EU importers of 3TG meet the OECD Guidance. The regulation directly applies to almost 1,000 EU importers and indirectly applies to 500 smelters and refiners of 3TG, based both inside and outside the EU (European Commission 2017). As the regulation only applies to 3TG, the aforementioned minerals will not be directly affected. However, additional minerals could be added to the regulation in 2024, when the EU executive is set to undergo a review of the bill.

The Chinese Chamber of Commerce for Metals, Minerals & Chemicals Importers & Exporters (2015) also introduced a framework to operationalize the OECD Guidance. The framework provides guidelines and tools to those Chinese companies that extract or use minerals in their products to help them identify, prevent and mitigate risks of conflict, human rights abuses and misconduct throughout the entire mining life cycle. The framework can be applied to all minerals, however, remains voluntary.

The U.S. Dodd-Frank Act is another prominent mechanism to promote responsible mineral supply chains, though with more limited geographic scope. Passed in 2010, Section 1502 of the Act requires U.S. publicly traded companies to assess and address any risks in the supply chains of 3TG that may originate from the DRC or neighbouring countries. While the scope of this legislation neglects the potential for other minerals and regions beyond sub-Saharan Africa to finance armed groups in mineral-rich fragile states, the legislation is a landmark example of public sector action against corruption, human rights abuses and violence in the mining sector.

In addition to these public sector mechanisms, civil society groups and private sector actors have also taken steps to ensure the responsible sourcing of minerals and metals. The Extractive Industries Transparency Initiative (EITI), for example, promotes the open and accountable management of oil, gas and mineral resources (EITI 2018). In addition, the London Metals Exchange (LME), the world's biggest market for industrial metals, has increased efforts to monitor and investigate cobalt sources. In July 2018, LME announced that it would require all companies that receive a minimum of 25% of their metal from ASM mines in the DRC to undergo a professional audit (Desai and Daly 2018). These moves demonstrate the growing awareness of the conflict-related risks associated with cobalt extraction in fragile states.

Industry schemes like the Responsible Minerals Initiative (RMI) are also contributing to the improved international governance of mineral supply chains. RMI is a coalition of more than 360 member companies aiming to improve the human rights conditions of mineral supply chains, by administering third-party audits of mineral sourcing in conflict-affected and high-risk areas (Sustainable Brands 2018). In 2016, the OECD began to assess five industry schemes, including RMI, regarding the schemes' alignment with the OECD Guidance and overall due diligence in sourcing practices. Although the official policies of the industry schemes were increasingly aligned with the OECD Guidance, the assessment found in its 2018 report that implementation of the due diligence processes lagged behind alignment (OECD 2018). Reporting and monitoring by civil society groups have pointed to sometimes shallow improvement in due diligence practices by mining companies and industries as well. The Responsible Sourcing Network, for example, assess companies' efforts to provide strong supply chain due diligence in their use of conflict minerals (Deberdt and Jurewicz 2017). The report found that, although more than 70% of assessed companies followed the OECD guidelines, most did so only superficially and few used the guidance in full (Deberdt and Jurewicz 2017).

The implementation gap is common across the various public sector, private sector and civil society actions to monitor and regulate the mineral supply chain, posing a risk to the responsible sourcing of green energy technologies and the minerals required to develop and deploy them. While legislation like Section 1502 of the Dodd–Frank Act and the upcoming EU Conflict Minerals Regulation represent significant milestones to responsible mineral governance, many are not yet applicable to minerals beyond 3TG. In the past few years, the conflict risks of cobalt sourcing have started to garner international attention. Subsequently, some civil society groups and private sector actors have started to incorporate responsible cobalt sourcing into their mandates. However, the supply chains and associated risks of other green conflict minerals like bauxite and alumina, lithium, nickel and rare earths are still largely unexplored and therefore largely absent from the current field of mineral supply chain governance.

6 Recommendations and Conclusions

Mining can be an inherently risky industry. Companies and small-scale miners alike constantly struggle with accessing uncertain deposits, managing volatile commodity prices, ensuring worker health and safety, and avoiding environmental catastrophe. Conflict adds another, significant risk to mining: it can interrupt operations, undermine the social license to operate, damage reputations and, at its most extreme, threaten the lives of those involved. Conflict and fragility are bad for business, communities and governments, and reducing conflict risks is in the interests of all stakeholders.

For minerals required to make the transition to a low-carbon economy, there are real risks of grievances, tensions and conflicts emerging or continuing, as has been made clear in the case studies and mapping exercise presented. For fragile states, weak governance, corruption and the inadequate implementation and enforcement of existing laws all work against ensuring that the benefits of mining accrue to the population and the country's sustainable development. Local voices are often left out of important decision-making, and meaningful engagement with communities does not always occur prior to the start of mining activities. Depending on the mineral, individuals can be subject to health and safety violations, human rights abuses, environmental risks and child labour.

There remains a lack of transparency across a number of key supply chains, including those for cobalt, lithium and bauxite. This opaqueness extends to the recycling industry, which will be increasingly important part of mineral and material provision in the future. Regulations and laws supporting increased transparency are not yet widespread enough to capture all relevant minerals, though important lessons can be drawn from international efforts to eliminate conflict from 3TG supply chains. The complexity of these supply chains, which include miners, traders, smelters, refiners, manufacturers, transporters and consumers, can be intimidating, but should not deter the international community from the important work that needs to be done to ensure they are conflict free.

Governments, the private sector and civil society should work together to ensure that the transition to a low-carbon economy, and the subsequent sourcing from mineral-rich fragile states, is conducted in a way which promotes peace, stability and sustainability. While some of the foundations are in place to ensure that these critical minerals do not emerge as conflict minerals in the coming decades, these must be extended and strengthened. In particular, communities should be engaged in a meaningful way across the mineral life cycle, transparency promoted, existing supply chain regulations should be expanded to apply to more minerals, and the implementation of these regulations improved.

To meet the aims of the Paris Agreement and the Sustainable Development Goals, it is imperative that we transition to a low-carbon economy. Green energy technologies like wind turbines, solar panels, EVs and improved energy storage will aid in this transition. However, the emergence or exacerbation of fragility, conflict and violence along the supply chains of the minerals needed to produce these technologies could threaten the overall "green" nature of this transition. In order to meet global goals around sustainable development and climate change mitigation, while contributing to lasting peace, the supply chains of these strategic minerals must be governed in a way that is responsible, accountable and transparent. Achieving this vision will require concerted effort from civil society, the private sector and governments.

References

Agence France-Presse (2015) Guatemala, wracked by 36 years of civil war. NDTV, 24 October 2015. https://www.ndtv.com/world-news/guatemala-wracked-by-36-years-of-civil-war-1235818

Allen, A (2017) World faces minerals shortages. Supply Manag. https://www.cips.org/supply-management/news/2017/april/world-faces-mineral-shortages/

Amnesty International (2014) Guatemala: mining in guatemala: rights at risk. Amnesty International Publications, London

Amnesty International & Afrewatch (2016) "This is what we die for:" human rights abuses in the Democratic Republic of the Congo power the global trade in cobalt. Amnesty International Ltd & African Resources Watch. https://www.amnestyusa.org/files/this_what_we_die_for_-_report.pdf

Amnesty International (2017) Time to recharge: corporate action and inaction to tackle abuses in the cobalt supply chain. Amnesty International. https://www.es.amnesty.org/uploads/media/Time_to_recharge_online_1411.pdf

Andrews T, Elizalde B, Le Billon P, Oh CH, Reyes D, Thomson I (2017) The rise in conflict associated with mining operations: what lies beneath? Canadian International Resources and Development Institute (CIRDI). https://cirdi.ca/wp-content/uploads/2017/06/Conflict-Full-Layout-060817.pdf

Arrobas DL, Hund KL, Mccormick MS, Ningthoujam J, Drexhage JR (2017) The growing role of minerals and metals for a low carbon future. World Bank Group, Washington, D.C. http://documents.worldbank.org/curated/en/207371500386458722/The-Growing-Role-of-Minerals-and-Metals-for-a-Low-Carbon-Future

Baystreet Staff (2017) The demand for lithium has companies scrambling from the Congo to Canada. Baystreet.ca, 8 November 2017. https://www.baystreet.ca/stockstowatch/2493/The-Demand-for-Lithium-Has-Companies-Scrambling-from-the-Congo-to-Canada

BBC News (2018a) DR Congo country profile. BBC News, 14 May 2018. https://www.bbc.com/news/world-africa-13283212

BBC News (2018b) Guatemala country profile. BBC News, 31 May 2018. https://www.bbc.com/news/world-latin-america-19635877

BloombergNEF (2019a) Electric vehicle outlook 2019. Bloomberg New Energy Finance, New York City

BloombergNEF (2019b) New energy outlook 2019. BloombergNEF, New York City

Bontron C (2012) Rare-earth mining in China comes at a heavy cost for local villages. The Guardian, 7 August 2012. https://www.theguardian.com/environment/2012/aug/07/china-rare-earth-village-pollution

Bosshard P (2018) Insuring coal no more: the 2018 scorecard on insurance, coal and climate change. 350.org. Center for International Environmental Law, ClientEarth, Consumer Watchdog, Foundation "Development YES - Open-Pit Mines NO", Ecologistas en Accion, Friends of the Earth France, Greenpeace, Indigenous Environmental Network, Market Forces

Bradsher K (2010a) After China's rare earth embargo, a new calculus. The New York Times, 29 October 2010. https://www.nytimes.com/2010/10/30/business/global/30rare.html

Bradsher K (2010b) China to tighten limits on rare earth exports. The New York Times, 28 December 2010. https://www.nytimes.com/2010/12/29/business/global/29rare.html?ref=rareearths

British Geological Survey (2015) Risk list. NERC. http://www.bgs.ac.uk/mineralsuk/statistics/risklist.html

Church C (2019) Is Vanadium the "Valyrian Steel" of the energy transition? IISD Blog, 16 April 2019

Deberdt R, Jurewicz P (2017) Mining the disclosures 2017: an investor guide to conflict minerals reporting in year four. Responsible Sourcing Network

Desai P (2018) Tesla's electric motor shift to spur demand for rare earth neodymium. Reuters, 12 March 2018

Desai P, Daly T (2018) Exclusive: London metal exchange aims to ban metal sourced with child labor. Reuters, 13 February 2018

Diaz-Struck E, Poliszuk J (2012) Venezuela emerges as new source of 'conflict' minerals. Int Consort Investig Journal. https://www.icij.org/investigations/coltan/venezuela-emerges-new-source-conflict-minerals/

Dietz K, Engels B (2016) Contested extractivism: actors and strategies in conflicts over mining. J Geogr Soc Berl 148(2–3):111–120. https://doi.org/10.12854/erde-148-42

Donnelly G (2017) Google just bought enough wind power to offset 100% of its energy use. Fortune, 1 December 2017. http://fortune.com/2017/12/01/google-clean-energy/

EITI (2018) The global standard for the good governance of oil, gas and mineral resources. The Extractive Industries Transparency Initiative. https://eiti.org

Environmental Justice Atlas (2018) Environmental conflicts in China. Environmental Justice Atlas, 27 February 2018. http://ejatlas.org/country/china

European Commission (2017) The regulation explained. European Commission: Trade, 13 December 2017. http://ec.europa.eu/trade/policy/in-focus/conflict-minerals-regulation/regulation-explained/

Forsythe M (2017) China aims to spend at least $360 billion on renewable energy by 2020. The New York Times, 5 January 2017. https://www.nytimes.com/2017/01/05/world/asia/china-renewable-energy-investment.html

Fox S (2014) History, violence, and the emergence of Guatemala's mining sector. Environ Sociol 1(3):152–165. https://doi.org/10.1080/23251042.2015.1046204

Hill D (2014) Central America's biggest nickel mine reopens amid violent clashes. The Guardian

Ho P, Yang X (2018) Conflict over mining in rural China: a comprehensive survey of intentions and strategies for environmental activism. Sustainability 10. https://doi.org/10.3390/su10051669

Hongqiao L (2016) The dark side of renewable energy. Earth Journal Netw. https://earthjournalism.net/stories/the-dark-side-of-renewable-energy

Huang X, Zhang G, Pan A, Chen F, Zheng C (2016) Protecting the environment and public health from rare earth mining. Rev Geophys 532–535. https://doi.org/10.1002/2016EF000424

Hudbay Minerals Inc. (2017) The facts: Hudbay's former operations in Guatemala. Hudbay. http://www.hudbayminerals.com/English/Responsibility/CSR-Issues/The-facts-Hudbays-former-operations-in-Guatemala/default.aspx#link3

Hunt P (2018) How nickel makes electric vehicles go green. Science|Business. https://sciencebusiness.net/news/how-nickel-makes-electric-vehicles-go-green

IIAG (2016) Democratic republic of Congo. Ibrahim Index of African Governance. http://iiag.online

Institute for Economics & Peace (2017) Positive peace report. Institute for Economics & Peace. https://reliefweb.int/sites/reliefweb.int/files/resources/Positive-Peace-Report-2017.pdf

International Energy Agency (2017) Global EV outlook. OECD/ IEA. https://www.iea.org/publications/freepublications/publication/GlobalEVOutlook2017.pdf

IPCC (2018) Summary for policymakers. World Meteorological Organization, Geneva, Switzerland

Jamasmie C (2017) 'Conflict minerals' entering tech supply chains from countries beyond Africa - report. Mining.com, 6 April 2017. http://www.mining.com/conflict-minerals-entering-tech-supply-chains-from-countries-beyond-africa-report/

Kassam A (2017) Guatemalan women take on Canada's mining giants over 'horrific human rights abuses'. The Guardian, 13 December 2017. https://www.theguardian.com/world/2017/dec/13/guatemala-canada-indigenous-right-canadian-mining-company

Klippensteins, Barristers, Solicitors (2018) The Lawsuits. Choc v. HudBay Minerals Inc. & Caal v. HudBay Minerals Inc. http://www.chocversushudbay.com/about/

Leotaud VR (2017) Thousands protest in Morocco against the "mines of death". Mining.com, 27 December 2017. http://www.mining.com/thousands-protest-morocco-mines-death/

Levin Sources (2017a) Hybrid electric, plug-in hybrid electric and battery electric vehicles. Levin Sources, Cambridge. http://www.levinsources.com/publications/green-economy-series-hybrid-electric-plug-in-hybrid-electric-and-battery-electric-vehicles-1

Levin Sources (2017b) Solar photovolatic and energy storage in the electric grid. Levin Sources, Cambridge. http://www.levinsources.com/assets/pages/Green-Economy-Series-Solar-Photovoltaic-and-Energy-Storage-in-the-Electric-Grid.pdf

Manberger A, Stenqvist B (2018) Global metal flows in the renewable energy transition: exploring the effects of substitutes, technological mix and development. Energy Policy 119:226–241. https://doi.org/10.1016/j.enpol.2018.04.056

MICLA (n.d.) Guatemala. McGill Research Group Investigating Canadian Mining in Latin America: http://micla.ca/countries/guatemala/

Naaeke J (2017) WBCSD launches initiative to create market for sustainable fuels. Ghana News Agency, 14 November 2017. http://www.ghananewsagency.org/science/wbcsd-launches-initiative-to-create-market-for-sustainable-fuels-124912

NASA (2018) The consequences of climate change. Global Climate Change Vital Signs of the Planet, NASA, 13 June 2018. https://climate.nasa.gov/effects/

Natural Resources Governance Institute (2017) 2017 resource governance index. Natural Resources Governance Institute. https://resourcegovernance.org/sites/default/files/documents/2017-resource-governance-index.pdf

NetworkNewsWire (2017) Cobalt perfectly positions as global cobalt demand surges. PR NewsWire, 17 November 2017. http://www.prnewswire.co.uk/news-releases/cobalt-perfectly-positioned-as-global-cobalt-demand-surges-658216703.html

Nickel Institute (n.d.) Nickel metal - the facts. Nickel Institute. https://www.nickelinstitute.org/NickelUseInSociety/AboutNickel/NickelMetaltheFacts.aspx

OECD (2016) OECD due diligence guidance for responsible supply chains of minerals from conflict-affected and high-risk areas, 3rd edn. OECD Publishing, Paris. https://doi.org/10.1787/9789264252479-en

OECD (2018) Alignment assessment of industry programmes with the OECD minerals guidance. Secretary-General of the OECD

Pavel CC, Lacal-Arantegui R, Marmier A, Schuler D, Tzimas E, Buchert M, Blagoeva D (2017) Substitution strategies for reducing the use of rare earths in wind turbines. Resour Policy 52:349–357. https://doi.org/10.1016/j.resourpol.2017.04.010

PBS News Hour (2011) Timeline: Guatemala's Brutal civil war. PBS News Hour, 7 March 2011. https://www.pbs.org/newshour/health/latin_america-jan-june11-timeline_03-07

Price, S (2015) A legacy of shame: Canadian mining companies leave behind decades of violence in Guatemala. Intercontinental Cry Magazine, 30 December 2015. https://intercontinentalcry.org/a-legacy-of-shame-canadian-mining-companies-leave-behind-decades-of-violence-in-their-wake/

RCS Global (2017) The battery revolution: balancing progress with supply chain risks. RCS Global Industry. http://www.rcsglobal.com/the-battery-revolution-balancing-progress-with-supply-chain-risks/

Reuters (2018) Cobalt to be declared a strategic mineral in Congo. Reuters, 14 March 2018. https://www.reuters.com/article/us-congo-mining-cobalt/cobalt-to-be-declared-a-strategic-mineral-in-congo-idUSKCN1GQ2RX

Reuters (2019) U.S. dependence on China's rare earth: trade war vulnerability. Reuters, 27 June 2019

Rodriguez S (2019) Despite drought, Costa Rica's electricity stays clean - but not cheap. Reuters, 19 June 2019

Safehaven.com (2017) Cobalt prices to rocket as tech giants scramble for supplies. PR Newswire, 15 November 2017. https://www.prnewswire.com/news-releases/cobalt-prices-to-rocket-as-tech-giants-scramble-for-supplies-657690933.html

Schlanger Z (2017) Apple wants to try to "stop mining the Earth altogether" to make your iPhone. Quartz, 20 April 2017. https://qz.com/964862/apple-says-it-will-stop-using-rare-earth-minerals-to-make-iphones/

Shaw M (2017) Rare earths outlook 2018: diversifying supply and spotlight on NdPr. Investing News, 11 December 2017. https://investingnews.com/daily/resource-investing/critical-metals-investing/rare-earth-investing/rare-earth-outlook/

Sustainable Brands (2018) New approaches, tools for ethical sourcing of conflict minerals. New Metrics, Sustainable Brands, 15 March 2018

The Fund for Peace (2018) Fragile states index. The Fund for Peace. http://fundforpeace.org/fsi/data/

The Fund for Peace (2019) Fragile states index annual report 2019. The Fund for Peace, Washington, D.C.

Transparency International (2017) Corruption perceptions index. Transparency International. https://www.transparency.org/news/feature/corruption_perceptions_index_2017

Transparency International (2019) Corruption perceptions index 2018. Transparency International

U.S. Geological Survey (2018) Mineral commodity summaries. Reston, Virginia. https://doi.org/10.3133/70194932

UNFCCC (2018) The Paris agreement. United Nations Climate Change, 20 April 2018. https://unfccc.int/process-and-meetings/the-paris-agreement/the-paris-agreement

United Nations (2018) Sustainable development goals knowledge platform. United Nations, Division for Sustainable Development Goals. https://sustainabledevelopment.un.org/

United Nations Environment Programme (2009) From conflict to peacebuilding: the role of natural resources and the environment. UNEP, Geneva

US EPA (2017) Global greenhouse gas emissions data. United States Environmental Protection Agency, 13 April 2017. https://www.epa.gov/ghgemissions/global-greenhouse-gas-emissions-data

USA News Group (2017) Demand for high grade manganese growing with EV market. PR Newswire, 8 November 2017. http://www.prnewswire.co.uk/news-releases/demand-for-high-grade-manganese-growing-with-ev-market-656099213.html

Watts J (2017) Growing number of global insurance firms divesting from fossil fuels. The Guardian, 15 November 2017. https://www.theguardian.com/environment/2017/nov/15/growing-number-of-global-insurance-firms-divesting-from-fossil-fuels

Wilson T (2017) We'll all be relying on Congo to power our electric cars. Bloomberg, 26 October 2017. https://www.bloomberg.com/news/articles/2017-10-26/battery-boom-relies-on-one-african-nation-avoiding-chaos-of-past

WindEurope (2019) A new identify. Wind Europe. https://windeurope.org/about-us/new-identity/

Yan Z (2012) Plan to oversee Ganzhou rare earths industry. China Daily, 03 July 2012. http://europe.chinadaily.com.cn/business/2012-07-03/content_15545079.htm

The Impacts of the Energy Transition on Growth and Income Distribution

Giacomo Luciani

1 The Impacts of the Energy Transition on Economic Growth and Income Distribution

Two conflicting narratives are frequently heard in connection with the economic impact of the energy transition. The first has it that the transition is a great opportunity to revitalize economic growth and increase employment. The second, in contrast, estimates that objectives like reaching carbon neutrality by 2050, as pledged by the European Union, would be "too expensive." Which is right?

The question is supremely relevant for the political viability of the transition and the implementation of the Paris agreement. It is not by chance that environmentalists have consistently asserted that decarbonization is not only good for the environment, but also for the economy. This assertion promises benefits to all, including people living in regions that may not be greatly affected by global warming; and to contemporaries versus future generation, thus fundamentally improving the political appeal of the transition. If, instead, decarbonization entails an economic burden, and even accepting that this burden is likely much smaller than that which would derive from climate change, the question of the sharing of such burden, both internationally and intergenerationally, inevitably arises. As we know, there has, unfortunately, been substantial and persistent pushback from electorates in democratic countries towards decarbonization policies; and governments are not implementing measures capable of delivering the goals they are subscribing to. This must have something to do with perceived costs and benefits!

In the following pages, we attempt at disentangling the multiple contrasting effects that might be expected from the energy transition. It goes without saying that the net effect, resulting from the balance of such multiple contrasting effects, is extremely difficult or impossible to predict. It will surely very much depend on the specific

G. Luciani (✉)
SciencesPo PSIA, Paris, France
e-mail: giacomo.luciani@sciencespo.fr

© The Author(s) 2020
M. Hafner and S. Tagliapietra (eds.), *The Geopolitics of the Global Energy Transition*,
Lecture Notes in Energy 73, https://doi.org/10.1007/978-3-030-39066-2_13

characteristics of the economy facing energy transition, notably its current energy system, rate of growth of energy demand, available energy resources, and opportunities for decarbonization. All of these are extremely variable country by country. It will also greatly depend on the specific transition path pursued, and especially the intended speed of the transformation.

The question, it should be stressed, is not whether it is more appropriate to move speedily with the transition; it is not whether mitigation is preferable to adaptation. It is quite possible—indeed supported by majoritarian expert opinion—that adapting to climate change might be much more expensive that avoiding or limiting the same. Nevertheless, if mitigation has a net cost in terms of economic well-being, disposable income, and income growth, the question of how the burden should be distributed inevitably arises. Finding a consensus or compromise on burden sharing may be very difficult indeed, and the path to mitigation may therefore result politically too arduous.

The prominence of the issue of burden sharing is abundantly evidenced by the acrimonious debates in successive COPs, and the difficulty that the European Commission has experienced in pushing for the adoption of the goal of carbon neutrality by 2050. Some member countries of the EU have much "cleaner" electricity systems, thanks to greater reliance on hydro and nuclear. Other member countries significantly rely on coal and consequently generate much larger emissions per kilowatt/hour. The Commission has proposed the creation of a Just Transition Fund aimed at taking some of the burden from the shoulders of the worst emitters and transferring it to countries with cleaner electricity systems. But to what extent is it acceptable that countries that have invested in cleaner sources early on should be called to contribute to the cost of cleaning the systems of countries that have resisted doing so? (And, in some cases, even opted to abandon nuclear, thus worsening the problem?)

2   Three Definitions of GDP

The discussion of the economic impact of the energy transition that is proposed in this chapter is organized around the definition of gross domestic product (GDP). GDP can be looked at in three different perspectives: from the point of view of production generation, of production use, or finally of production (income) distribution.

- From the perspective of production generation, GDP is defined as the sum of all value added generated in an economy in a given time:

$$GDP = \Sigma \text{value added} + (\text{taxes} - \text{subsidies}) \text{ on products} \qquad (1)$$

  where value added is defined as the difference between all costs of production (excluding the remuneration of factors of production, capital, and labor) and the realized value of the final product (i.e., total sales revenue at producers' prices).

- Alternatively, GDP can be viewed from the point of view of the utilization of production, and in this case it is equal to consumption plus investment (gross capital formation or GCF), plus exports minus imports:

$$GDP = Consumption + Investment + Exports - Imports \qquad (2)$$

  Gross capital formation (investment) in turn is composed of substitution of obsolete production tools to maintain existing production capacity, plus addition of new tools to expand production capacity (the latter constituting net capital formation or NCF).
- Finally, GDP can be viewed from the point of view of the distribution of income, in which case it is equal to total wages plus total profits and interests, plus net government transfers (taxes minus income subsidies):

$$GDP = Wages + Profits + Taxes - Subsidies \qquad (3)$$

All of the above is very relevant for our discussion, because the energy transition has implications for value added; for the allocation of income to investment rather than consumption; to the distribution of income between wages and profits; for taxes and subsidies; and finally for foreign trade. All of these implications must be spelled out and considered to achieve a thorough understanding of the impact of the transition on GDP and its growth.

GDP is a frequently criticized indicator. Indeed, some of the weaknesses of GDP are well known: in particular, only products and services that are commercialized, i.e., paid for, are included (except for government services, which are included at cost). There are plenty of subjectively very valuable products and services that we enjoy, and play a very important role in determining the well-being of each of us, and at the same time are not necessarily paid for, hence not included in GDP. Also, lots of public services that are provided for free are included at cost, independently of the outcome. Hence, alternative measures have been proposed and sometimes used, such as the Human Development Index (UNDP 2019) or either Gross or Net National Income. Each alternative has merits and demerits, but in the end GDP remains the most widely used and the best measure of "what money can buy." It is quite possible that a decline in GDP will allow you to conduct a happier life than before, but your money will still buy less than it used to.

3   Pricing Emissions

It is a widely accepted starting point that the economic cause of global warming is a market failure due to the fact that the cost of emitting CO_2 and other greenhouse gases (GHGs) into the atmosphere is not borne by the emitter (Nordhaus 2013). No one has to pay for using the atmosphere, and rules for preventing corporations and individuals from emitting pollutants are mostly concerned with local or, at most, national atmospheric conditions. Until very recently, the emission of GHGs has not involved a cost for the emitter, thus creating a negative externality.

This interpretation assumes that in the absence of a cost for emissions, carbon-intensive technologies will be more attractive than clean alternatives. According to a point of view which is more and more frequently expressed, some clean alternatives—notably non-dispatchable renewables—are becoming cheaper and cheaper, and soon will be, or are already absolutely preferable to carbon-intensive technologies even in the absence of the imposition of a cost for emissions. Such statements frequently ignore systemic costs arising from growing penetration of non-dispatchable renewables beyond a certain threshold (variously estimated at 35–50%). But even ignoring the issue of systemic costs, if it were true that clean sources may become cheaper than fossil ones, the market would, we may say, be vindicated, and policies to promote clean technologies would not be needed, because the latter would prevail out of their own greater competitiveness. At most, the energy transition might be a matter of speeding up (at a cost) a process that is taking place anyhow. In the rest of the chapter, I assume that fossil sources remain mostly cheaper than clean ones: therefore, if no cost is charged for emitting GHGs, the market will not by itself bring about decarbonization and avoid climate change.

If the cause of excessive GHG emissions is a market failure due to negative externality, the remedy must be sought in correcting the functioning of the market by imposing the internalization of the cost of emitting GHGs. This is the well-established "polluter pays" principle, which translates into the need of imposing a cost on carbon emissions.

By definition, the emergence of a new cost associated with the production of goods reduces the value added which the economy generates. This is because, other things being equal, the new cost increases the total cost of production. As energy enters in the production of all goods, this means that all productive activities will be faced with an increase in production costs.

Furthermore, also the cost of utilization of a given good may increase (if emissions are linked to the utilization as well as production stage: e.g., producing an internal combustion engine car will cost more, and using it will also cost more). In this case, either the producer accepts a lower per unit sales price, which is unlikely, or demand for the product may decrease because of the increased cost to the consumer: in both cases the net result will be a reduction in total value added.

GDP is the sum total of all value added generated in an economy, hence introducing a cost for carbon must reduce it. This effect may be more or less important depending on the level of the newly imposed cost and the extent to which the economy depends on carbon-intensive production and/or consumption; but it will inevitably be there.

In fact, one may argue that the downsizing of GDP when the cost of carbon emissions is made explicit is the consequence of the failure of acknowledging this cost in earlier years, since the beginning of the industrial era. In this view, past estimates of GDP, that do not include externalities, are exaggerated, and the introduction of an explicit cost for carbon emissions is just a remedy to past miscalculation. Following this line of thinking, the World Bank has proposed a concept of adjusted national income, which estimates environmental depletion associated with value-added generation, and not included as production cost; and corrects national income accordingly (Lange 2018).

The matter is further complicated by the time lag between damage to the environment and the emergence of the economic cost of such damage. We suffer today from emissions released by past generations over longer than a century; and future generations will suffer because of our emissions. The economic damage that emitting a ton of CO_2 today entails will only be visible in the future, and depends on how much CO_2 has been emitted in the past. Therefore, in fact we cannot internalize the externality by imputing as cost the present value of the future economic damage caused by an additional unit of emissions, because we have no precise idea of what this cost might be. We are, rather, imposing a price on carbon emissions in order to solicit a market response and achieve a reduction or elimination of emissions. This price then represents the opportunity value to the potential emitter of emitting one additional unit (ton of CO_2): he will stop emitting only if the price is higher or equal to the benefit that he may derive from emitting one additional GHG unit.

The explicit addition of a previously hidden cost is the reason that most governments are reluctant to introduce carbon pricing, whether under the form of a carbon tax or of a price generated by an emission trading system. Governments frequently prefer to resort to regulation and administrative measures, whose cost is non transparent and not immediately predictable by those on whose shoulders it will fall. But this cost exists: it may manifest itself as a shift from a preferred technology to a less commercially attractive one, or accelerated obsolescence of the existing capital stock, but in all cases it will lead to a decline in value added, hence of GDP.

4   Carbon Prices Are a Tax

But how is a price imposed on carbon? It is out of acts of government introducing an emission trading system or a carbon tax (or a combination of the two). In one way or another, the imposition of a price for carbon emissions translates into revenue for the government, i.e., higher taxation. Although in theory the imposition of a price on carbon emissions is justified by the additional cost that these emissions are likely to impose on future generations, the proceeds do not accrue to some fund set aside for future generations, but to the government of today. This is important, because taxes on products appear with a positive sign in the GDP Eq. (1). If all that happens is that a price on carbon translates into higher tax revenue, the impact on GDP will be neutral: higher costs imposed on enterprises and/or consumers are compensated by higher revenue for the government. If this higher revenue is used to subsidize select productive activities (e.g., production of electricity out of renewable resources by way of feed-in tariffs, or purchase of solar PV panels through reimbursement of part of the cost) all that will happen is that the value added of producers paying the tax will be decreased to the benefit of producers receiving subsidies, and aggregate value added may not change.

This would be our conclusion if the purpose of carbon pricing were to raise revenue for the government. In fact, the purpose is to stimulate a change in the behavior of producers and consumers, i.e., induce them to change their production methods or consumption baskets, reduce emission, and ideally decarbonize completely. Thus, the real cost consists of the "distortionary" effect of the combination of taxes (either explicit or through an ETS) and subsidies. In most cases taxes on products are designed in such a way as to minimize distortionary effects on production: in our case the distortion is the desired outcome. The "distortion" may be for the better, but still entails deviating from maximization of producers' profits or consumers' satisfaction, hence a worsening of their condition.

5 Consumption or Investment?

Next, we need to discuss the impact of the transition on the allocation of income to consumption vs. investment, as described by GDP Eq. (2). From this point of view, the impact of the transition is a needed decline in consumption, and increase in investment.

All consumption of goods and services entails some demand for energy. Energy saving is unanimously identified as a key component of the necessary decarbonization process: we need to drive less, fly less, heat or air condition less, and so on. We may shift to more efficient machines (requiring additional investment) in order to maintain the same level of net service while reducing energy consumption (increasing energy efficiency), but very likely reduced energy consumption is part of the deal.

Improvements in energy efficiency are unanimously considered an essential component of the energy transition. The International Energy Agency estimates that 37% of the difference between the Stated Policies and the Sustainable Development Scenario must be contributed by improved energy efficiency (IEA 2019, p. 79). Whether this is genuinely increased efficiency (i.e., less energy use for unchanged level of service) or simply reduced energy consumption (i.e., acceptance of reduced level of service) is not clear. Some transition optimists believe that humankind will be able to achieve expanded level of service with reduced energy use through extraordinary improvement in efficiency, but this cannot at all be taken for granted.

This points to the need for more investment. There is no progress possible toward decarbonization that does not require some form of investment. True, the energy sector always stood out as relatively capital-intensive, meaning that investment would in any case be necessary to satisfy growing demand or improve efficiency, even if we were to continue with emitting GHGs into the atmosphere; however, the decarbonization agenda entails even higher investment.

There are two main effects at work. The first is accelerated obsolescence of the existing capital stock. Physical assets with decades or years of technical life left in them will become stranded. This affects Net Fixed Capital Formation, which, as said earlier, is Gross Fixed Capital Formation (i.e., investment viewed as the sum total of buildings and machinery added to the production process) minus the replacement of worn-out capital from the existing stock. Accelerated obsolescence means that more of the new fixed capital added in a year will simply compensate for the retirement of existing capital, instead of contributing to the enlargement of productive capacity.

The second effect is an expected increase in the capital/output ratio, i.e., increased capital intensity. As mentioned, the energy sector has been relatively capital-intensive even before the need to decarbonize became established, but in a decarbonizing world it will become even more so. Almost all renewable energy sources are characterized by high initial investment costs and low subsequent operational costs. The latter are mostly maintenance costs, not directly related to the volume of production, while marginal costs may be nil or irrelevant.

In some cases, additional investment for decarbonization may even negatively affect production capacity, rather than the opposite. Think for example of carbon capture associated to a fossil fuel-based power plant: the process of capturing CO_2 absorbs some of the electricity generated by the plant, so that by investing in carbon capture we are actually decreasing the net output of electricity from the plant. Of course, what is gained is the elimination or decrease of emissions, but the usefulness of the plant with respect to its main product, which is electricity, is decreased. Or consider the expected transformation of the car industry from internal combustion to electric engines: this requires huge investment on the part of the automobile manu-facturers for the introduction of new models, of distributors or municipalities for the installation of recharging stations, of final consumers for buying new vehicles—and the end result is a mobility service which is somewhat more limited (because of range limitations or recharging times) or at most equivalent to what they enjoyed previously.

The macroeconomic impact of additional investment requirements may be larger or smaller depending on whether the country in question experiences stagnant or increasing energy demand. Where energy demand is stable or declining (as in the EU), the required investment is more likely to be in substitution of existing production capacity. In contrast, in an environment of rapidly growing energy demand relative decarbonization may be achieved by focusing on clean solutions for incremental energy production capacity, and the increase in investment requirements relative to a scenario of business as usual might be much more limited. In fact, there may even be situations in which investment in renewable energy sources may be easier, because production units are smaller and economies of scale not as important as for traditional technologies. Greater modularity may facilitate spreading investment over time and reducing the financial burden.

The shift from consumption to investment may under certain conditions justify the assertion that decarbonization will enhance growth rather than the opposite. Whether this is the case depends on the initial condition of the economy. If the economy is distant from an equilibrium of full employment of available resources, and savings exceed investment for lack of opportunities or inefficient intermediation of the financial sector, the emergence of new investment opportunities, especially if strongly supported by clear and consistent government policies, may facilitate a shift toward greater utilization of resources. But how likely might such a scenario be?

It may be argued that the European economy has in fact been far from full employment equilibrium ever since the financial crisis of 2008. Expansionary monetary policies have failed to stimulate either consumption or investment. In our capitalist economies, investment is justified by the expectation of profit, which ultimately is supported by consumer demand. In recent years, consumer demand in Europe has remained subdued because of uncertainty, and savings have exceeded investment. In the US, the situation is altogether different.

The transition requires increased investment even though consumption may not increase, or even decrease. If resources can be channeled towards investment to implement the transition, in the context of departure from consumerism, it is indeed possible that the economy moves towards fuller employment of all resources, including labor, thus resulting in an acceleration of growth. But is this a realistic expectation?

The answer largely depends on institutions and policies. The latter in particular need to be predictable and strong enough to solicit the desired response from investors. The profitability of investment for the transition must be clearly established and consistently supported for investors to take the plunge.

But even if the required new investment projects are demonstrably profitable, the relevant investors might not be able to invest. A main example is investment for energy savings, which to a large extent depends on decisions to be taken by millions of final consumers: the investors in this case may face limitations in financing the investment, or be put off by the generally long recovery periods, or simply not be aware of all available opportunities. Even for corporate investors, their balance sheets may not be solid enough to underpin large financial efforts. Certainly, venture capital may be available, and shareholders are ready to pay high prices for the shares of companies that promise a bright future, in contrast with the equity of old energy companies, which commands low valuations notwithstanding the high dividends paid. But the impression is that so far not enough of available savings have been channeled toward investment for decarbonization. In fact, the opposite is likely true: the uncertainty surrounding many an economic activity in a decarbonizing future increases investors' risk and discourages long-term investment.

This is all the more true if we move from the national to the global level. Globally, many investment opportunities in cleaner energy sources are to be found in countries with dubious or precarious governance, presenting a risk profile, which few investors are willing to undertake. Global decarbonization ideally entails a massive shift of financial resources from the industrial to the emerging countries, because there demand for energy is growing faster, and the deployment of renewable energy sources would in many cases be easier.

But in fact, we see very little of this shift taking place: energy projects that reach final investment decision are more easily geared to the development of hydrocarbon resources than to investment in renewables. According to the IEA, in the period 2014–18 out of total global average annual energy investment of 2 trillion US dollars (at 2018 prices), 1 trillion, or 50%, went to fossil fuels without carbon capture and sequestration. Renewables for power generation and final uses attracted 435 billion dollars, and energy efficiency 238 billion (IEA 2019, p. 50). Implementation of the Agency's Sustainable Development scenario would require an increase of global average annual energy investment to upwards of 3.5 trillion dollar (IEA 2019, p. 93), with roughly halving of investment in fossil fuels and tripling of other investment functional to the transition. Institutional barriers to such massive expansion and redirection of investment should not be underestimated.

It is therefore not clear at all that the decarbonization drive may per se suffice to overcome the low-employment equilibrium which is found e.g. in the European Union or Japan. Considerably more muscular policies than those currently implemented would be needed to convince private actors to substantially increase investment. Given the limited fiscal space that most governments enjoy, increased investment cannot be supported primarily by government expenditure. The imposition of a high enough carbon tax, whose proceeds were channeled exclusively and rapidly into support for energy transition-related investment, may become a driver for private investment, but would entail depreciation of existing assets, and uncertainty negatively impacting on corporate and household propensity to invest.

The task of achieving zero or negative CO_2 emissions requires a massive shift from consumption to investment, i.e., a further increase in the rate of savings over GDP, rather than simply efficiently channeling existing savings towards investment. We normally expect economies that devote more of their GDP to investment and less to consumption to grow more rapidly, because investment adds to the capital stock and expands the production possibility frontier. But the energy transition entails investment that is predominantly in substitution of existing productive capacity, and might even decrease rather than expand existing capacity. Thus, the pro-growth effect of a shift of resources from consumption to investment might be undermined by the acceleration in the obsolescence of the existing capital stock and the increase in the capital/output ratio that the energy transition requires.

The speed of the transition plays an important role in determining whether growth will be supported or undermined. Fast transition requires a larger shift from consumption to investment, and faster obsolescence of the existing capital stock; it is therefore less likely to be conducive to faster GDP growth. In all advanced economies, growth is driven by consumption: to compress consumption and shift to a model of growth led by investment independently of an expected increase in aggregate demand is a huge political–institutional task. This is implicitly recognized by proponents of very ambitious policies arguing in favor of something close to wartime mobilization, but then the required societal acceptance is far from being guaranteed.

6 Exports and Imports

Equation (2) describing the destination of GDP also includes external trade: exports are a possible destination together with domestic consumption and investment; and imports are a possible alternative way to satisfy domestic consumption and investment and must therefore be subtracted from GDP.

Advocates of pioneering the energy transition frequently insist on the fact that early movers may acquire comparative advantage, which will support their exports in the future. Furthermore, all countries that are net importers of fossil fuels would see their imports relatively reduced. The energy transition is therefore depicted as potentially improving the trade position of a country, and in this way contributing to its economic prosperity.

This is a simplistic approach for more reasons than one. Firstly, it should be underlined that export are a destination of GDP alternative to consumption and investment, so increasing exports must be matched by decreasing consumption or domestic investment, something which may be difficult to achieve in a context in which consumption already must be compressed to make room for a significant increase in investment. Exports are useful as driver of economic growth when domestic aggregate demand is not sufficient to justify the existing pace of investment—as was the case in the early years of most economic booms, including China's. But the energy transition cannot be led by export demand—at most the cost of some machines might be reduced if economies of scale are available, and export demand on top of domestic demand facilitates the attainment of large enough production runs. Thus, we may say that the energy transition is facilitated in China (and in the rest of the world) thanks to the collapse in the cost of PV panels that Chinese producers have been able to achieve; but surely this has been due to massive domestic demand in conjunction to export opportunities. It is the low cost of production rather than being pioneers that facilitates exports.

As for decreased import demand for fossil fuels, this can indeed benefit growth (especially in energy import-dependent emerging countries), inasmuch as it frees resources which otherwise might need to be devoted to exports, and improves the solvability of the country. It may therefore facilitate attracting foreign investment, which is a crucial consideration for supporting the energy transition in emerging countries. However, it is not always the case that decarbonization will allow decreasing imports: in coal-producing countries such as China or India a needed shift from coal to natural gas may lead to increased rather than reduced demand for imports.

7 Income Distribution

So far, we have discussed the impact of decarbonization on the formation and use of GDP: we must now discuss income distribution, i.e., how decarbonization may affect the share of income accruing respectively to labor (wages) and capital (profit).

The distribution of income between labor and capital is determined by the capital/output ratio. As discussed earlier, the energy transition entails an increase in the capital/output ratio, because of a shift to more capital-intensive technologies and little net benefit of the required additional investment (i.e., more capital needed for the same output).

An increasing capital/output ratio automatically results in an increasing share of income accruing to capital, unless fully offset by falling returns on industrial investment or interest rates on borrowed capital. We do live in a world of historically low interest rates, but there is no evidence that corporations are ready to accept lower returns. In fact, the opposite is true, as the perception of risk has widely increased, and in the energy industry the perspective of decarbonization further increases risk. Thus, the increase in the capital/output ratio associated with the energy transition inevitably also determines a shift of income from labor to capital—i.e., a widening of inequality in income and wealth distribution.

It should be stressed that the energy transition in this case simply reinforces a trend that has been underway ever since the end of the Second World War, as argued by Piketty (2013). Thus, while we certainly cannot attribute exclusive responsibility for growing inequality to the energy transition, the fact that it adds to an unwelcome existing trend further hinders public acceptance.

This effect on income distribution must be compounded with the effect of the increasing cost of energy on different income groups. It is generally accepted that an increasing cost of energy has a regressive impact on income distribution because energy expenditure is a larger share of the budget of poorer households. In addition, households are expected to invest to minimize the added cost, e.g., in insulation of their homes or buying new electric vehicles, but the vast majority of households has no net savings and no borrowing power either. Thus, richer households can contain the added cost by engaging in investment, but poorer citizens simply must bear the brunt of the decarbonization agenda.

The above is true at the level of individual countries, but even more so at the global level. Although emerging countries may offer better opportunities for decarbonization—because it is easier to decarbonize where energy consumption is growing than where it is stagnant—and because of more favorable climate and environmental circumstances in some cases; nevertheless, the burden of a higher capital/output ratio will be felt universally. The poor in emerging countries are even less able to bear the burden of added energy costs—although at the extremely low level of consumption that they currently enjoy the difference may not be felt (if you rely on collecting wood for cooking and on a small generator for lighting). And the availability of investment finance is certainly critical for all emerging economies and chronically insufficient to meet all investment needs.

It is indeed difficult to see how global decarbonization may take place unless policies and institutions are put in place to facilitate the emerging countries' access to investment finance, which, however, also implies an added financial burden on the industrial countries and a further reason why income inequality in the latter may be expected to widen.

It is therefore not surprising that the energy transition agenda has been accompanied by demands for financial transfers from rich to poor countries; and proposals for the introduction of a carbon tax in industrial countries are accompanied by the suggestion that the proceedings should be entirely redistributed to citizens on an equal basis, so that poorer citizens may actually end up being better off.

There are, however, at least two major problems with coupling environmental and redistributive measures. The first is that the energy transition contributes to, but is not the only cause of growing income inequality. Why should the introduction of a citizens' income be funded in particular by the carbon tax? These two measures are logically separate and the only reason for coupling them is to facilitate the swallowing of the bitter pill—the carbon tax—with sugar coating—citizens' income. The second problem is that devoting the revenue from a carbon tax to redistribution, rather than in particular supporting investment functional to the transition, would reduce the effectiveness of the policy with respect to its environmental goal. It should be recalled that the carbon tax has the ambition of eventually disappearing when decarbonization will have succeeded—the sooner the better; it is therefore not an appropriate fiscal tool for addressing a problem that will remain long after decarbonization has succeeded.

8 Employment

It is normally asserted that the energy transition will generate millions of new jobs globally, and in this way benefit the countries engaging in it. The evidence proposed consists of an estimate of all jobs created directly or indirectly by transition-related projects.

It is certainly to be expected that the investment surge linked to the transition will generate jobs. At the same time, jobs will also be destroyed in some industries—this being a major deterrent from reducing reliance on domestically produced coal or discouraging the sale of IC vehicles. But we need to approach the issue at a macro rather than micro level, and ask whether a shift in the composition of aggregate demand from consumption to investment is likely to increase employment. The answer is not straightforward.

Consumption demand has been progressively shifting from goods to services. Some categories of services are indeed labor-intensive, and normally associated with the growing number of poor-quality low-paying jobs; at the same time, other services have witnessed a huge improvement in productivity thanks to the introduction of information technology, and the threat to employment arising from artificial intelligence is a major preoccupation. Jobs created by investment expenditure are on balance likely to require higher skills and be better paid than jobs in services, although generalizations are questionable in this case.

It should be underlined that, although employment creation is a constant preoccupation for politicians and governments, labor is a cost, which should in principle be minimized. There is constant tension between increasing productivity and full employment: the former should be maximized, preferably with no detriment for the latter, which is only possible if total production is growing in line with productivity. We have noted that the transition must be expected to lead to an increase in the capital/output ratio, which also means a decrease in the productivity of capital (output per unit of capital is the inverse of capital/output). Assuming that, other things being equal, employment will also grow for a given output is tantamount to saying that the productivity of labor (which is the ratio of output to employment, or output per worker) will also decrease. In other words, we are envisaging a decline in both the productivity of capital and of labor, i.e., a poorer world.

Of course, productivity of capital and labor may increase in other industries and compensate for the loss of productivity in energy, but it is important to recall that increased employment in the energy industry, especially in countries where production is stagnant, is not per se a positive outcome.

The jobs created argument in support of the energy transition may very well turn out to be correct, but it is not clear that, other things being equal, it should be viewed as a net benefit. It does inevitably hide a decrease of productivity, which negatively affects total GDP and GDP per capita.

9   Concluding Remarks

If, as some claim, the energy transition were a win–win development, why should it be so difficult to implement? We need an energy transition and the cost of failing to make rapid progress in this respect may well be much higher than the cost of the energy transition itself, but there is a net cost to the energy transition both in terms of total available GDP and of its division among rich and poor.

If we want to make progress with the energy transition, it is necessary to acknowledge its cost and seek agreements on the division of the burden. Agreements are needed at the international level, between rich and poor countries, as well as domestically (within each country or the EU) between rich and poor citizens. For much too long proponents of the environmental agenda have bypassed this need and relied on the slogan that the transition is not only better for the environment, it is also better for the economy: unfortunately this may be true in the long term, but it is mostly not true in the short and medium term.

Governments have been searching for paths of lesser resistance, measures with limited costs, or non-transparent costs whose burden on each individual could not easily be predicted. This approach has largely failed: the share of fossil fuels on total global energy consumption has not decreased at all since the UN Framework Convention on Climate Change entered into force in 1994.

Imposing a sufficiently high price on carbon emissions is the only approach that is widely based enough to solicit the needed global response, but no government is ready to bite the bullet. The reason is clear: it is the one approach that most clearly would bring to the fore the cost of the transition (and hence create the greatest incentive to engage in it earlier rather than later). Governments keep on preferring ad hoc measures that provide only partial benefits and stimulate the search for loopholes.

The bipartisan US initiative to impose a significant carbon tax and redistribute the income of the same (CLC 2019) may be slowly gaining traction also in Europe, where a European Citizens' Initiative was launched in 2019 (EC 2019). However, the resistance of vested interests is strong and unlikely to fade away.

As is evident from our argument, the cost of the transition depends on the speed of it. Slowing down the energy transition will relatively minimize the losses connected to premature obsolescence of existing capital stock, facilitate the transfer of resources from consumption to investment, and allow technological progress to come up with better, more efficient solutions. It may also give time to our political systems to adopt policies for the redistribution of income and wealth—independently of the energy transition—which in recent decades have been largely undone in most democratic societies. Today we face underlying conflicts because of inequality and lack of social mobility; these are not caused primarily by the energy transition, but the latter ends up being the lightning rod that precipitates open conflict, so that the transition becomes politically unviable.

The reversal of the globalization process may lessen the pressure of international competition and its drag to the bottom, allowing governments greater latitude to engage in voluntaristic policies, including tackling inequality and climate change. But it could also lead to a loss of credibility for multilateral institutions, and lack of interest for a global coordinated approach, which is indispensible for avoiding catastrophic global warming. Global growth is likely to slow down anyhow, and with it the growth in energy demand. Possibly, this may give us more time to pursue the transition at a slower, less conflictual pace.

References

CLC (2019) Economists' statement on carbon dividends. Wall Street J
EC (2019) Commission Decision of 3.7.2019 on the proposed citizens' initiative entitled 'A price for carbon to fight climate change'; Brussels C (2019) 4973 final
IEA (2019) World energy outlook 2019, Paris
Lange (2018) Glenn-Marie, Quentin Wodon and Kevin Carey eds. The changing Wealth of Nations 2018. World Bank Group, Washington DC
Nordhaus (2013) William The Climate Casino. Yale University Press New York, London
Piketty (2013) Thomas Le capital au XXI siècle, Seuil Paris
UNDP (2019) Human Development Report 2019, New York

The Global Energy Transition and the Global South

Andreas Goldthau, Laima Eicke, and Silvia Weko

1 Introduction

The global energy transition, that is the full decarbonization of the world energy system until 2050, is attracting growing attention in global policy debates. While the geopolitics of the renewable transformation have attracted much recent attention (Goldthau 2017; IRENA 2019; O'Sullivan et al. 2017; Scholten 2018), the scholarly community has been surprisingly silent on what the low-carbon shift means for the Global South.

To be sure, there exists a host of writings on the development implications of renewables. A key theme here is improving energy access by decentralized electricity supply, notably in rural Africa or developing Asia (Alstone et al. 2015; Dagnachew et al. 2017). Other works center on the opportunity for developing countries to leapfrog centralized energy systems and instead make off-grid solutions the backbone of an economic development model (Kuzemko et al. 2016; Levin and Thomas 2016). Important reference points in this debate are the Sustainable Development Goals (SDGs), notably SDGs 3 (health and well-being) and 7 (centering on affordable and modern energy), which may be helped by a surge in renewable energy, prompting scholars to point to the co-benefits of the energy transition (Edenhofer et al. 2014; Helgenberger et al. 2017). Some works also zoom in on individual states (Baker et al. 2014; Power et al. 2016), highlighting the importance of the domestic political economy underpinning the energy transition.

That said, little scholarly attention is paid to the energy transition and the Global South beyond the fate of producer economies and the developmental state. Developing countries will be central, as 70% of the future energy demand is expected to come from non-OECD countries in 2040 (IEA 2018b), thanks to rapidly growing populations and economies. Among these, many developing countries in sub-Saharan

A. Goldthau (✉) · L. Eicke · S. Weko
IASS, Potsdam, Germany
e-mail: Andreas.Goldthau@iass-potsdam.de

© The Author(s) 2020
M. Hafner and S. Tagliapietra (eds.), *The Geopolitics of the Global Energy Transition*,
Lecture Notes in Energy 73, https://doi.org/10.1007/978-3-030-39066-2_14

Africa, South East Asia or Latin America are confronted with the imperative to foster economic development but at the same time lack the domestic capacity to ensure this is done in a sustainable way. China, India, and Brazil, by contrast, have large enough markets to make up for some of the potential risks for companies to do business there, and have proven capable of 'catching up' in many areas, including the (low carbon) energy domain.[1] The focus of the present analysis therefore is on the 'non-BRICS Global South', that is on developing countries other than the large emerging economies.

This chapter seeks to fill an apparent scholarly gap and offer a 'Global South perspective' by shedding light on the specific circumstances pertaining to countries of the Global South. More to the point, it argues that countries in the Global South may face a specific set of challenges in their efforts to embrace a low-carbon future (Sect. 2). Empirically, the chapter zooms into the trias of technology, finance, and trade, and suggests that there exist structural barriers and uncertainties that require the attention of scholars and policymakers (Sect. 3). Theoretically, it offers three different conceptual lenses on the low-carbon transition and the Global South, drawing on realist International Political Economy (IPE), critical IPE, and dependency theory (Sect. 4). A final section concludes.

2 A Global South Perspective on the Energy Transition

2.1 Innovation, Investment, and Low-Carbon Modes of Production

Judging by their cost curves, renewable technologies have entered the stage of market maturity. The unit costs of solar PV fell by around 90% over the past decade (IRENA 2019), and similar dynamics have unfolded in onshore wind turbines. This is largely a function of scale effects and a surge in global investment in renewable energy capacity. Low-carbon technologies are now cost-effective sources of power and have in many regions resumed a top place in power investment (IEA 2019).

That said, there exist significant differences in the distribution of global capital allocation (see Fig. 1). Northern America, Europe, and China are the top destinations of global investment in renewable energy. Within the developing world, the bulk of investment essentially goes to three countries: China, India, and Brazil. Other developing countries received a mere 12% of the total investment volumes of USD 280 billion in 2017 (UNEP/BNEF 2018). It is likely that this trend will continue, as the underlying investment decisions reflect public policies favoring renewables and a strong market pull stemming from sizeable economies. This lopsided investment pattern in renewables is in line with energy investment more generally. As the IEA

---

[1] This is evidenced by investment flows, among other (Murphy et al. 2015). Especially China should not be considered as a developing country due to its unique set of circumstances (Watson and Byrne 2012).

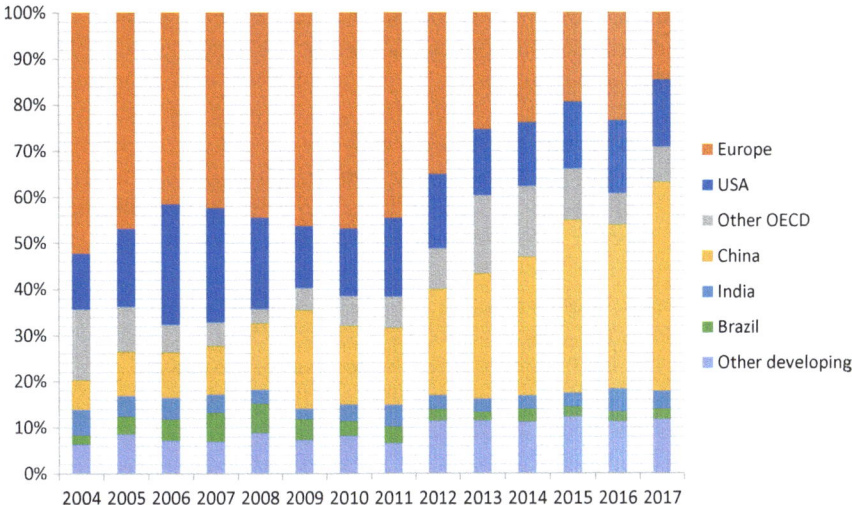

Fig. 1 Geographical distribution of investment in renewable energy. *Source* (UNEP/BNEF 2018)

estimates, middle-to low-income countries accounted for some 14% of overall investment in 2017, but represent 41% of the world's population. By contrast, high-income countries, representing less than 15% of the population, received more than 40% of investment volumes (IEA 2019).

To be sure, from a climate policy perspective it is highly desirable that investment in low-carbon energy sources is geared toward large (emerging) markets, as these will generate the bulk of GHG emissions increases in the near future. Yet, with a view to the SDGs, it is detrimental if (private) capital shies away from low-income economies, as it means perpetuating the energy access problem. Moreover, it is typically less developed countries (LDC) economies where most of the economic growth potential lies, given their generally young population (Gribble and Bremner 2012). Ensuring sustainable development in low-income countries would therefore require sufficient investment in low-carbon energy sources, setting them on a climate-friendly growth path. In fact, finance needs in mitigation technologies are found highest among developing nations (Tempest and Lazarus 2014).

A second challenge pertains to the fact that low-carbon technology remains concentrated in terms of ownership. Judging by the number of patents in the low-carbon technology domain, technology leadership essentially lies in the OECD world and China. Only few developing economies joined this rather exclusive club, and if at all, then only on a very selective basis. Brazil and Mexico, for instance, have become players in the field of biofuels, whereas Russia managed to catch up in the wind sector (IRENA 2019). The upshot is, however, that low-carbon technology innovation happens in the North, not the South.

In this context, observers have repeatedly pointed to the health sector for it exhibits structural similarities (Abbott 2009; Chon et al. 2018). Notably, intellectual property rights (IPRs) for drugs tend to be concentrated in OECD nations. IPRs are governed by stringent regimes protecting the right of the innovator to recoup significant upfront costs. In the health sector, fierce debates emerged around developing countries' access to essential medicines, and the question which good is more valuable: the private good of an IPR, or the public good of a population's health and wellbeing. Such arguments have been made also in the context of the UNFCCC, where representatives of the developing world raised the issue of IPRs as a potential impediment to their sustainable development and successful energy transition (Ockwell et al. 2010; Zhou 2019).

While it remains contested whether the pharma and the low-carbon tech sectors indeed are comparable, the more fundamental point here is a 'technology gap' (Castellacci 2011) facing many countries in the Global South, a cause of persisting underdevelopment and indeed poverty (Fofack 2008). Observers have therefore made the argument that efforts to foster access to low-carbon technologies are likely to generate a 'development dividend' in LDCs (Forsyth 2007). Note that this is about more than simply ensuring climate-friendly energy supply and energy access. It is about facilitating a low-carbon mode of production for any good or service a given country feeds into global value and supply chains. The rationale here is as simple as it is straightforward: as climate policies progress and become a determinant also in global trade, a low-carbon footprint ensures a country's products stay competitive in the global market. Short of access to low-carbon technologies, developing nations may therefore face the risk of being cut out of global trade relations through such mechanisms as carbon border adjustment, which has been lauded by economists (Akerlof et al. 2019) and planned at the EU level (Horn and Sapir 2019).

In this context, and as discussed in further detail below, countries rich in fossil fuels are believed to face a specific challenge. Predominantly found in the Global South, producer economies are typically locked in a resource-dependent economic model. Many qualify as rentier states and are trapped in a 'resource curse' that makes the extractive sector their dominant industry (Auty 1993; Ross 2012; Schwarz 2008). As research has shown, there exists great potential for solar and wind farming in the Gulf, Northern Africa or the Gulf of Guinea (IRENA 2018; Kruyt et al. 2009), world regions that at present are home to prime fossil fuel exporters. The low-carbon energy transition therefore presents these countries with the opportunity to embrace a new economic model based on industrial diversification. Yet leaving aside questions of political will (many resource-rich countries have shied away from diversifying their economies in the past), administrative capacity and technical expertise, it has been suggested that it will be hard for such states to leave the carbon-intensive development pathway that characterizes their socioeconomic system (Friedrichs and Inderwildi 2013). In other words, the development challenge goes beyond ensuring access to pertinent low-carbon technology and embedding them in national systems of value creation. It is about replacing a path-dependent, fossil-based rentier state with a fundamentally different, low carbon-based economic model.

2.2 The Limits of Adopting a 'Global South' Perspective

That said, there are clear limitations to what a 'Global South' lens can do for analyzing the global energy transition. It can only point out global trends and risks, and will not be able to account for the differences between Global South countries, which are clearly a heterogeneous group. Their differences in market size and investment risks, for instance, will mean that least developed nations clearly face more extensive risks and challenges than other developing countries, both in the realm of investment (Newell and Bulkeley 2017) and tech transfer (Ockwell and Byrne 2016). The degree to which Global South countries protect intellectual property rights may also impact their respective energy transitions, although this is a contested point and not one that can be addressed in detail here (see for example Abdel-Latif 2014; Ockwell et al. 2010; Pirrong 2014; Raiser and Bruhn 2017). Neither can a 'Global South perspective' address the national-level dynamics, which also shape energy transitions. Some challenges for developing countries in the energy transition are specific to their national political economies, such as governance and capacities (Jordana et al. 2006) and the interests of incumbent players (Baker et al. 2014).

3 Three Challenges for Countries in the Global South

Clean tech, finance, and trade are key elements for a successful transition to a low-carbon future. This section discusses these three elements with a view to identifying challenges and barriers for the case of the Global South.

3.1 Technology and Value Chains

A first challenge comes with the way value is captured in the global value chain (GVC) for low-carbon technology. Renewable energy magnifies the importance of technology for value creation due to the zero marginal cost problem: the resource such as the sun and wind is free, whereas the technology that converts energy and moves it to consumers earns money (see for example Overland 2019). Technology is made up of both hardware (equipment) and software, knowing to use and reproduce or improve upon hardware (Ockwell et al. 2010). Such technology is one of the most valuable parts of the value chain, which is a series of steps for a product or service to move from its conception to end use (Gereffi et al. 2005). In general, product development and design are more valuable than other activities such as resource extraction (Gereffi and Lee 2012). As global innovations systems scholars have elaborated, clean tech valuation differs between industries according to their standardization vs. customization, which is not static (Binz and Truffer 2017). Nevertheless, the value chains for low-carbon energy technologies may be categorized as producer-driven, which are 'capital, technology, or skill-intensive industries' with the potential for large, vertically integrated firms to control the production system (Gereffi 2014).

The value chains for low-carbon energy tech are quickly globalizing. Where solar PV was previously produced by vertically integrated firms within a country or region, its "production is now governed by multiple value chains made up of vertically specialized, and some integrated, firms, spanning across multiple countries" (Meckling and Hughes 2017: 227). This increased geographical scope presents the opportunity for technology diffusion and transfer,[2] which Bell (1990) conceptualizes as taking place in three flows: (A) the flow of capital goods and services; (B) skills and know-how to use and maintain technologies; and (C) skills and knowledge necessary to create technical change. While the first two flows may result in the diffusion of low-carbon energy technologies and green growth, only the third stream enables innovation. The prerequisite for this to take place is 'absorptive capacity' which involves high levels of human capital and is especially important for complex technologies (Bell's more recent work also focuses on low-carbon tech in particular, see Bell 2012).

However, not all countries will be able to benefit from globalizing value chains, and the Global South in particular risks being shut out. Many developing countries remain excluded from international technology flows (Glachant and Dechezleprêtre 2016). This is because international private developers see politically unstable countries as too risky of an investment; and in very poor countries or regions there is no market case for private industry to participate (Kirchherr and Urban 2018). The possibility for countries to escape infrastructural and technological carbon lock-in depends on the costs of moving away from high-carbon systems and options for alternatives (Seto et al. 2016). Therefore, in those countries which are currently not attractive for clean tech investments and do not participate in low-carbon tech value chains, existing technologies and infrastructure will resist changes. Some mechanisms exist to transfer low-carbon technology when private firms will not invest, led by international institutions (Abdel-Latif 2014; Ockwell and Byrne 2016; Rimmer 2019) and public–private partnerships (Chon et al. 2018). The extent to which such mechanisms succeed in transferring technology is unclear, but developing countries excluding the big three have remined below their potential (Edenhofer et al. 2014). This suggests that the carbon lock-in risk remains.

Countries which are attractive for investors in installing or producing low-carbon energy technologies may become a part of the global energy value chain and achieve co-benefits like better environmental quality and more jobs. However, in clean energy industries which are research- and technology-intensive, the most lucrative part of the value chain is tech development, which is centered in the US, EU, China, and Japan (Curran 2015; Nahm 2017). Most Global South countries lack the absorptive capacity to innovate on transferred hardware and therefore 'upgrade' within the value chain if they do not also receive 'software'. A key exception to this is China: its high

---

[2]We use the IPCC 2000 definition of technology transfer: "a broad set of processes covering the flows of know-how, experience and equipment for mitigating and adapting to climate change amongst different stakeholders" (Metz et al. 2000: 3), which is a commonly used definition.

local innovative capacity, combined with industrial policies such as local content requirements, helped its firms capture value and upgrade in the clean energy value chain (Gosens and Lu 2013; Schmitz and Lema 2015; Zhang and Gallagher 2016). Yet such industrial policies may not succeed in countries with small markets, poor regulatory design and coherence, or low industry/innovation potential, and has only seen moderate success even in large markets such as India (Johnson 2016). Whether or not the Chinese experience can be recreated in other Global South countries with the help of tech transfer initiatives remains unclear. Preliminary research on low-carbon tech transfer initiatives suggests that initiatives mainly focus on transferring hardware rather than software, which will help clean technology diffuse, but not encourage indigenous innovation capacities or widespread entry into value chains (see Fig. 2).

If Global South countries do not build up indigenous capacities, they may become dependent on the Global North for low-carbon technologies, which comes with two potential risks: increased trade litigation and political tensions, and monopolies. Work by Lewis (2014) has shown that substantial trade imbalances between renewable energy technology manufacturers and users resulted in increased international trade litigation and escalating trade tensions. While Lewis' work focuses on actors in the EU, the US, and China, this pattern may be exacerbated if Global South countries become technology users without upgrading within GVCs. North–South tensions may also be exacerbated in the international political arena, as a lack of access to low-carbon tech has already been a stumbling block for UN climate negotiations (Abdel-Latif 2014). Currently, many nationally determined contributions (NDCs) are conditional on climate finance and the transfer of low-carbon tech to meet their goals, and calls for low-carbon tech transfer have included discussions of compulsory

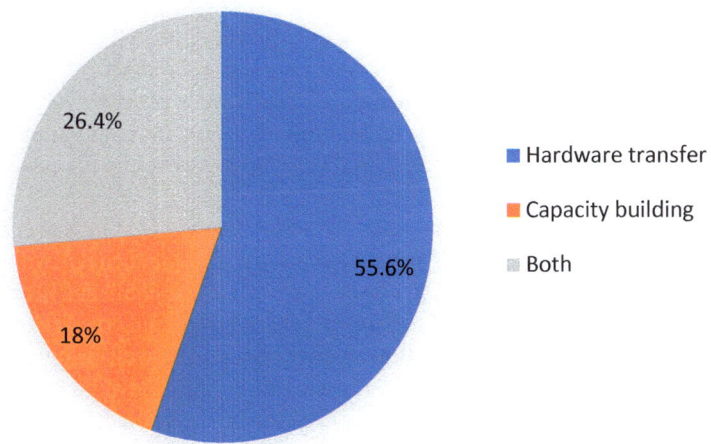

Fig. 2 Focus of global tech transfer initiatives. *Source* Own data

licensing which have been strictly opposed by the EU and the US (Pueyo et al. 2012). In addition, because low-carbon tech value chains are producer-driven, large, vertically integrated firms have the potential to monopolize production. Already authors have argued that the current system uses IPRs to prevent market entry and follow-on innovation (Baker et al. 2017; Raiser et al. 2017). Monopolies allow technology producers to distort markets, and this dynamic may be worsened if the Global South is increasingly dependent on low-carbon tech from a few powerful actors, rather than having developed indigenous technologies. This risk may be more pronounced for some technologies such as wind power, while solar PV may be less likely to develop captive value chain governance (Binz and Truffer 2017).

3.2   Financial Risk and Path Dependency

A second challenge pertaining to the global energy transition and the Global South are financial transition risks facing particularly resource-rich countries. Today, many developing countries of the Global South find themselves at the crossroads between trying to leapfrog carbon-intensive industrial production, and replicating the carbon lock-in that characterizes the developed countries' rise to industrial powerhouses. As many countries only start shaping their future national energy systems, today's decisions matter: the infrastructure built now will be determining these countries' energy mix for decades to come (Seto et al. 2016). Yet, carbon lock-in is not only contingent on domestic factors, but also external actors. A case in point is Chinese overseas investments in fossil fuels, which amounted to USD 128bn in comparison to only USD 32bn in renewables between 2000 and 2018 (Gallagher et al. 2018; Wright 2018). Such investments create path-dependent positive returns in fossil infrastructure which has the potential to delay the adoption of low-carbon technologies and the deployment of renewables despite their economic viability for decades (Unruh 2000; Unruh and Carrillo-Hermosilla 2006). This may seriously impede low-carbon future development pathways of the recipient countries of such investments. Leaving such a lock-in is possible, but comes with high transaction costs as it requires changing long-established infrastructure, rules, and (economic and political) institutions (Seto et al. 2016).

This point extends to another aspect, which is the significant exposure of carbon locked-in countries to several elements of financial risk. One such risk which has been widely discussed is that of stranded assets. In essence, the challenge here consists in the looming devaluation of fossil fuel assets held on public and private balance sheets. As the 2015 Paris Agreement to limit global warming to well below 2 °C implies two-thirds of known fossil fuel reserves need to remain unexploited (IEA 2018b), much of the oil, gas, or coal currently held in the books will not be monetized. The question therefore is not whether, but when these assets will be devalued and become 'stranded'. Estimates go up to $100 trillion losses in fossil fuel assets' value until 2050 (CitiGPS 2015). Modeling suggests that the resulting global 'carbon bubble' could lead to losses comparable to the 2008 financial crisis (Mercure et al. 2018). As the IEA estimates, oil and gas producer states bear the risk of some USD 7 trillion losses in income in a Paris scenario: Nigeria risks a decrease of about 500 billion in income, Saudi Arabia could lose almost USD 2 trillion and the United Arab Emirates faces the risk of some USD 900 billion in foregone revenue (IEA 2018a).

A related question pertains to the distribution of such financial risks: that is, who will be hit first and with which impacts. It has been argued that in the OECD, fossil fuels are mainly held and exploited by publicly traded companies, which typically have lower reserve-to-production ratios and more flexible business models as compared to their state-owned peers (NOCs) elsewhere. Moreover, the majority of crude and gas reserves are in state hands. This implies a higher degree of vulnerability and indeed risk for countries of the Global South and their NOCs (Krane 2017). Investment in fossil fuel based energy infrastructure in the face of rising energy demand in many developing countries further enhances this risk (Seto et al. 2016).

Short of decisive transition management, some countries might come under twin pressure: they may lose resource rents whose redistribution ensures social stability, but in some cases also domestic support for autocratic rule; at the same time, they may face deteriorating terms of trade as their exports might decrease in relation to imports and hence mounting pressure on their currencies. At worst, the financial risk of stranded assets may become a political risk of 'stranded nations' (Manley et al. 2017) with possible implications for the stability of the international system. The global energy transition therefore flips the logic of 'resource curse' (Karl 1997) on its head: it no longer is the abundance of economic rents, but rather them withering away faster than countries can adapt that creates problems for resource-rich states. Still, the implications might look similar to the ones observed today, especially poor economic performance (Sachs and Warner 1995), undemocratic tendencies (Ross 2001), and possibly even civil wars (Collier and Hoeffler 2004) (Fig. 3).

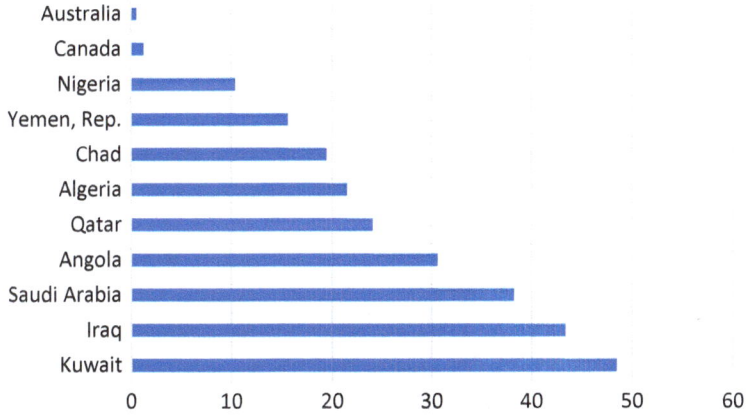

Fig. 3 Oil rents as percentage of GDP, average 2007–2017. *Source* World Bank, World Bank Open Data (https://data.worldbank.org/indicator/NY.GDP.PETR.RT.ZS), selected countries

3.3   Trade

Finally, the global energy transition is susceptible to challenging incumbent global trade patterns in energy and electricity. A key question here is whether global trade in energy will increase or decrease. Underpinning this question is the fact that the global energy transition in principle enables every country to produce renewable energy, yet at different efficiency and cost levels. One scenario therefore suggests countries will become more self-sufficient and global energy trade will falter (Scholten 2018b). This could give importing states the upper hand in trade and also reduce pressure on trade routes of fossil fuels such as the Strait of Hormuz, which have repeatedly been the site of conflict. A contrasting view holds that trade volumes in energy will stay the same or even increase (Schmidt et al. 2019). Technology innovation in renewable fuels and Power-to-X are seen as increasingly enabling global trade in renewable energy. Different production conditions, but also different acceptance levels create price differences among countries, driving international trade. Some fossil fuel infrastructure such as gas pipelines might see second use cases in, e.g., hydrogen trade. This could reduce the risk of stranded investment and therefore also hedge the possible financial losses of decarbonization (Schmidt et al. 2019). Depending on which scenario will unfold, it will impact on the relations between today's energy net-importing countries, transit countries, and net exporting countries. Countries of the Global South will face the challenge to reposition themselves in this new, emerging energy trade order.

What is more, the global energy transition is likely to particularly impact regional integration dynamics in electricity trade. Regional electricity trade is expected to increase as part of the energy transition, as money can be made from cross-border balancing of fluctuations in renewables supply (Bahar and Sauvage 2013; Criekemans 2018: 46). This results in more interconnected cross-border electricity grids. Observers suggested that this gives an edge to countries dominating and owning regional grids, the most efficient producers and balancing states. The control over regional grid infrastructures such as power lines, storage, or software will become vital for national security and for projecting influence and power (Criekemans 2018: 47). Some have also suggested that regional integration might happen around powers centers in such networked grids (Goldthau et al. 2019). For instance, China heavily invests in the strategic build-up of grid infrastructure as part of its great Belt and Road Initiative. The Desertec project, planned to connect Europe to large North African solar plants, has failed so far, by contrast. As existing evidence from the EU and other world regions suggests, political stability, and mutual trust are, in addition to administrative capacity, crucial preconditions for effective governance structures of cross-border spanning grids—a challenging task for some countries and regions of the Global South.

A last element and indeed significant uncertainty lies in carbon-related trade measures. In a recent elite study we conducted among international energy experts and energy investors, 87% confirmed that the global energy transition will raise the importance of carbon content in products. Asked for the key driving forces behind

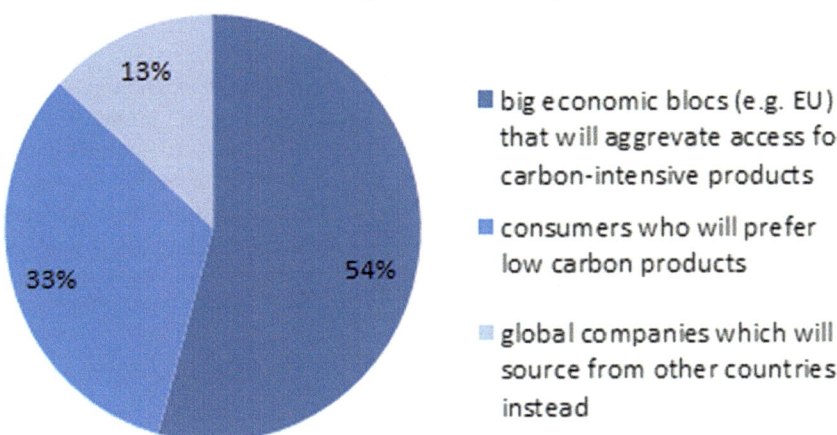

Fig. 4 Elite perceptions on drivers of decarbonization. *Source* Own data

this, 54% trade regulations by big economic blocs as much more important drivers than consumer preferences (33%) or companies (13%) (Fig. 4).

In fact, recent discussions in the EU have started to use trade regimes as a potentially powerful instrument to foster a global energy transition. A case in point are carbon border adjustments, which put a levy on carbon-intensive imports (Mehling et al. 2019). Such measures could level the playing field for domestic and foreign producers against the backdrop of more progressive climate policies in the EU, help to export the EU's decarbonization agenda, and strengthen the internationalization of own efficiency standards (Keohane et al. 2015). This discussion ties into a broader agenda of climate clubs, as championed by Nobel prize-winning William Nordhaus. Economic modeling and theory suggests that such a climate club, by effectuating small trade penalties on non-participants, is able to foster a large, stable coalition with high effects on carbon reductions (Nordhaus 2015). Once introduced, carbon border adjustments or even a coalition of climate clubs could have strong implications for trade relations with countries of the Global South, as not only direct trading partner countries but also large companies would most likely establish carbon content and efficiency requirements along their production cycle (see also Vandenbergh 2007). This goes back to the point made above, to the effect that a successful domestic energy transition would become prerequisite of a country's participation in global value and supply chains.

4   Three Conceptual Lenses on the Emerging Geopolitics of the Energy Transition

A 'Global South perspective' on the global energy transition needs to go beyond merely describing the possible structural barriers pertaining to technology, finance, and trade. So how can we make sense of the three core aspects outlined above in more conceptual terms? We use three different conceptual lenses to interpret current dynamics pertaining to the low-carbon transition and the Global South, drawing on realist IPE, critical IPE, and dependency theory. Each of these reflects a specific strand in current policy debates on the global energy transition, and offers a distinct take on the technology, finance, and trade in that context.

4.1   Realist IPE

The first lens broadly trades under the rubric of realist IPE. Essentially relying on a state-centered perspective on International Political Economy (Gilpin 1987), realist IPE firmly links economics and high politics by positing that state interests are central to the shape and function of international economic relations. States strive for economic security, a goal which is considered equal to military security. In its nationalist reincarnation, realist IPE views economics as a source of power and supremacy, giving rise to a mercantilist approach in global political economy (Hamilton 1791; List 1841). An important aspect here is the balance of trade, which economic nationalists tend to view as an indicator of a country's relative strength or weakness vis-a-vis competitors on the international stage. Because state interests and economics are linked, the latter can become a function of the former, or vice versa, thus giving rise to the patterns of 'trade following the flag' or 'the flag following trade' (Baru 2012).

In the context of the global energy transition and the Global South, the important point here is that technology, trade or finance are considered a means to an end, rather than sectors of their own. More to the point, from a realist IPE point of view, low-carbon solutions cannot be assumed to globally diffuse thanks to free market forces, sufficient demand pull, and falling unit costs. Rather, they may be available depending on whether it is in the strategic interest of the very states these technologies originate from, and also of the organized incumbents in the recipient countries. Given their status as tech laggards, countries in the Global South may therefore risk becoming politically dependent on the goodwill of the major green technology patent leaders—China, the US, the EU, Japan, or Korea. Likewise, trade as such cannot be considered to happen on a global-level playing field. Instead, the flows of goods and services may be a function of strategic trade policies, facilitated by preferential regimes or prevented by select tariff or non-tariff barriers. Similar assumptions can be made regarding financial flows, whose allocation may be informed by states' desire to maintain a surplus in their current account balance, rather than a deficit, thus perpetuating the lopsided global distribution of RES investment (see Fig. 1).

In short, such as lens suggests that all three aspects may become subject to 'an economics agenda for neo-realists' (Moran 1993).

From a realist IPE lens, the export of low-carbon products (like solar PVs) and services may help improving the trade balance and the current account. In turn, this means that keeping control over the technological know-how is imperative to ensure this effect to happen. As Meckling and Hughes (2017) have shown, for instance, states may take action to ensure certain parts of the renewable value chain remain on their territory. Or take renewable fuel trade, which may pick up thanks to advances in technology (Schmidt et al. 2019): from an economic nationalist perspective, it in principle does not make much difference whether it is fossil or green molecules that are imported, as both deteriorate the trade balance. It is therefore not inconceivable that states will try and curb imports of renewable fuels, not for reasons related to climate skepticism but because it is not considered in the national interest. This will result in global markets developing at a much smaller scale than technological advances would suggest or allow.

Moreover, as realist IPE comes with a mercantilist edge, there is no such thing as open and free trade. Energy technology transfer may be subject to strategic trade policies rather than business opportunity, and may even fall prey to geopolitical scheming (see also Goldthau et al. 2019). As the trend toward a retreat of the liberal order that started to unfold in the twenty-first century weakens multilateral frameworks such as the global trade regime, this may further exacerbate growing economic nationalism, also in the energy domain. For the Global South, this essentially implies that the Global South may fail to adequately profit from advances in low-carbon technology and be excluded from participating in renewable value chains.

4.2 Critical IPE

A second lens offered here is critical IPE, which promises insights into interactions between the state, markets, and society, zooming in on the underlying power structures. An obvious starting point here is Karl Polanyi (1957), whose term of a 'Great Transformation' has gained renewed attention in the context of the post-fossil economy (Fraser 2017; WBGU 2011). At the most fundamental level, the deep decarbonization implied by the Paris Agreement will arguably indeed bring about profound change. The global energy transition is therefore likely to reshape not only interstate relations or international trade along the lines discussed above, but equally so the socioeconomic structure. And yet, there has been surprisingly little analysis of the energy transition from a critical international political economy perspective until now (see Newell 2019 for an exception).

To be sure, Polanyi's work centered on the rise of market capitalism in nineteenth-century Great Britain. But some of the core concepts may offer useful starting points for the dynamics underpinning the global energy transition. His analysis of the rise of the market society, for instance, reminds us of the risk of 'disembedding' economic activity from the broader society, and of allowing this activity to develop its own distinctive laws. While the global energy transition arguably is about more than mere economic activity, elements of it clearly resonate with Polanyi's critique of subordinating nature and society to separate systemic logics. This extends to the notion of a 'double movement' between the competing dynamics of the expansion of market forces and societal protection. This concept could be put to work for analyzing dilemmas of the global energy transition: on the one hand, there exists the market-based globalization, creating social, and environmental externalities such as climate change. On the other hand, we witness the societal measures for protection, such as climate change mitigation and adaptation policies. Sometimes these policies aimed at ensuring human security and the habitat might in turn themselves foster the expansion of market dynamics, e.g., through the creation of carbon markets or REDD+ certificates, which in turn are mainly produced in countries of the Global South.

A rich tradition in Polyanian-inspired IR theory (see Dale 2016) may, further, inspire thinking about the implications for the international relations—or even the geopolitics—of the unfolding global energy transition. Critical development studies in particular offer important ways to conceptualize the 'marginalized others' (Inayatullah and Blaney 2016) in this context, that is countries of the Global South that may remain outside of what may emerge the dominant paradigm of a new 'climate capitalism' (Newell and Paterson 2010).

Antonio Gramsci's theory on hegemony (2011) may be another promising approach. A key focus of Gramsci's work rests on the historical-materialist factors underpinning large-scale transformations. In contrast to realist IPE, which by and large views states as black boxes, it is the intra-state power dynamics between contrasting forces that are central to the analysis, making the state a function of social relations. The supremacy of a ruling group is understood as hegemony, which is characterized by a mixture of coercion and consent, requiring the formation of selective compromises between rulers and ruled (Gramsci 2011). From a Gramscian lens, the global energy transition and the emerging climate capitalism may be well viewed as entailing an element of cultural (and indeed discursive) hegemony. Not only is it the Global North that arguably set the global policy agenda pertaining to the energy transition. It also defines the approach for addressing the climate challenge, which remains compatible with the incumbent capitalist paradigm. Such a perspective could also add to our understanding of the contemporary contestations surrounding the deep shifts facing the society's very material basis that is the energy system. Especially Gramsci's conceptualization of the state and its internal dynamics can guide analyses of the state's role as a transition manager in this process (Newell 2019).

A Gramscian approach may also add value to analyzing global dynamics. For instance, questions of access to technology and international finance may translate into vulnerabilities for countries in the Global South and therefore become subject to additional contestation. As Newell (2019) argues, the jury is out whether new actor constellations arising in a renewable energy world will lean towards 'trasformismo', referring to co-optation into a dominant coalition, in which potentially conflicting ideas become assimilated and incorporated (Cox 2016) rather than a great transformation of deep structural change in Polanyi's sense. It is also not inconceivable that it will give rise to a new counter-hegemony of emancipatory movements, rebelling against the globally unjust distribution of benefits, risks, and losses within the globalized production system of a global energy transition. Emerging debates around the imperative of a 'just transition' accounting for developing countries, marginalized groups, and economically vulnerable communities mirror this normative dimension underpinning the low-carbon shift.

4.3 Dependency Theory

A final lens on energy transitions and the Global South is dependency theory, which posits that a country's economic development potential is determined by its relative position within the economic structure of a world system. A key analytical category is the distinction between core and periphery within this system. Dependency theorists posit that given their peripheral status, some countries of the Global South may be structurally unable to catch up in terms of technological development. Whereas some transfer of technology occurs on the back of the influx of foreign capital, innovation processes remains concentrated in the center (Vernengo 2016). In the context of the global energy transition, the concentration of renewable technology patents can be seen as indicative for the perpetuation of the established OECD 'core' in the global economic system, complemented by the emergence of a select number of new core countries, notably China.[3] The mere deployment of solar panels across the world does not do break the principle logic of replicating dependency relations of old.

This point extends to trade. Dependency theory would suggest that the core exports mainly advanced products, whereas the periphery delivers raw products and natural resources in return. In the energy sector, many countries in Africa and Latin America have indeed long served as exporters of oil, gas, or other resources, and may in part also resume that position when it comes to rare earth materials that are considered crucial for renewable energy technologies and electric vehicles. The OECD and China, in turn, are set to dominate the export of renewable technologies as well as related products and services, thanks to their role as innovation leaders. The underlying pattern here reflects some of the theory's structural assumptions on the typical international division of labor between center and periphery countries, perpetuating dependency relations. Importantly, resource exports are generally subject to higher price volatility than manufactured goods (Becker 2008), as illustrated by the notorious oil price fluctuations. The Prebisch-Singer effect also suggests that the terms of trade tend to deteriorate for primary commodity exporters, as the relative demand

---

[3] In terms of technological development, lessons may be drawn from the recent Chinese experience. China heavily relied on state support and steering coupled with domestic content requirement for foreign investment, thus ensuring technology transfer. What in the 1960s would have qualified as state-assisted import substitution industrialization (ISI) has proven a model for the 21st century and gave rise to technology leadership in one of the most important sectors going forward.

for resources tends to shrink as the world income increases (Engel's law) (Prebisch 1949; Vernengo 2016). It is therefore doubtful whether the peripheral Global South stands to gain relative power over industrialized countries thanks to being rich in rare earth materials. Instead, the pattern suggests they will remain in a position of underdevelopment.

Finally, it has been argued that structural dependence in finance may constitute an even stronger limitation of development and growth than technology dependency (Becker 2008; Tavares 1985), an argument that extends to the global energy transition. Renewable energy investment prominently relies on foreign capital in many countries. This fact gains special importance in phases of economic growth, when investments rise. The availability of foreign currency limits domestic accumulation processes (Becker 2008). Capital inflow is highly dependent on economic trends within the center. In phases of prosperity, the willingness to invest in the 'riskier' periphery is low. During times of little investment opportunities and excess liquidity (as we are currently facing at zero interest rates in Europe), investment in peripheral states picks up (Becker 2008). In turn, foreign debt rises, which in the past was seen as strengthening the creditors' hand over the debtors (e.g., through global financial agencies such as the IMF and the World Bank). This suggests that renewable energy investment in the Global South will remain a function of the economic cycles in the center—the Global North.

5   Conclusion

The global energy transition is likely to generate important benefits for countries in the Global South. At the same time, it will also throw up new questions and give rise to novel challenges. As this chapter argued, these challenges are likely to center on the trias of finance, technology, and trade. In each of these three aspects, there exists the potential for outright 'energy transition risk' for countries that have yet to catch up economically. To be sure, individual sectors such as the financial industry have started to take these risks seriously. By definition, however, individual sectors adopt stovepipe approaches, which presents a call on the scholarly community to make sense of the bigger picture.

This call is yet to be heeded. While some important scholarly work has been done, for instance, on the prospects of oil producer economies, much more effort is needed to comprehensively grasp the implications of the global energy transition for the Global South. As the discussion in this chapter suggests, there exists a serious chance of some of the Global South either not gaining access to low-carbon energy technologies, or of becoming technology-dependent. While China, officially still considered a developing country, has been able to 'upgrade' its place in the global low-carbon tech value chain, most of the Global South only sees technology diffusion. This may be enough to deliver important co-benefits in the shape of health effects, the empowerment of women, or decentralized energy access for local communities. It will not be enough to make developing countries climb up the economic ladder and industrialize on a low-carbon model—the idea that underpins the promise of leapfrogging. What is more, the poorest and least stable countries which either do not manage to participate in global (low carbon) value chains or fall out of it thanks to various carbon regimes ringfencing consumer blocs, run the risk of not transitioning. Such an uneven transition brings its own set of geopolitical consequences (Eicke et al. 2019).

In terms of theory, these challenges offer great potential for further analytical advancement. This chapter has suggested three approaches to make sense of trends and possible risks, but there may be more. Admittedly, the present bias towards left-leaning or outright Marxist approaches—dependency theory and critical IPE—leaves open flanks for criticism. This, however, is intentional. Liberal theories, often the basis of cost-related energy technology diffusion arguments, rely on equally strong normative foundations and indeed assumptions that are hard to maintain in the very real-world setting presented by the global energy transition. To drive the point home again: while there is good reason to believe that solar PVs will help the poor across the planet thanks to falling unit costs, it would be false to ignore the underlying structural imbalances that have emerged over centuries of uneven global economic development. For the energy transition to become a fully global phenomenon, it is arguably these imbalances that need to be addressed.

References

Abbott KW (2009) Innovation and technology transfer to address climate change: lessons from the global debate on intellectual property and public health. In: Intellectual property and sustainable development series. International Centre for Trade and Sustainable Development (ICTSD), Geneva

Abdel-Latif A (2014) Intellectual property rights and the transfer of climate change technologies: issues, challenges, and way forward. Clim Policy 15(1):103–126

Akerlof G, Greenspan A, Maskin E, Sharpe W, Aumann R, Hansen LP, McFadden D, Shiller R, Baily M, Hart O, Merton R, Shultz G (2019) Economists' statement on carbon dividends. Wall Street J

Alstone P, Gershenson D, Kammen DM (2015) Decentralized energy systems for clean electricity access. Nat Clim Chang 5:305–314

Auty RM (1993) Sustaining development in mineral economies: the resource curse thesis. Routledge, London, UK, and New York, NY

Bahar H, Sauvage J (2013) Cross-border trade in electricity and the development of renewables-based electric power

Baker D, Jayadev A, Stiglitz J (2017) Innovation, intellectual property, and development: a better set of approaches for the 21st Century. AccessIBSA

Baker L, Newell P, Phillips J (2014) The political economy of energy transitions: the case of South Africa. New Polit Econ 19(6):791–818

Baru S (2012) Geo-economics and strategy. Survival 54(3):47–58

Becker J (2008) Der kapitalistische Staat in der Peripherie: polit-ökonomische Perspektiven. Journal für Entwicklungspolitik XXIV(2–2008):10–32

Bell M (1990) Continuing industrialisation, climate change and international technology transfer. University of Sussex SPRU, Brighton

Bell M (2012) International technology transfer, innovation capabilities, and sustainable development. In: Ockwell DG, Mallett A (eds) Low-carbon technology transfer: from rhetoric to reality. Routledge, pp 21–47

Binz C, Truffer B (2017) Global innovation systems—A conceptual framework for innovation dynamics in transnational contexts. Res Policy 46(7):1284–1298

Castellacci F (2011) Closing the technology gap? Rev Dev Econ 15(1):180–197

Chon M, Roffe P, Abdel-Latif A (2018) The cambridge handbook of public-private partnerships, intellectual property governance, and sustainable development. Cambridge University Press

CitiGPS (2015) Energy darwinism II. why a low carbon future doesn't have to cost the earth. CitiGroup, London

Collier P, Hoeffler A (2004) Greed and grievance in civil war. Oxf Econ Pap 56(4):563–595

Cox RW (2016) Gramsci, hegemony and international relations: an essay in method. Millenn: J Int Stud 12(2):162–175

Criekemans D (2018) Geopolitics of the renewable energy game and its potential impact upon global power relations. In: Scholten D (ed) The geopolitics of renewables. Springer, London, pp 37–74

Curran L (2015) The impact of trade policy on global production networks: the solar panel case. Rev Int Polit Econ 22(5):1025–1054

Dagnachew AG, Lucas PL, Hof AF, Gernaat DE, de Boer H-S, van Vuuren DP (2017) The role of decentralized systems in providing universal electricity access in Sub-Saharan Africa—A model-based approach. Energy 139:184–195

Dale G (2016) In search of Karl Polanyi's international relations theory. Rev Int Stud 42(3):401–424

Edenhofer O, Pichs-Madruga R, Sokona Y, Minx JC, Farahani E, Kadner S, Seyboth K, Adler A, Baum I, Brunner S, Eickemeier P, Kriemann B, Savolainen J, Schlömer S, von Stechow C, Zwickel T (2014) Climate Change 2014. Mitigation of Climate Change. Working Group III contribution to the fifth assessment report of the intergovernmental panel on climate change. Cambridge University Press, New York

Eicke L, Weko S, Goldthau A (2019) Countering the risk of an uneven low-carbon energy transition. IASS Policy Brief. Potsdam: Institute for Advanced Sustainability Studies

Fofack H (2008) Technology trap and poverty trap in Sub-Saharan Africa. World Bank, Washington, DC

Forsyth T (2007) Promoting the "Development Dividend" of climate technology transfer: can cross-sector partnerships help? World Dev 35(10):1684–1698

Fraser N (2017) Why Two Karls are better than one: integrating Polanyi and Marx in a critical theory of the current crisis. In: Working Paper der DFG-Kollegforscher_innengruppe Postwachstumsgesellschaften 1/2017

Friedrichs J, Inderwildi OR (2013) The carbon curse: are fuel rich countries doomed to high CO_2 intensities? Energy Policy 62:1356–1365

Gallagher KP, Zhongshu L, Mauzerall D (2018) Estimating Chinese Foreign investment in the electric power sector. In: GCI Working Paper 003(10/2018)

Gereffi G (2014) Global value chains in a post-Washington consensus world. Rev Int Polit Econ 1–29

Gereffi G, Humphrey J, Sturgeon T (2005) The governance of global value chains. Rev Int Polit Econ 12(1):78–104

Gereffi G, Lee J (2012) Why the world suddenly cares about global supply chains. J Supply Chain Manag 48(3):24–32

Gilpin R (1987) The political economy of international relations. Princeton University Press, Princeton

Glachant M, Dechezleprêtre A (2016) What role for climate negotiations on technology transfer? Clim Policy 17(8):962–981

Goldthau A, Bazilian M, Bradshaw M, Westphal K (2019) Model and manage the changing geopolitics of energy. Nature 569(7754):29–31

Goldthau A (2017) The G20 must govern the shift to low-carbon energy. Nature 546:203–205

Gosens J, Lu Y (2013) From lagging to leading? Technological innovation systems in emerging economies and the case of Chinese wind power. Energy Policy 60:234–250

Gramsci A (2011) Letters from prison. Columbia University Press, New York

Gribble JN, Bremner J (2012) Achieving a demographic dividend. Popul Bull 67(2):1–15

Hamilton A (1791) Report on the subject of manufactures. Philadelphia

Helgenberger S, Gürtler K, Borbonus S, Okunlola A, Jänicke M (2017) Mobilizing the co-benefits of climate change mitigation: building new alliances—Seizing opportunities—Raising climate ambitions in the new energy world of renewables. In: COBENEFITS IMPULSE (Policy Paper). Institute for Advanced Sustainability Studies (IASS), Potsdam

Horn H, Sapir A (2019) Border carbon tariffs: giving up on trade to save the climate?. Bruegel

IEA (2018a) Outlook for producer economies 2018. In: What do changing energy dynamics mean for major oil and gas exporters?. International Energy Agency (IEA), Paris

IEA (2018b) World energy outlook. International Energy Agency (IEA), Paris

IEA (2019) World energy investment 2019, Paris

Inayatullah N, Blaney DL (2016) Towards an ethnological IPE: Karl Polanyi's double critique of capitalism. Millenn: J Int Stud 28(2):311–340

IRENA (2018) Renewable power generation costs in 2017. Abu Dhabi

IRENA (2019) A new world. In: The geopolitics of the energy transformation. Masdar City

Johnson O (2016) Promoting green industrial development through local content requirements: India's National Solar Mission. Clim Policy 16(2):178–195

Jordana J, Levi-Faur D, Puig I (2006) The limits of europeanization: regulatory reforms in the spanish and portuguese telecommunications and electricity sectors. Governance 19(3):437–464

Karl TL (1997) The Paradox of plenty. In: Oil booms and petro states. University of California Press, Berkeley

Keohane N, Petsonk A, Hanafi A (2015) Toward a club of carbon markets. Clim Change 144(1):81–95

Kirchherr J, Urban F (2018) Technology transfer and cooperation for low carbon energy technology: analysing 30 years of scholarship and proposing a research agenda. Energy Policy 119:600–609

Krane J (2017) Climate change and fossil fuel: an examination of risks for the energy industry and producer states. MRS Energy & Sustain 4:E2

Kruyt B, van Vuuren DP, de Vries HJM, Groenenberg H (2009) Indicators for energy security. Energy Policy 37(6):2166–2181

Kuzemko C, Lockwood M, Mitchell C, Hoggett R (2016) Governing for sustainable energy system change: politics, context and contingency. Energy Res Soc Sci 12:95–105

Levin T, Thomas VM (2016) Can developing countries leapfrog the centralized electrification paradigm? Energy Sustain Dev 31:97–107

Lewis JI (2014) The rise of renewable energy protectionism: emerging trade conflicts and implications for low carbon development. Glob Environ Polit 14(4):10–35

List F (1841) Das nationale system der Politischen Ökonomie. Cotta'scher Verlag zu Stuttgart, Stuttgart

Manley D, Cust J, Cecchinato G (2017) Stranded nations? The climate policy implications for fossil fuel-rich developing countries. In: OxCarre Policy Paper 34

Meckling J, Hughes L (2017) Globalizing solar: global supply chains and trade preferences. Int Stud Quart 61(2):225–235

Mehling MA, van Asselt H, Das K, Droege S (2019) Beat protectionism and emissions at a stroke. Nature 559(321–324)

Mercure J-F, Pollitt H, Viñuales JE, Edwards NR, Holden P, Chewpreecha U, Salas P, Sognnaes I, Lam A, Knobloch F (2018) Macroeconomic impact of stranded fossil fuel assets. Nat Clim Change 8:588–593

Metz B, Davidson OR, Martens J-W, van Rooijen SNM, Van Wie McGrory L (2000) Methodological and technological issues in technology transfer. A Special Report of IPCC Working Group III. Cambridge University Press, Cambridge

Moran TH (1993) An economics agenda for neorealists. Int Secur 18(2):211–215

Murphy K, Kirkman GA, Seres S, Haites E (2015) Technology transfer in the CDM: an updated analysis. Clim Policy 15(1):127–145

Nahm J (2017) Renewable futures and industrial legacies: wind and solar sectors in China, Germany, and the United States. Bus Polit 19(1):68–106

Newell P (2019) Trasformismo or transformation? The global political economy of energy transitions. Rev Int Polit Econ 26(1):25–48

Newell P, Bulkeley H (2017) Landscape for change? International climate policy and energy transitions: evidence from sub-Saharan Africa. Clim Policy 17(5):650–663

Newell P, Paterson M (2010) Climate capitalism: global warming and the transformation of the global economy. Cambridge University Press, Cambridge

Nordhaus W (2015) Climate clubs: overcoming free-riding in international climate policy. Am Econ Rev 105(4):1339–1370

O'Sullivan M, Overland I, Sandalow D (2017) The geopolitics of renewable energy. In: Working paper. Belfer Center for Science and International Affairs, Cambridge, MA

Ockwell D, Byrne R (2016) Improving technology transfer through national systems of innovation: climate relevant innovation-system builders (CRIBs). Clim Policy 16(7):836–854

Ockwell DG, Hauma R, Mallett A, Watson J (2010) Intellectual property rights and low carbon technology transfer: conflicting discourses of diffusion and development. Glob Environ Change 20:729–738

Overland I (2019) The geopolitics of renewable energy: debunking four emerging myths. Energy Res Soc Sci 49(March):36–40

Pirrong C (2014) The economics of commodity trading firms. Trafigura, Houston

Polanyi K (1957) The great transformation: the political and economic origins of our time. Beacon Hill, Boston, MA

Power M, Newell P, Baker L, Bulkeley H, Kirshner Joshua, Smit A (2016) The political economy of energy transitions in Mozambique and South Africa: the role of the Rising Powers. Energy Res Soc Sci 17:10–19

Prebisch R (1949) El desarrollo económico de América Latina y algunos de sus principales problemas. CEPAL, Santiago de Chile

Pueyo A, Mendiluce M, Naranjo MS, Lumbreras J (2012) How to increase technology transfers to developing countries: a synthesis of the evidence. Clim Policy 12(3):320–340

Raiser K, Naims H, Bruhn T (2017) Corporatization of the climate? Innovation, intellectual property rights, and patents for climate change mitigation. Energy Res Soc Sci 27:1–8

Rimmer M (2019) Beyond the paris agreement: intellectual property, innovation policy, and climate justice. Laws 8(1)

Ross ML (2001) Does oil hinder democracy? World Polit 53(April):325–361

Ross ML (2012) The oil curse: how petroleum wealth shapes the development of nations. Princeton University Press, Princeton

Sachs JD, Warner AM (1995) Natural resource abundance and economic growth. In: Development discussion paper. Harvard Institute for International Development, Cambridge, 517a

Schmidt J, Gruber K, Klingler M, Klöckl C, Camargo LR, Regner P, Turkovska O, Wehrle S, Wetterlund E (2019) A new perspective on global renewable energy systems: why trade in energy carriers matters. Energy Environ Sc

Schmitz H, Lema R (2015) The global green economy. In: Fagerberg J, Laestadius S, Martin B (eds) The triple challenge for Europe: economic development, climate change, and governance. Oxford University Press, pp 119–141

Scholten D (ed) (2018a) The geopolitics of renewables. Routledge, London

Scholten D (2018b) The geopolitics of renewables—An introduction and expectations. In: Scholten D (ed) The geopolitics of renewables. Springer, London, pp 1–36

Schwarz R (2008) The political economy of state-formation in the Arab Middle East: Rentier states, economic reform, and democratization. Rev Int Polit Econ 15(4):599–621

Seto KC, Davis SJ, Mitchell RB, Stokes EC, Unruh G, Ürge-Vorsatz D (2016) Carbon lock-in: types, causes, and policy implications. Annu Rev Environ Resour 41:425–452

Tavares MC (1985) A retomada de hegemonia Americana. Revista de Economia Política 5(2):5–16
Tempest K, Lazarus M (2014) Estimating international mitigation finance needs: a top-down perspective. Stockholm Environment Institute, Stockholm
UNEP/BNEF (2018) Global trends in renewable energy investment 2018. Frankfurt
Unruh GC (2000) Understanding carbon lock-in. Energy Policy 28(12):817–830
Unruh GC, Carrillo-Hermosilla J (2006) Globalizing carbon lock-in. Energy Policy 34(10):1185–1197
Vandenbergh MP (2007) The new wal-mart effect: the role of private contracting in global governance. UCLA Law Rev 54:913
Vernengo M (2016) Technology, finance, and dependency: Latin American radical political economy in retrospect. Rev Radic Polit Econ 38(4):551–568
Watson J, Byrne R (2012) Low-carbon innovation in China: the role of international technology transfer. In: Ockwell D, Mallett AG (eds) Low-carbon technology transfer: From rhetoric to reality. Routledge
WBGU (2011) Welt im Wandel: Gesellschaftsvertrag für eine große Transformation. Hauptgutachten. Wissenschaftlicher Beirat der Bundesregierung für Globale Umweltveränderungen (WBGU), Berlin
Wright C (2018) Chinese overseas investments in fossil fuel 100x bigger than renewables since Paris. climatetracker.org
Zhang F, Gallagher KS (2016) Innovation and technology transfer through global value chains: evidence from China's PV industry. Energy Policy 94:191–203
Zhou C (2019) Can intellectual property rights within climate technology transfer work for the UNFCCC and the Paris Agreement? Int Environ Agreem 19:107–122

Governing the Global Energy Transformation

Maria Pastukhova and Kirsten Westphal

An effective and efficient governance is key for the global energy transformation. We argue that the process under the Paris Agreement, its 'rulebook' and the nationally determined contributions (NDCs) will have to be accompanied by focused and tailored governance mechanisms in the energy realm. The energy sector itself is key to limiting global warming to two degrees centigrade compared to the preindustrial level, because it is responsible for over two-thirds of global greenhouse gas emissions. Yet, neither the energy transition nor energy governance start from scratch. Energy governance is already happening on many levels: the local, the national, the regional and the global. These multi-level governance structures are necessary to enable, facilitate, and accelerate the energy transition(s) on the ground. They have to be adapted, however, to the changing and transforming energy world as we argue in the conclusions.

In a first step, we conceptualize the notion of 'energy transition' and relate it to the concept of 'energy transformation'. We argue that it is necessary to firstly move beyond the normative and target-driven idea(s) behind 'transition' and to secondly bring in the systemic aspects of energy transformation. Moreover, energy security, economic efficiency, sustainability and climate neutrality have emerged over time as the guiding paradigms, forming a strategic quadrangle, as opposed to a strategic triangle, traditionally used to define energy security. In a second step, we present an overview of the current international energy governance system where multilayered governance structures have developed over time. We argue that the existing architecture is stemming from the past and is neither fit for governing the energy transition, nor even reflecting the proccesses underway in todays' world. In a third step, we highlight that the energy *transformation* has and will have tremendous techno-economic, socio-technical and political (Cherp et al. 2018) effects that have both internal and external dimensions. Moreover, the transformation comes with (geo)political effects as it changes the political economy of energy on all levels: the global, the regional, the

M. Pastukhova · K. Westphal (✉)
Stiftung Wissenschaft und Politik (SWP), Ludwigkirchplatz 3-4, Berlin 10719, Germany
e-mail: kirsten.westphal@swp-berlin.org

© The Author(s) 2020   341
M. Hafner and S. Tagliapietra (eds.), *The Geopolitics of the Global Energy Transition*,
Lecture Notes in Energy 73, https://doi.org/10.1007/978-3-030-39066-2_15

national and the local. In the final step, we look at ways forward. We argue that it is necessary to preserve existing multilateral institutions and to strengthen them. Moreover, we assume that governance approaches towards and inside regions will have to be re-shaped or even created from scratch. We conclude that the crumbling of the global liberal order and the crises of multilateralism are complicating the approach to a better governance of the energy transition on the global level. Moreover, we witness the emergence of illiberal tendencies in the Western democracies as well. Climate and energy are playing into the polarization of societies as the two topics emerged as a major cleavage and a conflict line. We emphasize that a just and inclusive energy transition, both on national and international levels, is necessary to keep countries and the world on a sustainable energy transformation path. The challenge faced by the planet is indeed systemic.

1 Energy Transition—Lost in Conceptualization?

If governance of the energy transition is to be exercised effectively and efficiently, a common understanding of 'energy transition' seems to be helpful and necessary. Nowadays, 'energy transition' is a concept widely accepted and operationalized by national governments, regional and international bodies and non-governmental organizations alike. Although the term "Energiewende"[1] has been first introduced in the early 1980s by the German Ökoinstitut (Krause 1980), it hasn't found its way into the vocabulary of policymaking until the twenty-first century. Yet, when Germany published the "Energy Concept for environmentally sound, reliable and affordable energy supply" (BMWi and BMU 2010), which was readapted after the nuclear accident in Fukushima by the 2011 Energy Concept and the related package on the "*Energiewende*", its English pendant '*energy transition*' has become the international buzzword for a shift towards cleaner and more sustainable energy systems.

As omnipresent and relevant the term 'energy transition' might be nowadays, it is remarkably difficult to grasp, not least because of the lacking conceptual clarity and uniformity. The lack of both a comprehensive definition and a theoretical framework to support the concept of energy transition is not only lamentable from a scientific point of view: the resulting lack of common understanding among (inter)national actors also incapacitates the development of functioning international governance mechanisms to address this global issue.

The main reason for this lack of conceptual integrity lies in the broadly preferred focus on the "toolbox", that is, the single components or tasks of the energy transition. Energy transition is most commonly defined extensionally (see Fig. 1), e.g. through its components such as the increasing share of renewable energy sources in the total energy mix (IRENA, IEA), energy efficiency (EU, IPEEC, IRENA), phase-out of fossil fuels (IRENA, EU) and nuclear energy (German Fed. Gov.), electrification of

---

[1]German for "Energy Transition".

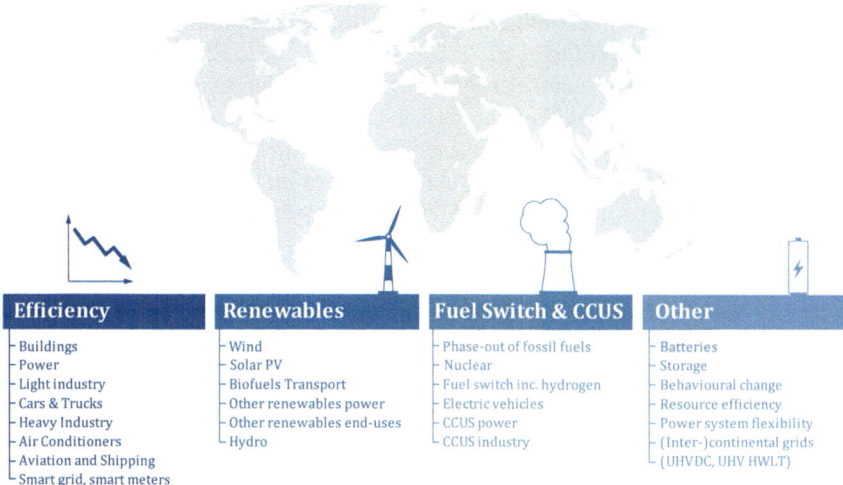

Efficiency	Renewables	Fuel Switch & CCUS	Other
⊢ Buildings	⊢ Wind	⊢ Phase-out of fossil fuels	⊢ Batteries
⊢ Power	⊢ Solar PV	⊢ Nuclear	⊢ Storage
⊢ Light industry	⊢ Biofuels Transport	⊢ Fuel switch inc. hydrogen	⊢ Behavioural change
⊢ Cars & Trucks	⊢ Other renewables power	⊢ Electric vehicles	⊢ Resource efficiency
⊢ Heavy Industry	⊢ Other renewables end-uses	⊢ CCUS power	⊢ Power system flexibility
⊢ Air Conditioners	⊢ Hydro	⊢ CCUS industry	⊢ (Inter-)continental grids
⊢ Aviation and Shipping			⊢ (UHVDC, UHV HWLT)
⊢ Smart grid, smart meters			

Fig. 1 Components of energy transition(s). *Source* IEA (2019a, b) World Energy Outlook, Authors'
analysis

the transport sector (IEA, EU), development of carbon capture and storage technologies (Norway, Saudi Arabia). An important observation at this point is, that the set
of these components differs among countries, regions and organizations according
to their respective agenda. In other words: the global community lacks a uniformly
agreed energy transition agenda.

The respective policy approaches are guided by a set of paradigms, the central
one often being *energy security*. Traditionally, in the EU and in the OECD countries,
energy security has been defined through the strategic energy triangle, consisting of
the three objectives of security of supply, sustainability and economic efficiency. Yet,
there has always been the issue of prioritization of these objectives, given that there
are not only synergies but also trade-offs between the policies addressing them. On
the other side, the World Energy Council highlights that countries face an energy
trilemma of addressing security of supply, ecological sustainability and energy justice
simultaneously. The different wording chosen by the World Energy Council illustrates the variety of notions associated with the paradigms across the globe. While
security aspects have been at the heart of energy governance since the emergence of
the first energy institutions, the economic aspects have been gradually added after-
wards, whereby their definition differs largely across the world. In the OECD countries, economic efficiency, competitiveness and affordability are prevailing notions.
In other parts of the world, energy equity and energy justice underpin the economic,
or better socio-economic angle of energy policies.

The concept of energy transition is pervasively used as a normative one, it is
also often tailored to fit certain policy objectives or to underpin specific measures
and steps. Therefore, international fora such as the G20 or the World Energy Council
make the case for *multiple* energy transitions, i.e. structural shifts in the energy

system of each country according to its respective goals and economic and resource potential (G20 2019; WEC 2014).

It is obvious that the various positions of countries in energy trading (influenced by their world market share/their position as a net importer/net exporter), in the globalized economy (trade surplus/deficit), with regard to their respective degrees of economic and social development (population growth/industrialization/urbanization) (Bradshaw 2010) as well as to the state of the energy system and the level of access to modern energy supplies determine the weighing of objectives and the prioritization of energy policy goals. With climate change mitigation, adaption and resilience added to the set of objectives, this diversity of priorities has proven to be a heavy burden and at times an obstacle for energy and climate governance beyond national levels (ibid.). Multilateral initiatives aiming to shape energy relations are in general hampered by widely scattered interests, which exacerbate the already considerable existing uncertainties. As a result, states have pursued very different pathways in energy governance: In the OECD area, it has been above all a matter of safeguarding prosperity; the post-socialist states have had to deal with the after-effects of the Soviet era, including the task of socio-economic transformation and a search for a new position in the global economy. The 'resource curse' and rent-seeking patterns have determined the energy dilemma of the energy-rich countries, while the question of sufficient access to energy has occupied the energy-poor countries (ibid.).

In addition to the traditional paradigms, sustainable development and growth have become key concerns. Since the second decade of the 2000s, the two key objectives of security of supply and climate protection have been accompanied by the goals of sustainable development and a fair and equal supply of energy worldwide, above all promoted by the United Nations. At that time, the United Nations also began to take an active stance on sustainable energy with its *Sustainable Energy For All* (SE4All) initiative. In the same vein, the Millennium Development Goals were translated into the Sustainable Development Goals (SDGs) in 2015. Goal 7 is to ensure access to affordable, reliable, sustainable and modern energy for all by 2030. Sustainable development is very much connected to the issues of energy justice and energy poverty, but also to environmental protection, and more specifically, protection of water, soil and air. In the same year of 2015, the Paris Agreement on Climate Change was signed. According to the Paris Agreement, countries' ambitions on NDCs have to progress in 5-year cycles. According to Art 2.1 of the Agreement, the NDCs should be formulated in line with the goal to keep global warming well below the 2 degrees Celsius compared to pre-industrial levels and pursue efforts to limit temperature increase to 1.5°. The Paris Agreement, the consecutive Conferences of Parties (COPs) and reports by the United Nations Framework Convention on Climate Change (UN FCCC) have added a sense of urgency to the issue of climate protection, but at the same time possibly aggravated the dilemmas in addressing all four objectives even further. Local air, soil and water pollution as part of sustainable development are in many countries a major driver and mitigating climate change comes as a *transformation dividend* (Goldthau et al. 2018), rather than as a policy goal on its own. Although sustainable development has become a major underlying theme, e.g. also in the International Energy Agency and its World Energy Outlook(s), and climate is often subsumed under 'sustainability', we argue that both, climate and sustainability constitute paradigms in their own right.

Therefore, we suggest that a *strategic energy quadrangle* rather than a triangle is informing energy policies across the globe. Energy security, economic efficiency, sustainable development and climate change mitigation, adaptation and resilience form four major angles or baskets, to which countries associate very different notions. At the same time, however, these four angles significantly overlap and create and numerous synergies to exploit and lift.

To summarize, when it comes to energy transition governance, countries differ in terms of their starting points, their path dependencies and their future pathways as well as their ambitions. In view of this diversity, the Paris Agreement, its rulebook as well as the bottom-up mechanism of nationally determined contributions (NDCs) are important and necessary, but not sufficient preconditions to steer energy transition towards climate neutrality nor appropriate to govern energy transition(s) to meet the other objectives.

Indeed, if we are talking not about one, but multiple energy transitions, defining them through their respective components makes a lot of sense, since such a definition can be easily operationalized by, say, national policy makers. However, in order to enable global governance and international cooperation mechanisms on this issue, there must be an understanding of energy transition every stakeholder can identify itself with. Although different stakeholders propose different measures and elements, there is indeed one common element such a definition can be based on: the characteristics of the future energy system they deem necessary and aim for are the same. All major stakeholders, some explicitly (G20 2019; BMWi 2015, p. 3; G20 2019, p. 1; MOFA Japan 2018), and some implicitly (IEA 2019a, b; EC 2015; IRENA 2018; national governments, e.g. the PRC's government[2]) define sustainability, environmental safety, economic efficiency and security of supply to be the central goals and the end-state to which the process of energy transition should lead. A future energy system with these characteristics is indeed universally aspired—the Sustainable Development Goal 7 on Energy, that is, *access to affordable, reliable, sustainable and modern energy for all* (UN 2019), has been adopted by governments of 193 countries.

Moreover, instead of energy *transition,* talking about energy *transformation* reflects the necessary systemic nature shifts in the energy system. Sometimes both concepts are used interchangeably: in IRENA's report "Global energy transformation—roadmap to 2050" energy transformation is a means to achieve energy transition, which is conceptualized as the end-state itself: "The challenge that policy makers around the world face is how to accelerate the transition. Fully delivering the energy transition will require a transformation in how we view and manage the energy system. Transitioning in a few decades from a global fossil-fuel powered energy system, built-up over several hundred years, to one that is sustainable, will

---

[2]As stated, for instance, in Xi Jinping's speech at the 19th National Congress of the Communist Party of China (Xi 2017).

require a much greater transformation than current and planned policies (the Reference Case) envisage" (IRENA 2018: 68). As in several recent academic studies, in one if its newest reports "Geopolitics of energy transformation", IRENA uses the term "energy transformation" intentionally instead of energy transition, to point out the broader implications a transition to low-carbon energy sources brings with it (IRENA 2019).

Against the above said, we suggest having an *intensional*[3] *definition* of energy transition that is formulated as follows: *a policy-driven process which involves systemic shifts towards (a) sustainable and climate-friendly, economically efficient and secure energy system(s)*. The measures and building blocks of such a transition will differ from country to country. Yet, there should be a governance system behind these national efforts, to pave the way, facilitate, enhance and accelerate the energy transition(s).

2 The Status Quo of Energy Governance and the Institutional Landscape

The existing energy governance landscape began emerging in the second half of the last century and has developed over time. It is sketchy and fragmented. Within this landscape, there are very few multilateral institutions that tackle energy issues in a comprehensive way (see Fig. 2). This is the result of (1) the different positions and roles of countries in the international energy system and (2) the diverging national priorities in energy policies regarding the strategic quadrangle of energy.

The traditional organizations such as the Organization of Oil Exporting Countries (OPEC), the Gas Exporting Countries Forum (GECF) or the International Atomic Energy Agency (IAEA) focus on specific energy sources, respectively oil, natural gas and nuclear energy.

Whereas OPEC and GECF are providing platforms for dialogue and cooperation among producing countries, the International Energy Agency (IEA) was formed by the OECD countries as an organization of energy consumers and primarily in response to the first oil crisis of 1973–1974. The IEA has been dealing with different energy sources ever since, albeit it has always had a pillar on oil crisis management and prevention. The IEA has been adjusting its role constantly to the new energy and climate realities. However, its membership structure, restricted to the OECD countries, came under increasing criticism as non-OECD countries like China and India have become powerful energy market players. In face of the changing dynamics, an association process has been currently under way with major non-OECD energy

[3]An intensional definition provides the meaning of an expression by specifying necessary and sufficient conditions for correct application of the expression. An intensional definition should be distinguished from an extensional definition, which merely provides a list of those instances in which the expression being defined is applicable. Cook, Roy T. "Intensional Definition". In *A Dictionary of Philosophical Logic*. Edinburgh: Edinburgh University Press, 2009. 155.

Scope	Mandate	Forum	Year of inception
Multilateral	Comprehensive	UN Energy	2004
		UN SE4All	2012
		Sustainable Development Goals	2015
		Energy Charter Track (European Energy Charter, Energy Charter Treaty, International Energy Charter/International Energy Charter Conference)	1991, 1994, 2015
	Specific	International Atomic Energy Agency (IAEA)	1957
		International Energy Forum (IEF)	1991
		Carbon Sequestration Leadership Forum (CSLF)	2003
		REN-21	2004
		Global Bioenergy Partnership (GBEP)	2005
		International Renewable Energy Agency (IRENA)	2009
Plurilateral (restricted membership)	Comprehensive	Group of Eight/ Seven (G8/7)	1975
		G20	2009
	Specific	Organization of the Petroleum Exporting Countries (OPEC)	1960
		International Energy Agency (IEA)	1974
		Gas Exporting Countries Forum (GECF)	2001
		International Partnership for Energy Efficiency Cooperation (IPEEC)	2009
		Clean Energy Ministerial	2009
		International Solar Alliance (ISA)	2015
Corporative		Global Energy Interconnection for Green and Low-Carbon Development (GEIDCO)	2016

Fig. 2 Institutional landscape: selected Governance Fora. *Source* Westphal 2015 based on Lesage et al. 2010, updated

powers. In 2020, the IEA comprises 30 member states, 8 association countries and 2 countries in accession. Though being a display of IEA's adaptability, the association process is certainly an attempt to maintain the existing order.

The creation of the International Renewable Energy Agency (IRENA) in 2009 meant a significant advance, both in renewable global governance as well as with regard to multilateralism (see also Roehrkasten 2015). IRENA got a clearly defined mandate to "be the global voice and knowledge base for the use of renewable energy, to serve as a forum for international technological cooperation, and to advise the member states on these matters". (Roehrkasten and Westphal 2013; Roehrkasten 2015). The specific focus has been on renewables. IRENA has also been looking into the geopolitical implications of an energy transformation (IRENA 2019).

As Fig. 2 shows, there are some organizations and fora that deal with specific energy sources or encompass a particular group of countries. This overview contains institutions on the global level, whereas regional organizations that specifically focus on energy (such as the Latin American Energy Organization (OLADE)) or have energy in their portfolio (United Nations Economic Commission for Europe (UNECE), European Union (EU), United Nations Economic and Social Commissions for Western Asia, for Asia and the Pacific and for Latin America and the Caribbean (UNESCWA, UNESCAP, UNECLAC), etc.) are not included here. Not included is also the World Trade Organization, which has played an important role in setting the rules for trade generally, but not in the energy sector (with an exception of energy services). The European Energy Charter in 1991 and the Treaty in 1994 were an attempt to translate similar rules into the energy trade, transport and investment. Yet, from today's perspective, it can be said that it is very doubtful whether the Energy Charter Process can be revived and modernized in a way to provide a 'rule book' or a 'code of conduct' for international energy trade, transport and investments, despite the 2015 signed International Energy Charter.

As Fig. 3 shows, very few existing institutions equally address energy policy objectives in an institutionalized manner.

At the end of the 2000s, there was a strong impetus to strengthen the coordination among the existing governance mechanisms and organizations. The initial idea was to better integrate the new powers such as China and India, and to have an outreach to the regions. The outreach and association process of IEA as well the International Platform of Energy Efficiency Cooperation (IPEEC) under the umbrella of IEA resulted from initiatives of the Group of Eight (G8). It was the G8 that reacted to the fact that energy governance did neither reflect the energy landscape any longer nor the changes in global politics in general. In an increasingly multipolar world, energy governance (Lesage et al. 2010) became a matter of steering committees and clubs, first and foremost of the G7/8 and G20. The G7 transformed back into an exclusive OECD-club with the crisis in and over Ukraine, when Russia was excluded from the process in 2014. The Group of 7 carried on with the agenda of tackling climate change and energy security (with the primary focus on natural gas).

In 2009, the G20 emerged as the new 'club' to primarily address the financial crises. The G20 began to work on energy matters under the US presidency in 2009, when G20 members declared their intent to phase out harmful and inefficient fossil

Institution	Energy Security	Sustainability	Climate Neutrality	Competitiveness/Affordability
UN Energy	Energy Access (to affordable, reliable, sustainable and modern energy for all)	Transition to Sustainable Energy	Mitigating Impacts from Climate Change through SDG7	—
UN Sustainable Energy for All (UN SE4All)	Energy Access (Raising share of RE)	Sustainable Development	Empowering action towards the achievement of Paris Agreement's goals	Equality
International Atomic Energy Agency (IAEA)	Nuclear Safety	Nuclear Safety	Promotion of nuclear as low-carbon source	—
International Energy Forum (IEF)	Producer and consumer dialogue Data collection	—	—	Market Transparency and Data Collection
International Energy Agency (IEA)	Coordination between members and associated countries strategic stocks; Market observation	Sustainable Development Scenario	Energy Outlook with reference to climate targets	Outlook and Market Analysis
International Renewable Energy Agency (IRENA)	Promotion of RE as a contribution to a diversified energy mix	Promotion of RE as win-win	Promotion of RE as win-win	Vague
Group of Eight/Seven (G8/7)	Ad hoc and delegation	Ad hoc	Ad hoc	Ad hoc
G20	Ad hoc	Ad hoc	Ad hoc	Phasing out inefficient fossil fuels
Organization of the Petroleum Exporting Countries (OPEC)	Security of demand Price stability	—	—	—
Gas Exporting Countries Forum (GECF)	Security of demand	Promotion of gas as a comparable "clean" fossil fuel	Promotion of gas as a comparable "clean" fossil fuel	—
International Partnership for Energy Efficiency Cooperation (IPEEC)	EE enhancing energy security	Positive effects	Positive effects	Positive effect on energy bill
Energy Charter Conference	Promotion of rules for trade, transit, investment	Certain standards	—	Promotion of rules for trade, transit, investment

Fig. 3 Global energy institutions, their mandate and activities. *Source* Based on Westphal 2015, updated

fuel subsidies (Van de Graaf and Westphal 2011). This new focus was also intended as an answer to the financial and economic crises as the member countries committed themselves to a resilient, sustainable and green recovery.

The G20 is perfectly positioned to steer global energy transition. Along with the G7 countries Canada, France, Germany, Italy, the UK and the US, the G20 includes the European Union (EU), Argentina, Australia, Brazil, China, India, Indonesia, Mexico, Russia, Saudi Arabia, South Africa, South Korea, as well as Turkey. It comprises countries that are of utmost importance for a successful energy transition and includes major energy producers, consumers and key players in existing international institutions. Also, in terms of climate policies, the G20 countries would make a huge difference, if acted together, as they account for 81% of global emissions (in comparison, G7 accounts for 25%). Last but not least, the G20 includes all permanent members of the UN Security Council, and major financiers of principal international organizations.

The G20 has constantly stepped up its voluntary cooperation in energy-related areas such as subsidies, market transparency and price volatility, international energy collaboration, energy efficiency, energy access and renewable energies. The G20 summits provide countries with an opportunity to meet on an equal footing and to exchange national views and standpoints on energy topics, a major step forward being made in 2015 with the first G20 Energy Ministers Meeting that took place under the Turkish Presidency. Yet since then, every new presidency set its own priorities, which hampered continuity as an important precondition for efficient and effective energy governance. Energy ministers have met each year since 2015, except for 2017 under the German Presidency. Overall, though the G20 unites a representative group of industrialized countries and new powers that can have an impact in their respective regions, the members have very distinct and diverse policies and perspectives. This limits the role of the G20 when it comes to global energy governance and is also the reason for the group's focus on less controversial issues, such as energy efficiency. As a result, the G20 has only partly lived up to its potential as a steering committee (Van de Graaf and Westphal 2011). In its current form, the G20 builds on the principle of voluntariness and on "soft" modes of steering, such as agenda setting, coordination among G20 members, information exchange and the steering of international organizations (ibid.). At the same time, the G20 has moved international energy governance up on its policy agenda, has rhetorically connected energy and climate policies and has enlarged its focus to sustainable development.

Since 2009, the G20 has continued to exchange on and monitor the progress towards phasing-out of fossil fuel subsidies: in 2010 the IEA, OPEC, OECD and the World Bank published reports tracking fossil fuels subsidies (IEA/OPEC/OECD/WB 2010, 2011). In 2013, the G20 endorsed a methodology for voluntary peer reviews "on inefficient fossil fuel subsidies that encourage wasteful consumption" (G20 2013, paragraph 94). Since 2013, the G20 has been addressing energy issues more comprehensively. At the 2014 G20 Summit in Brisbane, Australia, the G20 endorsed the G20 Principles on Energy Collaboration. The Chinese Presidency in 2016 continued this initiative to make energy institutions more inclusive and effective under the title "Global Energy Architecture".

The G20 affirmed its support for the SDG target Number 7 and pledged to increase the share of renewable energy substantially by 2030. At the core of the G20 action on renewable energy is the toolkit of voluntary options, developed by IRENA. The following five options are presented as particularly beneficial for the G20 action: (1) in-depth and country-specific analyses of renewable energy costs and reduction potentials, (2) exchange good practice examples on enabling national policy frameworks, (3) development of renewable energy-specific risk mitigation instruments, (4) country-specific assessment of renewable energy technology potential and development of roadmaps and (5) support the sustainability indicators and further actions of the Global Bioenergy Partnership (GBEP), in close cooperation with IRENA and IEA Bioenergy.

In addition, G20 members decided to explore the potential for increased regional infrastructure connectivity and cross-border investment to enable greater levels of investments in renewable energy, and to continue the support for international cooperation, including capacity building for developing countries and encouraging the use of existing cooperation platforms. In 2019, energy transformation has been officially put on the agenda at the ministerial meeting on "Energy Transitions and Global Environment for Sustainable Growth" in Japan (G20 2019).

Today's fragmented energy landscape increasingly amplifies the contours of a multipolar world. It is clear that in its current state, energy governance is far from being comprehensive, efficient and effective to steer a global energy transition. Moreover, in the current geopolitical environment the efforts to strengthen theglobal cooperation and work on global public goods seem more and more futile. At the turn of the new decade of the 2020s, personal ambitions of politicians determine politics. These are less directed to multilateral negotiations rather than to bilateral tit-for-tat zero-sum games. The volatility of personal relationships among the world leaders, where selfish nation-first policies dominate, increasingly compromises the stability and continuity of international relations.[4]

In the same vein, the global geopolitics around climate have changed fundamentally: Since the adoption of the Paris Agreement in 2015, the consensus among major powers has faded. COP 25 in Madrid in 2019 ended without a clear statement on raising the ambitions of the nationally determined contributions (NDCs), with the communiqué being watered down by the US, Brazil and Australia. The 'NDC explorer' shows, that the absolute majority of the countries put renewable energy first, while carbon capture, utilization and storage (CCUS) technologies are hardly mentioned. Even when taking into account that the NDCs have been produced under time pressure and that they may not be the best-grounded pledges, massive political and financial gaps are obvious and the ambitions are far too low (Pauw et al. 2019). This altogether makes energy (transition) governance—albeit granular, selective and regional—more important than ever.

---

[4]We owe these thoughts to Carlos Pascual and his inspiring presentation in December 2019 in Berlin.

3   The Energy Transitions and Their Geopolitical Impact

Policymakers at all levels face the Herculean task of making energy systems more sustainable and climate friendly. Moreover, at the same time, they have to ensure the supply of, e.g. fossil fuels, for the transitional period without perpetuating the existing energy system (Westphal 2012). If one looks to the horizon 2050, in which the world aims to become carbon-neutral, the energy supply worldwide should be structured in such a way that the expected nine to ten billion people on earth have access to modern, affordable and sustainable energy supplies without further destroying the livelihoods of present and future generations (ibid.).

We assume that energy transition pathways differ and depend on countries' respective preferences and imagined energy futures (see Chapters in this volume). Thus, we defined energy transitions as an *intensional* policy-driven process which involves systemic shifts towards (a) sustainable and climate-friendly, economically efficient and secure energy system(s).

There is no single and simple solution to transitioning the energy system(s) in line with these paradigms, as stated by the IEA in its World Energy Outlook of 2019.

The IEA's WEO 2019 has been very clearly stating that there is no silver bullet at hand, but a combination of technologies ranging from energy efficiency, renewables, fuel switch, nuclear, CCUS, etc. as well as—not least—behavioural change are needed to put the world on track (see Fig. 4).

The systemic nature of energy transitions in general has been pointed out before, most notably in works on historical energy transformations (ex. Smil 2010; Kander et al. 2014, etc.). In its core, previous energy transitions have been transitions from one energy source (wood, coal, oil, electricity) and one type of energy converter (manpower, animal power, steam engine) to another. In all of these cases, major

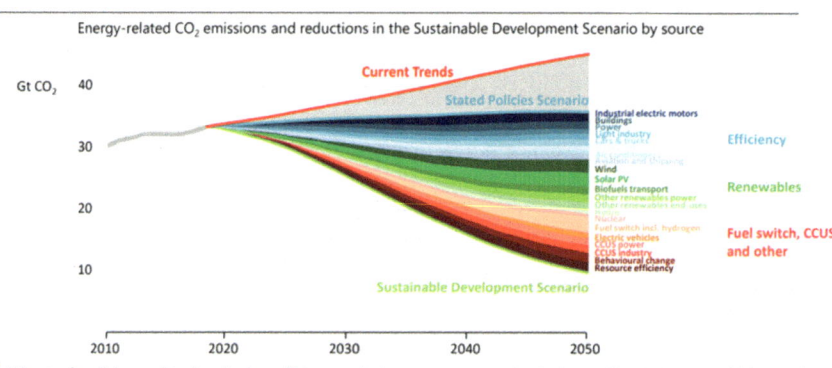

Fig. 4 No single or simple solutions to each sustainable energy goals. *Source* IEA (2019a, b), World Energy Outlook. All rights reserved

inventions of leading technologies such as the steam engine, electric lightning, etc., have kick-started processes of transition. However, all these transitions have been also accompanied by the profound and irreversible shifts on societal, ideological, political and economic levels (ibd.). In this regard, the energy transition that is taking place nowadays is no exception: it involves a shift to new, low-carbon energy sources. Yet, this time, the range and speed of the transition is and needs to be different. Moreover, the range of measures that will have to be deployed is enormous. As a result, the scale of socio-economic and political changes to be expected from the energy transitions happening around the world is not comparable to the historic cases. In other words, we have to think about energy transformation, in a way Karl Polanyi described the "Great Transformation"—neither national nor global energy systems are discrete elements. They are closely intertwined with politics, as well as with economic and social systems. A transition to a low-carbon economy doesn't just change the energy system. It has massive knock-on and distributional effects, causing re-allocation of resources both nationally and internationally.

How to think about the New World (IRENA, 2019) and the difficult, painful, but promising transition phase? (1) The energy transition(s) come with various *structural shifts* that create new patterns of energy supply and demand, investment and data flows, new infrastructure systems and new power balances. (2) The new system will be more electrified, digitized, demand-side driven and distributed. This requires large infrastructure to adapt, to modernize or to be developed, depending on the respective countries. (3) Today's energy system rests on individual sectors (i.e. electricity, buildings, transport, industry), each characterized by a dominant mix of (fossil) fuels (Goldthau et al. 2018). In the system of the future, the sectors (electricity, industry, heating and cooling, transport and mobility) will be coupled by the use of electricity and synthetic/decarbonized molecules and liquids. (4) As a consequence of these changes in the system, a relocation of production and demand as well as a reconfiguration of energy spaces will take place. (5) In the new energy world, the value is no longer generated primarily from the fossil fuel resource such as coal, oil or gas, but rather at the stage of conversion into end-user energy/services (ibid). In other words, more and more value will be created downstream of the energy supply chain and in services (e.g. lightning, heating, cooling, etc.). As a consequence, profits will be generated by the availability and use of low-carbon technologies.

Energy transformation does not only recalibrate energy value chains. It also (re)configures energy spaces, which are shaped by infrastructure, production chains, and industrial clusters. Energy infrastructure can be viewed as an "infrastructured" geography of "long durée" that shapes spaces and even creates its own "ecology" and topography (Högselius 2013). This is particularly true for electricity grids and their different shapes (central, decentralized) as well as sizes (local, national, trans/continental). The spatial effects of the energy transition(s) result from techno-economic change, e.g.in the shape of local micro grids or region-spanning super grids, such as those promoted by China's Belt and Road Initiative. Connectivity will be newly defined, knocking on existing interdependencies, alleviating old sensitivities and vulnerabilities, but also creating new ones. The interconnectedness of two critical infrastructures, the electric grid and the internet creates specific challenges and hybrid threats.

If we assume that the energy transition has tremendous political, economic and social effects, the interaction with international political and geographical factor is evident (Ivleva and Tänzler 2019). Geopolitics can be understood as dynamics that stem from the interaction of geographical factors and international politics (ibid.; Scholl and Westphal 2017). In international politics energy is (intended to be) used as a tool and means to influence political outcomes, achieve foreign policy goals and as a lever to project power (Ivleva and Tänzler 2019). The geopolitics of energy transformation constitute a governance challenge in its own right. There is a growing body of literature on the energy transition having a geo-economic and geostrategic character (Bradshaw 2014; Scholten 2018; Goldthau et al. 2018; Bazilian et al. 2019; IRENA 2019).

Importantly, the very notion of energy security will change along with the transformation of energy systems: To be more precise, in the oil-centered world of the past, national security and the issue of import dependencies were at the heart of energy security. In the new energy landscape, where electrification is a major trend, the stability of the power grid will be the defining feature of energy security. Not surprisingly, energy is and will remain at the heart of national sovereignty and/or statecraft, as Daniel Yergin's definition of energy security as "adequate, reliable supplies of energy at reasonable prices in ways that do not jeopardize major national values and objectives" hints to (Yergin 1988). Energy has always proven to be a major policy field, involving a strong role of the state, as it is closely related to its traditional tasks of providing prosperity, security and stability. The energy transition offers new opportunities to shape the energy system in a new vein, which is in line with national values and interests, while also providing and protecting global commons and goods.

While traditional geopolitics is related to power relations, the energy transition implies power shifts and alters the political economy on the national and the international levels. It creates winners and losers (IRENA 2019; Overland et al. 2019). Petrostates and coal-exporting countries are repositioning themselves in the international system as their major assets become de-valued (IRENA 2019). At the international level, fossil fuel producers are vulnerable to the fundamental changes caused by the energy transitions. Resilience to the energy transition effects depends on the percentage of fossil fuel rents in the GDP and the diversification of their economy (IRENA 2019). If petrostates such as Russia, Saudi Arabia or Iraq, etc., are confronted with declining oil rents, their socio-economic model and political systems come under severe pressure. Fossil fuel exporters are not only faced with a devaluation of their natural resources, but increasingly face fundamental challenges to their economic and social systems, since the resource wealth is part of the social contract. This in turn affects the political stability and economic growth in these countries. Yet, it is fair to say that the US shale gas and tight oil revolution has already caused a landslide by creating a new situation of energy abundance, shaking up the position of energy rich countries and diminishing their respective rents. In a sense, the US shale revolution has already anticipated certain effects for the petrostates.

Transit countries such as Ukraine, Morocco or Tunisia that gain rents from their midstream part in fossil fuel supply chains, will also feel the effects of energy transitions. Obviously, the energy transformation will have knock-on effects along the whole fossil fuel value chain making the exporting and transit countries to losers of an energy transformation. Evident winners are major importing countries, which will be able to produce more energy from renewables locally and at home or in cross-border cooperation within 'grid communities' (Scholten 2018), formed by political choice and not due to geological circumstances. At the same time, renewable technology leaders are emerging (IRENA 2019; Goldthau et al. 2018), gaining more and more political weight and a central place in the global markets. Hence, energy governance has to tackle the geopolitical ramifications of energy transformation and aim for a transition that is as smooth as possible. In this respect, the notion of a 'just transition', energy justice or evenness, is key for the global energy transition, in particular with regard to the Global South (see Goldthau in this volume).

Finally, the energy transformation has profound and even disruptive structural effects. At the national level, it inherently entails structural ruptures and puts stress on the incumbent energy system. Incumbent utilities like the German companies E. On and RWE lost significantly in their market capitalization and/or changed their asset base, which also split their renewable branches. In the socio-technical realm, a paradigm shift will have to take place, with the end consumer and/or the community moving into focus. Consumers are becoming key actors as both, consumers and producers ("prosumers") of sustainable energy, which requires a behavioural change beyond energy saving and efficient energy use. The EU, for instance, has paid a tribute to this paradigm shift by focusing on the end consumer in its 'Clean Energy Package for all Europeans' of 2017. In general, this paradigm shift from supply to the end consumer has three dimensions. First, consumer behaviour is critical to the success and speed of an energy transition: consumers have to take up responsibility, be empowered and become to a certain extend 'owners' of the transition. Second, there is the social dimension of access, availability and affordability of (modern) energy: among others, governance measures are needed to deal with e.g. (temporary) price increases. Last, but not least, structural ruptures such as a 'coal phase out' are part of the decarbonization. This requires states to mitigate social risks, e.g. ones expected from the coal phase-out negotiated in Germany in 2019 (BMWi 2019). Germany's decision to phase out coal by 2038 was achieved by a societal consensus after a long negotiation process and backed by a set of structural policies in coal regions. These policies translate into a long phase-out over almost 20 years, depending on the age (and technology used) of the respective power plants. Other countries might have the chance to leap-frog this technology, or might have to close down coal-fired power plants before their respective lifespan ends(as is likely to be the case for some countries in Asia) (IEA 2019a, b). In any case, the socio-economic dimension of the energy transformation cannot be underestimated.

As mentioned above, energy transition affects societies. Energy transition has a considerable effect on labor markets. Germany is a case in point where the narrative of creating green jobs has not really (or only temporarily) delivered. When feed-in tariffs triggered the diffusion of PV in Germany, the 'solar valley' in Sachsen-Anhalt boomed.[5] Yet, as China began to expand its solar panel manufacturing industry, many involved in the same sector in Germany lost their green jobs. Socio-technical dynamics have to be closely analysed and the respective policy measures developed: In Germany, traditional energy sectors are covered by unions, primarily the respective labor union IGBCE (IG Bergbau, Chemie, Energie), whereas workers in new industries such as solar and wind energy are not organized in trade unions and therefore don't have the same kind of support.

This particular challenge to tackle the socio-economic effects of the energy transition is thus increasingly debated as the "just energy transition". The focus on a just transition is inextricably linked to the question of "who wins, who loses, how and why". It is imperative to ask this question both in relation to the existing distribution of energy (e.g. "who lives with the side effects of extraction, production and generation?"), and with regards to the ongoing energy transformation (e.g. "who will bear the social costs of decarbonizing energy sources and economies?") (Newell and Mulvaney 2013, p. 3). This, in turn, necessitates addressing the issues of distribution and access to political power, natural, social and economic resources, and the political economy behind socio-technical energy transitions (Goldthau and Sovacool 2012). Attention has to be paid to the interrelation between a just energy transition and the speed of decarbonization, though. In the beginning, mitigating social effects can be an impediment to moving ahead with the rapid decarbonization, but a sound social consensus is needed as a stabilizing element to transform the energy system. The creation of green jobs can serve as a catalyst and is even more important than social measures to compensate for income losses and job cutbacks. These social aspects are moving to the political core of many Western societies, where, i.e. resulting from such movements as "Fridays for Future", the energy-climate cleavage has started to influence the politics and polarize the societies.

4   Conclusions and Recommendations

The message this chapter can't emphasize enough is that the energy transition(s) will play out differently in countries and regions, but they all will have a huge impact on all levels: the global, the regional, the national and the local. Moreover, while the targets and paradigms are in place, creating the institutional framework fit to steer the transitions' pathways remains an open issue. As there is no simple and single solution (see Fig. 4), there is no one-size-fits-all approach to governance. Instead, what is needed is a flexible multilevel architecture which is (1) reflecting new connectivities and energy spaces, (2) enabling, promoting and diffusing new technologies and know-how and (3) adapting the institutional and regulatory framework to the changes that come along with the transformation of the energy system(s) (see Fig. 5). There is a need for a better global and regional, and a good governance on the national and

---

[5]https://www.mdr.de/sachsen-anhalt/dessau/bitterfeld/solar-valley-solarzellen-photovoltaik-chronologie-q-cells-solibro-100.html.

Policy Directions for Global Energy Transition on Governance	Needed policies and cooperation mechanisms	National	Bi-/Plurilateral	Regional	Global
Reflecting New Connectivities and Energy Spaces	Development of cross-border infrastructure and grid planning		◆	◆	
	Management of cross-border infrastructure		◆	◆	◆
	Creating New Markets (e.g. for raw materials etc.)	◆	◆	◆	◆
	Consumer-producer Dialogue	◆			◆
Institutional and Regulatory Framework	Coalitions of the Willing		◆	◆	◆
	Preserving and Reforming the World Trade Organization (focus on IPR and patents)				◆
	Greening of International Financial Institutions	◆	◆	◆	◆
	Technical and regulatory dialogue	◆	◆	◆	◆
	Dialogue on critical infrastructure resilience	◆			◆
	Exchange on Energy Forecasts and Common Modelling Exercises		◆	◆	◆
	Stepping up Climate Disclosure Mechanisms	◆		◆	◆
Enabling, Promoting and Diffusing New Technologies and Know-How	Best Practices and Know How Exchange (i.g. „Patents")	◆	◆	◆	◆
	Tandems between Global North and South in promoting specific technologies		◆		◆

Fig. 5 Policies and cooperation mechanisms for global energy transition governance. *Source* Authors' analysis

local level. Particularly in today's world of nation-first policies and geo-economics, it is imperative to establish, maintain and improve the multilateral energy landscape.

Moreover, the governance task is not only to move forward with the energy transition, and to transform the system, but also to deal with the geopolitical aspects which the energy transition brings about. Energy transformation comes with a cost, but the costs of doing nothing are higher, even if less immediate and more diffuse. The call on governance is evident, because it is assumed that this will make the transition(s) faster, smoother and more even. The transition period is assumed to have systemic transformational effects on political systems, economies and societies, and thus can be messy, disruptive and conflictive. Furthermore, the energy transition(s) are not taking place in a vacuum but have the potential to add to the geo-economic rivalry. Moreover, technology leadership and control over mineral resources can add to the struggles over political authority and power.

The global environment of great power rivalry and the crises of multilateralism are clearly complicating the global energy transformation. There is less political will to work together to create and preserve global common goods than in the past. Under these circumstances preserving the existing institutions such as WTO, IRENA and UN Energy that build on 'altruism' and are aiming at a level-playing field is of the utmost importance. If the global consensus is shaky, it will remain important to act plurilaterally, in clubs, coalitions and alliances. The existing governance structure (Figs. 2 and 3) can and has to ensure the functioning of today's energy system without perpetuating it. The format of Clubs, the comprehensive institutions of IAEA and IEA will have a role to play. This is equally true for IRENA, UN Energy and SE4All, which are directed to changing the energy system(s) along the paradigms of the strategic energy quadrangle. In order to ensure that the measures undertaken (see Fig. 4) not only contribute to diversification, but in fact *transform* the system, the efforts in governing energy efficiency and renewable energy have to be stepped up. Aside from the global level, new governance structures will have to be developed or adapted at the regional level.

Without an aspiration of being comprehensive, we recommend the following focus areas.

Governing Energy Regionalization and Connectivity. The energy transition will reshape regions, but also create webs and routes within and between them (connectivity). So far they are largely un- or only partly governed. Yet, the existence of large infrastructure (e.g. power grids) running across regulatory spaces will require new norms, rules and standards which deal with interoperability of systems and cross-border management of flows (Scholl and Westphal 2017; Overland et al. 2016; Balmaceda and Westphal forthcoming). Energy regionalization takes place without a recognition of existing jurisdictions and polities. The question of who defines the rules of the game in the "infrastructured" space or across production and value chains is very acute. New governance schemes are necessary, also to prevent regulatory fault lines from feeding into geopolitical conflicts. This has systemic, structural and spatial implications of a transboundary nature that have repercussions on the regional and global level. Among them are the global shift of investment and financial flows due to the changes in energy and technology markets; the emergence of new geographies of demand and supply; where the digitization comes with its own risks and challenges, as both energy, IT and telecommunication sectors are connected by super infrastructures, which are highly critical to modern societies. The cascading effects in case of a crisis or a 'black-out' demand for specific resilience measures to be taken in smart/super grid communities. This issue is already very tangible in the EU, where electricity security and grid stability have gained utmost importance. The creation of synchronized grid communities that include Turkey, and soon the Baltics and Ukraine, come with their own governance challenges, not least of them being connected to cyber security.

A Common Set of Global Rules. One immediate blind spot to address in global energy governance is the lack of a code of conduct and/or a set of common rules. Both are needed in order to create a level playing field as well as transparent and functioning markets. The more technology-driven the energy system will be, the more important will rule-setting organizations such as the WTO become. Patents and intellectual property rights will remain important in order to make profits from innovation, but at the same time, solutions will be needed to provide access to important technologies for developing countries.

Investments. One of the key challenges for a sustainable energy transition is to get the investments right and right in time. Under the current price regime of low energy prices and in an era of abundance of energy sources, price signals to turn away from fossil fuels will be too weak or simply lacking. Investments into production sites and infrastructure predetermine and cement path dependencies given the long lead times and lock-in effects they create. Policy measures and regulatory frameworks will be key in the transition toward a sustainable global energy system. Institutions will play a central role in incentivizing and realizing the big shifts in technology, as well as creating and capturing the value and creating new business models. This is also related to the question of who will finance the necessary infrastructure. Therefore, 'shifting the trillions' and getting the financial and taxation framework right, is of a paramount importance. In many countries with high renewable energy and energy efficiency potential, the cost of capital is too high. New power grid infrastructure, renewable energy facilities, development of energy efficient buildings and appliances, restructuring transport sector, etc., require huge sums of infrastructure investments, great coordination efforts, and a stable regulatory framework to realize the shift. The unprecedented oil price slump triggered by the Covid-19 pandemia in the first quarter of 2020 is burning capital which will be missing in the energy sector and for the transformation as a whole.

Technology-specific governance schemes and mechanisms. Moreover, there are many issues around specific measures and technologies to be addressed by specific governance mechanisms. The table above is not considered comprehensive. Yet, it aims to visualize the complexity of the tasks faced by the global community. The instrumental and operational level aims to grasp the multitude of the technologies, components and tools needed to bring forward energy transition, among them the deployment of low-carbon energy sources, Power-to-X and synthetic fuels, but also the new approaches to power grid design (e.g. smart grids or super grids) as well as energy efficiency, sector coupling, and phase-out of fossil fuels. These elements and tools are employed altogether or in part, based on the political agenda and preferences of the respective countries—their efforts on the operational are therefore highly heterogeneous.

A consumer-producer dialogue. One of the key mechanisms to ensure a successful energy transition is an enhanced and effective consumer-producer dialogue. Such a dialogue is particularly important for depletion strategies and gradually phasing-out of fossil fuels. Moreover, this dialogue can create new partnerships to produce, trade and transport 'de-carbonized' molecules and fuels (hydrogen and Power-to-X) and coopt fossil fuel producers. The International Energy Forum (IEF), whose primary focus is consumer-producer dialogue, is not delivering on that. It seems that it would be more fruitful to move ahead with the 'clubs' and 'coalitions of the willing' to e.g. gradually develop hydrogen markets.

Get the institutions right. Steering energy transition on the global level requires enhanced technical and regulatory dialogue as well as a continuous exchange on best (and worst) practices among countries, regions and communities. There is no lack of targets, but the major challenge is to create effective incentives, frameworks and regulations to implement and accelerate the energy transformation. In addition, exchange and cooperation on a knowledge- and database on energy (including exchange on energy forecasts and common modelling exercises) must be developed, to provide transparency needed for an efficient energy transition governance. For these modes of inter- and transnational cooperation, the multilateralism of energy governance institutions should be kept up and preserved. Coordination and coherence among the existing institutions, clubs, and coalitions of the willing must be strengthened, while governance structures where cooperation on energy transition can take place in a level-playing field and is not driven by geo-economic rivalry have to be developed. For this, countries and other relevant stakeholders have to abandon the traditional perception of the quadrangle of energy security, economic efficiency, climate and sustainable development as mutually exclusive dilemmas, and put synergies from coordinating these four elements to the forefront instead.

Tandems between Global North and South are imperative to efficiently addressing the issues of energy poverty and just energy transition globally and to making sure that no region is left behind. With regard to the deployment of renewables, 'tandems' may be a way forward as pursued in the G20 with phasing-out fossil fuel subsidies. Healey and Barry (2017) rightfully highlight an "increasing inequality—of income, wealth and resource ownership" in general, and rising "inequality of access to safe and affordable energy" as well as "energy poverty" as major challenges. Energy transition regimes must address inequalities in power and injustices across entire socio-energy systems. The issue of energy justice must be incorporated into all governance mechanisms. This can be done by paying tribute to the eight principles of energy justice: availability, affordability, due process and information, responsibility, sustainability, intra-generational and inter-generational equity (Sovacool and Dworkin 2015).

Education, research and information. At the end of the day, the benefits of energy transition have to be communicated and pushed forward in the public debates: for instance, human security gains as improved air and water quality are among the 'dividends' of energy transition (Goldthau et al. 2018). Education and information are key in addressing societies in general and in particular increasingly polarized societies.

Approach climate and energy security through the lens of public goods. Energy transition has the potential of re-localizing the economy around human-scale enterprises rooted more closely in the communities they serve. Internally, energy transition should be 'democratized' as entailing a shift towards empowerment and ownership, transforming end-users into "prosumers" in the true sense of the word. Normatively, the aim should be a reconfiguration of "transition arenas" from spaces for 'coalitions of frontrunners' towards more open spaces for deliberation, dialogue and participation (Barry et al. 2015). For this, approaching energy and climate security through the lens of global public goods as opposed to a strategic national interest is important.

We conclude that a just energy transition on all levels, but certainly on the national and the international level, is necessary to keep countries and the world on a sustainable energy transition path. It is essential for the international community to stick to the Paris Agreement in order to keep up the level of ambition nationally. But also vice versa, the international ambition cannot be sustained without an enduring social consensus in major countries.

References

Barry J (2015) Low carbon transitions and post-fossil fuel energy transformations as political struggles: analyzing and overcoming 'carbon lock-in', pp 3–23

Bazilian M, Bradshaw M, Goldthau A, Westphal K (2019) How the energy transition will reshape geopolitics. Nature 569(7754):29–31. https://doi.org/10.1038/d41586-019-01312-5

Bradshaw MJ (2010) Global energy dilemmas: a geographical perspective. Geogr J 176(4):275–290. https://doi.org/10.1111/j.1475-4959.2010.00375.x

Bradshaw MJ (2014) Global energy dilemmas. In: Energy security, globalization, and climate change. Cambridge, Malden, Polity

Cherp A, Vinichenko V, Jewell J, Brutschin E, Sovacool B (2018) Integrating techno-economic, socio-technical and political perspectives on national energy transitions: a meta-theoretical framework. Energy Res Soc Sci 37:175–190. https://doi.org/10.1016/j.erss.2017.09.015

European Commission (2015) Communication from the commission to the European parliament, the Council, the European Economic and Social Committee, the Committee of the Regions and the European Investment Bank: a framework strategy for a Resilient Energy Union with a Forward-Looking Climate Change Policy. European Commission. Brussels (COM(2015), 80 final). Available online at https://eur-lex.europa.eu/legal-content/EN/TXT/?uri=COM:2015: 80:FIN. Accessed 20 Dec 2019

Federal Ministry for Economic Affairs and Energy (BMWi) (2015) Making a success of the energy transition: on the road to a secure, clean and affordable energy supply. https://www.bmwi.de/Redaktion/EN/Publikationen/making-a-success-of-the-energy-transition.pdf?__blob=publicationFile&v=6. Accessed 20 Dec 2019

Federal Ministry for Economic Affairs and Energy (BMWi) (2019) Commission "Growth, Structural Change and Employment". Final Report (orig. Germ.: Kommission "Wachstum, Strukturwandel und Beschäftigung". Abschlussbericht). https://www.bmwi.de/Redaktion/DE/Downloads/A/abschlussbericht-kommission-wachstum-strukturwandel-und-beschaeftigung.pdf?__blob=publicationFile&v=4. Accessed 20 Dec 2019

Federal Ministry of Economics and Technology, Federal Ministry for the Environment, Nature Conservation and Nuclear Safety (2011) Energy concept for an environmentally sound, reliable and affordable energy supply. https://cleanenergyaction.files.wordpress.com/2012/10/german-federal-governments-energy-concept1.pdf. Accessed 20 Dec 2019

G20 (2019) G20 Karuizawa innovation action plan on energy transitions and global environment for sustainable growth. In: Ministerial meeting on energy transitions and global environment for sustainable growth. Karuizawa, Japan. https://www.meti.go.jp/press/2019/06/20190618008/20190618008_02.pdf. Accessed 20 Dec 2019

G290 (2013) G20 Leaders' Declaration. St. Petersburg. http://www.g20.utoronto.ca/2013/2013-0906-declaration.html. Accessed 20 Dec 2019

Global Commission on the Geopolitics of Energy Transformation (2019) A new world. In: The geopolitics of the energy transformation. International Renewable Energy Agency (IRENA), Abu Dhabi, United Arab Emirates

Goldthau A, Sovacool BK (2012) The uniqueness of the energy security, justice, and governance problem. Energy Policy 41:232–240. https://doi.org/10.1016/j.enpol.2011.10.042

Goldthau A, Keim M, Westphal K (2018) The geopolitics of energy transformation. governing the shift: transformation dividends, systemic risks and new uncertainties. In: SWP Comment (42). https://www.swp-berlin.org/fileadmin/contents/products/comments/2018C42_wep_EtAl.pdf. Accessed 20 Dec 2019

Healy N, Barry J (2017) Politicizing energy justice and energy system transitions: fossil fuel divestment and a "just transition". Energy Policy 108:451–459. https://doi.org/10.1016/j.enpol.2017.06.014

Högselius P (2013) Red gas. Russia and the origins of European energy dependence. 1st edn. Palgrave Macmillan (Palgrave Macmillan transnational history series), Basingstoke, New York

IEA/OPEC/OECD/World Bank (2010) Analysis of the scope of energy subsidies and suggestions for the G20 initiative. Joint report, prepared for submission to the G-20 Summit Meeting Toronto (Canada). https://www.oecd.org/env/45575666.pdf. Accessed 20 Dec 2019

IEA/OPEC/OECD/World Bank (2011) Joint report by IEA, OPEC, OECD and World Bank on fossil-fuel and other energy subsidies: An update of the G20 Pittsburgh and Toronto Commitments. Prepared for the G20 Meeting of Finance Ministers and Central Bank Governors (Paris, 14–15 October 2011) and the G20 Summit (Cannes, 3-4 November 2011). https://www.oecd.org/env/49090716.pdf. Accessed 20 Dec 2019

International Energy Agency (IEA) (2019a) Clean energy transitions programme. In: Accelerating clean-energy transitions in major emerging economies. https://www.iea.org/areas-of-work/programmes-and-partnerships/clean-energy-transitions-programme

International Energy Agency (IEA): World Energy Outlook (2019b) IEA, Paris. https://www.iea.org/reports/world-energy-outlook-2019

International Renewable Energy Agency (2018) Global energy transformation: A Roadmap to 2050, International Renewable Energy Agency. Abu Dhabi, United Arab Emirates. https://www.irena.org/-/media/Files/IRENA/Agency/Publication/2018/Apr/IRENA_Report_GET_2018.pdf. Accessed 20 Dec 2019

Ivleva D, Tänzler D (2019) Geopolitics of decarbonisation: towards an analytical framework. Adelphi. Berlin. https://www.adelphi.de/en/publication/geopolitics-decarbonisation-towards-analytical-framework. Accessed 20 Dec 2019

Kander A, Malanima P, Warde P (2014) Power to the people. In: Energy in Europe over the last five centuries. Princeton University Press, Princeton

Krause F (1980) Energie Wende. Washstum und Wohlstand ohne Erdol und Urana. S Fischer Verlag, Frankfurt a M

Ministry of Foreign Affairs of Japan (2018) 'Evolving energy diplomacy—Energy transition and the future of Japan. https://www.mofa.go.jp/files/000383102.pdf, Accessed 20 Dec 2019

Newell P, Mulvaney D (2013) The political economy of the 'just transition'. Geogr J 179(2):132–140. https://doi.org/10.1111/geoj.12008

Overland I, Scholl E, Yafimava K, Westphal K (2016) Energy security and the OSCE: the case for energy risk mitigation and connectivity, SWP Comments2016/C 26, May 2016

Overland I, Bazilian M, Uulu TI, Vakulchuk R, Westphal K (2019) The GeGaLo index: geopolitical gains and losses after energy transition. Energy Strategy Rev 26:100406

Pauw WP, Castro P, Pickering J, Bhasin S (2019) Conditional nationally determined contributions in the Paris agreement: foothold for equity or achilles heel? In: Climate policy, pp 1–17. https://doi.org/10.1080/14693062.2019.1635874

Roehrkasten S (2015) Global governance on renewable energy. In: Contrasting the ideas of the German and the Brazilian Governments. Springer Fachmedien Wiesbaden, Wiesbaden

Roehrkasten S, Westphal K (2013) IRENA and Germany's Foreign renewable energy policy. aiming at multilevel governance and an internationalization of the energiewende? In: Working paper FG08. https://www.swp-berlin.org/fileadmin/contents/products/arbeitspapiere/Rks_Wep_FG08WorkingPaper_2013.pdf. Accessed 20 Dec 2019

Scholl E, Westphal K (2017) European energy security reimagined. mapping the risks, challenges and opportunities of changing energy geographies. In: SWP research paper (4/2017)

Scholten D (2018) The geopolitics of renewables, no (61). Springer International Publishing, Cham

Smil V (2016, 2010) Energy transitions. In: History, requirements, prospects. Santa Barbara [etc]: Praeger

Sovacool BK, Dworkin MH (2015) Energy justice: conceptual insights and practical applications. Appl Energy 142:435–444. https://doi.org/10.1016/j.apenergy.2015.01.002

United Nations (2019) Sustainable energy for all: SDG 7. https://sustainabledevelopment.un.org/sdg7. Accessed 20 Dec 2019

van de Graaf T, Westphal K (2011) The G8 and G20 as global steering committees for energy: opportunities and constraints. Global Policy 2(4):19–30. https://doi.org/10.1111/j.1758-5899.2011.00121.x

van de Graaf T, Lesage D, Westphal K (2010) Global energy governance in a multipolar world. Routledge (Global environmental governance), London, New York

Westphal K (2012) Globalising the German energy transition. SWP Comments, 14. Berlin: Stiftung Wissenschaft und Politik

Westphal K (2015) International energy governance revisited: fragmented landscapes, diverging dilemmas, and emerging (dis)orders. In: Knodt M, Piefer N., Müller F. (Hrsg.), Challenges of European external energy governance with emerging powers, Ashgate, Farnham, pp 289–306

World Energy Council (2014) Global energy transitions: a comparative analysis of key countries and implications for the international energy debate. http://wec-france.org/DocumentsPDF/donnees/Global-Energy-Transitions-2014.pdf. Accessed 20 Dec 2019

Xi J (2017) Speech at the 19th National Congress of the Communist Party of China. Peking. http://www.xinhuanet.com/politics/19cpcnc/2017-10/27/c_1121867529.htm. Accessed 20 Dec 2019

Yergin D (1988) Energy security in the 1990s. Foreign Aff 67(1):110. https://doi.org/10.2307/20043677

Setting Up a Global System for Sustainable Energy Governance

Vladimir Zuev

1 Energy Geopolitics: From Security Above Anything to Sustainability Among Everything

The global energy landscape is currently shaken by tectonic shifts.[1] We witness dramatic changes in energy geopolitics, the formation of the global system of energy governance, a huge wave of massive technological innovations, global markets are undergoing a radical transformation embracing a fast multiplication of new sources of energy, new products, new producers, and suppliers, coupled with the development of the vast and sophisticated infrastructure and an increasing efficiency in energy use. Each and every component of the system is touched upon by a wind of change that brings about the contours of a new energy global order.

One core element of the current transformation is evidently the transition from fossil fuels to renewables or rather a rapid rise in renewables usage, especially in those parts of the world which used to be poor in possession of traditional fossil fuels. This transformation will mean a radical shift in the focus of energy geopolitics. Some aspects of this shift have been already mentioned in several studies (Overland 2019). We'd like to outline one more aspect of the new energy geopolitics that seems important for this analysis.

Western countries used to be heavily dependent on oil imports from the Middle East (oil crises of 1973 and 1979), or on gas imports from Russia (Hassanzadeh et al. 2014) (the gas crises between Russia and Ukraine in 2006 and 2009) (Sharples 2016), and on the constantly rising oil prices that reached a peak of $150 a barrel before the global financial and economic crisis. Nowadays, we can register a new phenomenon

---

[1]The study was supported by the World Economy and International Politics Faculty of the National Research University Higher School of Economics, with a contribution from Frolova K., Bukanova D., Ermolaev D.

V. Zuev (✉)
National Research University Higher School of Economics (NRUHSE), Moscow, Russia
e-mail: vzuev@hse.ru

M. Hafner and S. Tagliapietra (eds.), *The Geopolitics of the Global Energy Transition*, Lecture Notes in Energy 73, https://doi.org/10.1007/978-3-030-39066-2_16

that we can call a "reversed dependence." This time, developing countries producing an abundance of fossil fuels are becoming increasingly dependent on consumers from the major developed and emerging economies. The higher the level of renewables will be in the future energy mix the less fossil fuels will be needed globally. If that happens (and there is a high chance it will), this may mean a terrific blow to the developing countries' energy incomes, affecting many economies and re-shaping geopolitical influences.

The oil-producing countries themselves come to realize all the more, the necessity to diversify their economies away from fossil fuels dependence (OPEC 2019), which could be yet another factor for sustainable economic development. For energy-exporting developing countries that rely on revenues from a limited range of natural resources, the need to reorient their economies is also growing because of an emerging regulatory framework on climate change. Achieving diversification is considered vital for the long-term economic sustainability of their economies. The notion of economic diversification suggests a strategy to transform the economy from using a single resource, or a relatively narrow set of income sources, into multiple sources of income or a considerably broader variety of new and emerging economic sectors. Such a diversification pathway or strategy may be driven by various motivations, with the key objective being to boost economic performance along with sustainable growth.

Energy security, perceived above all as an important element of military and political security, has been one of the top priorities of each and every government of the international community (Blumer et al. 2015). The rapid growth of renewable energy sectors is providing room for a different way of geopolitical thinking, focusing more on economic and sustainability aspects of energy production within, or outside of the national economies, rather than focusing primarily on military and political aspects of energy security at a time of being fully dependent on energy imports (Kelanic 2016; Barnes and Jaffe 2006). The shift can also be traced by using the Energy Trilemma Index of the World Energy Council that is aimed to help countries to formulate better policy through balancing energy security, equity, and environmental sustainability. In other words, the more energy self-sufficient the countries become (most of the renewables can be produced practically everywhere), the better the chances are for sustainable energy development.

It is not only the rise in renewables that is the key to understanding the new system. It is not a mere transition from one energy source to another. In each sector, energy production and consumption become different: more efficient, reliable, safe, affordable and available to everybody, eco and climate-friendly. The bottom-up market and technology-driven transformation of the energy sector, multiplication of energy production sites in each and every corner of the world, have completely revolutionized the way we consume and produce energy. By combining renewable energy, digital technologies, and advanced materials, supported by appropriate infrastructure, the world can modernize the energy system and reduce the flow and the waste of primary resources. Overall, energy usage is becoming, or at least should become sustainable.

2   Energy Governance Institutions—A Key to Sustainable Transformation

The market and tech developments represent a basis that is a necessary precondition for the introduction of the new principles into the energy governance. In order to push the economic development in line with the wishful sustainable scenario, one has to have an appropriate system of governance at hand. According to the International Renewable Energy Agency (IRENA) roadmap to 2050, "energy transformation" is possible due to *digitalization, education, and regulation.* We can agree that education is important and not alone in this sector, but in each and every field. Education for new energy generation is imperative. We could add that not only digitalization, but the technological progress in general (batteries, panels, etc.) is also vital for the development of renewables. And the governance challenge that we describe in this chapter, is for sure critical for this transformation.

A sustainable development goal on Energy, formulated as an *access to affordable, reliable, sustainable, and modern energy for all*, has been adopted within the United Nations by governments of 193 countries (UN 2015). The UN set up the foundations, basics for the legal framework of the energy sector development, according to the principles of sustainability and climate protection, as in the UN Convention on climate change, supported by concrete mechanisms provided for in the Kyoto protocol and in the Paris agreement on climate change. The UN has the largest country coverage, as the number of participants is close to totality. However, consensus is hard to reach, decisions are difficult to implement, and big deals are rarely finalized. That is the reason other institutions take their turn to fulfill the mission of global energy governance. We will take a brief look at some of the most important of them for the reasons of the suggested analysis.

The G20 stands next to the UN in terms of the scope and coverage of the sector, as its members represent about 85% of global GDP. The G20 as an informal institution is quick to react on the burning issues in energy governance, and energy was in focus of most G20 Summits of the recent decade as a high-profile issue. Energy efficiency has been a long-term priority for the G20 as it contributes to the optimum utilization of energy resources. G20 members agree that increased collaboration on energy efficiency can drive economic activity and productivity, strengthen energy security, and improve environmental effects. Energy security, economic efficiency, and environmental safety came to be fully integrated into the G20 sustainable energy development concept.

At the G20 Global Summit on Financing Energy Efficiency, Innovation and Clean Technology in Tokyo, Japan, June 2019, the CEOs of major investment funds and senior financiers joined G20 policy makers, deciding how to close the world's energy efficiency investment gap. The Summit was organized during the 2019 Japanese Presidency of the G20 in conjunction with the Ministerial Meeting on Energy Transitions for Sustainable Growth. The finance industry debated, with G20 government delegations, on the scope of finance for innovation required to boost the world's USD 240 billion annual energy efficiency investment market up to one trillion-dollars

(G20 2019a). Another priority was how to improve the world´s emerging sustainable energy finance markets. One measure envisaged, is the so called "*green tagging*"—the attachment of energy performance and data to financial performance and data. Increased transparency of the energy performance of banks' assets through their accelerated tagging could be an additional factor to achieve energy transformation. Green tagging can also serve as an instrument to inform regulators about the impact of energy efficiency on financial activity. At the G20 Global Summit on Financing Energy Efficiency, over 150 high-level delegates concluded a declaration on improving the energy performance of asset investments by financial institutions. They also set up a mode for the implementation of the 2017 G20 Energy Efficiency Investment Toolkit that could support transformation inside member economies. The G20 Energy Efficiency Finance Task Group has developed tools to enable 122 private banks and six public financial institutions, bringing USD 4 trillion for energy efficiency activities. Ministers adopted the "G20 Karuizawa Innovation Action Plan on Energy Transitions and Global Environment for Sustainable Growth," which will enhance cooperation at national, regional, and international levels (G20 2019b). They also agreed to the initiative "Research and Development 20 for Clean Energy Technologies (RD20)," to promote international collaboration among research institutes in G20 countries.

Though these are mostly recommendations, they are aimed at the point. The G20 Energy Efficiency Leading Program (G20 2016) called for the broadening and deepening of private sector engagement, including through the establishment of a Private Sector Energy Efficiency Investment Platform. Using best practices for energy performance through networks of leading financial institutions is another instrument of the transformation, like the UNEP FI Energy Efficiency Finance Platform. There are many more G20 initiatives, but the ones already mentioned, demonstrate the level of the G20's engagement to the global sustainable energy governance.

OPEC has a special role in global energy governance, as it has an extended mandate and can fix oil production quotas, thus affecting oil prices that can affect all energy prices indirectly. Its goals are well known, and though officially the goals are set to keep prices stable, reasonable, and to reduce price volatility, in reality, they are aimed more to protect the interests of producers in order not to suffer too much from the relatively low price levels, the way these prices happen to be most of the time since the global financial crisis of 2007–09. The deficit of impact on the global production levels, and accordingly on prices, was partially offset by the OPEC+ 11 non-OPEC countries agreement (now +10 as Equatorial Guinea became an OPEC Member in May 2017). Had this been done 5 years earlier, the effect would have been much greater.

Currently, the USA, not being a member of OPEC and not a part of the OPEC+ deal, has become a new important actor on the global energy markets, undergoing a spectacular transformation from a net importer to a net exporter of oil and gas after the shale revolution. The USA alone is capable to compensate for the reduction in the oil production quotas of all OPEC and 11 non-OPEC countries altogether, which downgrades the regulative effects of OPEC's action on oil production and oil prices. Drone attacks, or the USA—Iran 2020 military tensions, or coronavirus threat from China, do influence more, the volatility of oil prices than OPEC's actions. The meaning of these consequences is that governing the global security issues has become a factor more important for oil price volatility than energy-related governance per se. Thus, the role of the global institutions like the UN or the G7 could be no less important for global energy governance than the role of specialized energy institutions.

The transformation of the OPEC cooperative framework is very illustrative in a way that it brings together more countries to decide upon the global oil agenda. OPEC+ format becoming permanent is already making OPEC more globalized. Bringing China into the permanent cooperation frame is another change in this direction. Three High-level Meetings of the OPEC-China Energy Dialogue proved to be promising. The third Meeting provided a platform for knowledge exchange, and contributed to the deepening of energy dialogue in general between China and OPEC. Another illustrative change in OPEC's activities consists of the fact that the sustainability agenda becomes an all the more sensitive and important topic in its work. For OPEC Member Countries and other energy-exporting developing countries that rely on revenues from natural resources, the imperative need to reorient their economies is growing, owing mainly to an emerging stringent regulatory framework on climate change and a sustainable development agenda. Achieving diversification is proclaimed vital for the long-term economic sustainability of the OPEC economies (OPEC 2010, 2019). It remains to be seen how far a strategy to transform the economy from using a single resource, or a relatively narrower set of income sources, into one based on multiple sources of income, or a considerably broader variety of new and emerging economic sectors, will go. For us, it is important to note that such a diversification strategy may be driven by sustainable development concepts.

Energy markets become all the more globalized, like in the case of gas markets that used to be either national or regional connected by the pipe-lines infrastructure. The fast spread of the LNG facilities has led to the fast globalization of the gas markets. As the International Energy Agency notes in its World Energy Outlook, liquified natural gas will surpass pipeline gas as the main way of gas trade over long distances by 2030 (IEA 2019). Within the next 30 years, LNG's share in total gas demand is projected to rise from 20% in 2018 to 40% in 2040.

Global governance response to these fast changes is slow, and appears to be so far, inadequate. One can't claim that the *Gas Exporting Countries Forum (GECF)* becomes kind of the new OPEC for the gas markets. Though already in the Doha Declaration, adopted at the first GECF Summit in November 2011, member-states agreed upon the need for fair pricing with respect to a balanced distribution of market risks between gas producers and consumers. However, no robust mechanism to safeguard fair pricing was provided for. The idea of gas prices indexation to oil prices, or support of the long-term gas contracts, put forward at the 2nd Summit in 2013 in Moscow, came into conflict with the fast development of the global gas market with spot sales booming. There is simply no need and it is not realistically feasible to arrange the gas production in a similar way to oil production.

It is interesting to note that the fossil fuel-producing countries put a special emphasis on making evident a link between sustainable development and fossil fuels' continued production increase to the energy-consuming countries. At the 4th Summit of GECF in Bolivia in November, 2017, the participating countries focused on the promotion of natural gas as an environmentally friendly type of fuel, and on the need for using the potential of natural gas for the implementation of the UN approved Sustainable Development Goals and the Paris Climate Agreement. At the 5th GECF Summit in Malabo (Equatorial Guinea) in November, 2019, the Declaration itself was under-titled "Natural Gas: Energy for Sustainable Development," to highlight the need to use gas as the core source of energy for Africa. The same focus on sustainability has been noticed through OPEC's activities (see above). It becomes clear that the future of fossil fuels will depend to a large extent, among other factors, on the regulators' ability to make fossil fuels usage, eco-friendly, and sustainable. This task is becoming more ambitious under additional pressure from the fast rise in renewables usage, that will serve as an extra point of reference to the efficiency of the traditional energy sources.

Renewables surge can be largely accredited to the regulators' deliberate policies to push forward sustainable energy networks. An evident example of this kind could be the EU energy policies. *The International Renewable Energy Agency (IRENA)* is another global intergovernmental organization case that supports countries in their transition to a sustainable energy future, and serves as a platform for international cooperation, facilitating technology transfer on renewable energy. IRENA promotes the widespread adoption and sustainable use of all forms of renewable energy, in the pursuit of sustainable development, energy security, and low-carbon economic growth. The adoption of the Agenda for Sustainable Development and Sustainable Development Goals (SDGs), and the Paris Agreement on Climate Change in 2015, provided a powerful global signal for a transition to sustainable energy. Since then IRENA membership doubled to 154 countries and 26 more in accession, with around 1100 governmental representatives, demonstrating the Agency's rise in global significance for energy governance. The instruments of cooperation are innovative, compared to traditional energy institutions, the way renewables are innovative. The work is based on designing a centre of excellence for energy transformation, making heard the "Global voice of Renewables" and spreading around the "advice and support" initiatives, creating a Network Hub.

The paradox critical assessment of the IRENA's work could be expressed in a way that it is too much centered upon renewables. The global goal, officially formulated and advanced by IRENA, consists of supporting and fostering the "energy transit" and turning the energy system *from the one based on fossil fuels to another one that enhances efficiency and is based on renewable energy.* The *International Energy Agency's* (IEA) work, as well as the task of the Agreement on the International Energy Program signed in Paris in November, 1974 are similarly envisaging to reduce dependence on oil and to develop alternative energy sources.

However, in traditional energy sectors from coal to atomic energy, countries, and producers do also improve efficiency and security. For instance, in the OPEC Long-term strategy, adopted 2019, it is stated that "OPEC supports the development and promotion of technologies that advance the environmental performance of oil, and advocates the continuous improvement in standards for exploration and development activities" (OPEC 2010, 2019). Thus, these other traditional energy sectors could also be integrated into the concept of sustainable development, especially having in mind the realistic assessment that their importance will be still great for decades ahead. Though, it is natural that priority in IRENA's or in the IEA work is given to renewables, denying that much of a role for other sources of energy in the future system could be unnecessary discouraging for many countries and sectors in their efforts toward sustainability.

Many energy organizations that seemed to be far from the sustainable energy concept, are doing a lot, building alliances and shifting their priorities toward sustainable development. For instance, since 1996, IAEA joined the Uranium Group (since then it is called Joint NEA-IAEA Uranium Group). The Nuclear Energy Agency (NEA) itself was set up within the OECD framework, to deal with nuclear power safety. Recently, the IAEA and NEA under the Uranium Group started to coordinate their work to better meet the climate change targets, to advance research and technologies on small modular reactors, upgrading safety standards, and embracing human capital into the concept of nuclear energy development. All of these changes fall well into the concept of sustainable energy governance. Thus, the traditional energy institutions like OPEC, GECF, IAEA are in the process of active adaptation of their respective agendas, to the sustainable development goals that have to be further encouraged by the international community.

3 Setting Up a Global System of Sustainable Energy Governance

Global, regional, and national institutions set up a frame and a vector for the energy sectors' transformation and a top-down policy driven decarbonization, affecting both market conditions and fostering tech innovations, predetermining to a high extent the future energy landscape. We put into focus, global governance institutions and the way they act and should act to make energy transition sustainable. National energy systems are no longer isolated, they are becoming all the more interconnected and interdependent, thus making global governance—an objective imperative. With time, the necessity for coordinated action will be felt as more of a necessity. To make Global Energy Governance (GEG) ready to meet the challenges of the day is becoming an important global policy task. Which mechanisms, at a global level, will help to generate the major changes in a sustainable way? What is missing to make the energy sector sustainable? What model should be chosen to move forward at a global institutions level?

Analyzing the activities of the international energy organizations, we can find a lot of evidence of intensive cooperation links between them either, on ad hoc or on permanent basis, to support the argument of an emerging global energy network. Some cases are listed below.

The United Nations has its special place in the center of the system, setting up the concepts, the goals and institutions, providing a legal basis for energy governance by concluding conventions and agreements, and providing a framework for cooperation between all the major international energy organizations. Advancing the Sustainable Development Goals had a special milestone meaning for the global energy governance, with a particular emphasis to achieve the 7th Goal: "Ensure access to affordable, reliable, sustainable and modern energy for all." Numerous expert discussions in different formats, like at the Conference of Parties (COP) of the United Nations Framework Convention on Climate Change (UNFCCC), do have a huge capacity-building potential for energy governance.

The G20, G7, OECD, IEA assure the global coordination for sustainable energy. They work together on concepts and ways to achieve the transformation. Since the Pittsburgh G20 Summit in 2009, the IEA has actively contributed to all energy-related activities of the G20, including those on energy security, energy data, market transparency, renewable energy, energy access, energy efficiency, and phasing-out fossil fuel subsidies.

The G20 Summit's work on financing energy efficiency in June 2019 in Tokyo, was a very illustrative evidence of an intense cooperation between the global general competence organizations and specialized energy institutions. This work was coordinated by the *UN Environment's Finance Initiative (UNEP FI)* and the *International Partnership for Energy Efficiency Cooperation (IPEEC)*, as co-hosts of the G20 Energy Efficiency Finance Task Group ("EEFTG"). The experts, negotiators, and industry representatives were brought together under the Japanese Presidency of the G20 with the CEOs of major financial institutions, to find a common way forward in energy governance (UNEP Finance Initiative 2019).

The *World Energy Council* is another place where national committees from 100 countries and about 3,000 energy-related organizations, work together promoting the sustainable supply and use of energy. The World Energy Congress, which takes place every three years, is a strategic place for many countries where trends in the rapidly changing energy sector are discussed. In 2019, the congress was held in Abu Dhabi and the Council issued a Report on the "World energy scenarios," that showed global energy pathways to 2040 in line with the sustainable scenario.

June 7, 2017, IRENA and the State Grid Corporation of China (SGCC), the world's largest utility company, agreed to enhance cooperation on advancing the energy transition. The International Atomic Energy Agency (IAEA) and IRENA cooperate in the area of energy planning. Collaboration between the IAEA and IRENA was formalized by a Practical Arrangement in 2016. In 2019, the International Energy Agency and IRENA enhanced cooperation between the two organizations by signing a Memorandum of Understanding. The IEA plays an important role in the global energy debate, and co-operates with a broad range of international organizations and forums. It hosted a number of multilateral organizations at its headquarters in Paris, including the Clean Energy Ministerial Secretariat and the Energy Efficiency Hub. The IEA was also the facilitator for the Bio-Future Platform.

Each year, the IEA, the International Energy Forum (IEF), and OPEC, work together in a joint Symposium on Energy Outlooks, which becomes an important part of their working program. The symposium gathers senior analysts and delegates from energy producing and energy-consuming countries, bringing together oil companies, banks, and experts, to discuss the IEA World Energy Outlook and OPEC's World Oil Outlook. This dialogue is leading to greater convergence in the baseline data. During the 8th IEA-IEF-OPEC symposium, which was held in 2018 in Riyadh, Saudi Arabia, IEA Executive Director Dr. Birol, who began his career at OPEC before joining the IEA, emphasized that a major dialogue between the IEA and OPEC is critical to ensuring global energy security in an environmentally sound and economically sustainable way (IEA-IEF-OPEC 2018).

Sustainable energy agenda became a priority for many organizations that are not directly or solely linked to energy. For instance, the Sustainable Development Working Group of the *Arctic Council* of the eight Arctic states is committed to promoting sustainable development in the Arctic and improving the living conditions of Arctic communities in general. The spreading of green energy in the Arctic region will become one of the main vectors for the Arctic Council work in the coming years. Protecting the Arctic has a special meaning, therefore, energy projects in the Arctic should be agreed upon and supervised more rigorously from a sustainable energy perspective.

The Food and Agriculture Organization (FAO) is one of the IAEA's partners. Since 1964, the two organizations govern together the Joint FAO/IAEA Division of Nuclear Techniques in Food and Agriculture. The cooperation envisages, among other things, common targets, joint programming, co-funding, and coordinated management.

Global and regional organizations are in close cooperation on sustainable energy issues. In 2011, the IEA and ASEAN formally recognized their ongoing cooperation in energy-related activities by signing a Memorandum of Understanding focused on information-sharing, training, and capacity-building on key energy priorities in the region such as stable and affordable energy supply, power sector development and market integration, the ASEAN Petroleum Security Agreement, and energy efficiency. Another illustrative example of this kind could be the 2018 Agreement between the IEA and the African Union, for a strategic partnership toward a more secure, sustainable and clean energy future for countries across the African continent, through a memorandum of understanding. Eradicating energy poverty is a priority for the IEA, and the agreement will play a vital role in stepping up efforts to achieve secure and sustainable energy for all.

In October 2015 the IEA signed a Statement of Intent with the APEC Energy Working Group at the APEC Energy Ministers Meeting in Cebu, Philippines. This statement builds upon many years of extensive cooperation, and seeks to expand collaboration in areas including energy security, energy data and statistics, renewable energy, fossil fuel subsidy reform, energy market analysis, and capacity building.

Among the new lines of cooperation between global and regional institutions in the sphere of energy, we can mention the recent developments, when for the last year, the GECF and the EAEU have been discussing the models of cooperation. An agreement has been reached on sending, in the year 2020, an invitation to the

Eurasian Economic Commission, to take part in the upcoming GECF Summit. These organizations also discussed eventual spheres of cooperation to create a sustainable and transparent energy market with an option to sign a memorandum of cooperation.

We could further continue listing the evidence of intensive cooperation links between the global and regional energy organizations. Having looked through and having analyzed the above-cited cases, it becomes clear that a solid energy cooperation base has been established within a comprehensive format. There are far-reaching goals and concepts proclaimed. Formal and informal energy institutions do actively work and cooperate with each other in an intensive way to form a new energy system, sustainable and viable with a just place for everybody and access to energy for all. Having made the analysis of the activities of the major energy sector institutions, and taking into consideration the increased level of interaction between them, we can arrive to a conclusion that *a system of global energy governance is being actively formed.* We see more coordination at an international level between the policy regulators, more coherence, a higher role of global institutions in policy priorities formulation in general. We can also conclude that this system of *global sustainable energy governance*, has all the chances to be no less solid and important than recognized systems of global trade or financial governance.

4   Looking Forward to Sustainable Energy Governance

One method to advance new ideas within global energy governance may consist of bringing for consideration some of the existing mechanisms from the best regional (EU) and global practices (trade and finance). To what extent these can be applied at a global energy governance level remains to be seen. We made a first attempt of this kind in this book.

It appears that the *European Union* provides the best *model* so far, for regional and global energy governance (see chap. 2). The EU has moved in a tremendously dramatic and successful way from the common energy policy, intensified after the Lisbon Treaty, to the Energy Community and Energy Union concepts, all the way up to the recent Green Energy Initiative. The European Union's clean energy package sets up an ambitious target of reaching 32% for renewable energy's share in the total final energy consumption, by 2030. It was made with the support of IRENA, which is another demonstration of the intensity and importance of interlinks between the institutions within the system of the global energy governance. The European Green Deal re-announced in December 2019 by the European Commission, seems to be so far, the best comprehensive plan to achieve sustainable development (European Commission 2019a). The EU concept is coming as a global benchmark—a certain guide "how-to advance" a transformation to a prosperous, inclusive, and sustainable economy on the basis of clean energy concept.

It is not only the EU that can provide useful experience of implementing good governance procedures for the global energy governance. About three decades ago, there was another spectacular development at the global trade governance level.

Although few drawbacks can currently be registered within this system. Another governance case, which provides a useful reference for us, can be seen by the global financial governance that has been built up within a decade after the global financial crisis. After the global trade and financial governance surge, today comes the turn of the consistent global energy governance system. Both global trade and financial governance so far, are more advanced in the variety of instruments and more efficient in governing the system compared to global energy governance.

Assessing what is missing from the global energy governance initiatives, could be made by a comparative analysis between a respective global energy institution with the energy policies in the EU, and the ones in global trade and global financial governance according to a set of criteria. Before inventing a bike let's have a look whether we can make good usage of the already existing one. Some of the most important points for good governance are mentioned below.

Availability of the appropriate vision, *ideas, concepts, goals*. These are well advanced in the system of the global energy institutions (WEC, IEA, IEF, G20). The concept of sustainable energy is well advanced in the UN SDGs and in the agenda of major global institutions—G20, OECD, IEA, International energy forum, GECF, IAEA, OPEC, IRENA. In this chapter, we provided concrete examples of the good goals of different institutions which match in a good way with the concept of sustainable development. The focus within these concepts is made on structural shifts that create new patterns of energy investment, production, supply, and consumption, facilitated by technological transformation of economy, data flows, and new infrastructure systems.

Progress in establishing a solid *legal basis* for energy governance was evident (many conventions and agreements already signed). However, the international legal platform for energy governance could be further improved, mainly in the direction of more binding commitments from the stake-holders. Even in the area of atomic energy and IAEA activities, where binding commitments are badly needed, the reality remains that the nuclear security regime "is still a patchwork of voluntary, non-binding, non-transparent national commitments, ad hoc bilateral and multilateral initiatives, and vague legally binding measures that provide no specific standards that states must follow" (FMWG 2013).

Some important agreements, like the European Energy Charter Treaty, were supposed to set up the sound legal basis for cooperation between the energy producing and energy-consuming countries. The Russian Federation planned to be part of this order and signed the Charter. Unfortunately, the Charter was not ratified by Russia. Time went by and the cooperation environment changed dramatically. In May 2019, the European Commission adopted a proposal to modernize the Energy Charter Treaty (European Commission 2019b). The Commission recommended that the Energy Charter Treaty provided stronger provisions on sustainable development and on the energy transition, in line with recent agreements. In the coming years the necessity to either reconsider the existing legal order or to set it up anew, will only grow.

Monitoring procedures become a special issue for energy governance. With some exceptions, monitoring mechanisms within the system of global energy governance are either not provided for at all, or not well organized. For instance, OPEC sets quotas for oil production, but there is no OPEC body to monitor compliance with the fixed quotas. It remains under the responsibility of the countries concerned. Under the OPEC+ agreement there is no clear-cut monitoring mechanism at all. The same situation of the weak structures for monitoring is typical for the global emissions quotas. Thinking about the nuclear security, the 20/20 Commission of Eminent Persons recommended that binding agreements should "...give the IAEA a precise mandate to confirm that these (nuclear security) standards are being implemented" (IAEA 2008). One main achievement of the IAEA was the verification of Iran's nuclear program, re-confirmed in 2019, that allowed the achieving of a diplomatic breakthrough from 2015. A joint comprehensive plan of action followed, the implementation of which remains again under the control of the IAEA. Even with these instruments in place, suspicions and mistrust remain (D. Trump's Administration claims against Iran in the beginning of 2020 could be provided as proof). This controversial case is highly revealing. Irrespective of the US President's Administration's discontent at the results of the work of IAEA inspectors in Iran overall, monitoring could be considered as sufficiently reliable, as all the sites were under control of international inspectors with appropriate reports at hand. Working further on the reliability of monitoring procedures remains a challenge for sustainable energy governance.

Access to finance is critical for the success of energy governance. Proclaimed initiatives and goals should be supported by an appropriate financial contribution. The EU experience in encouraging the development of renewables by an extensive financial contribution, both in direct and indirect way, is a good illustration of this success story (see Chap. 2). So far, the common financial funds for energy systems' transformation and adaptation are rare to find, though the traditional fossil fuel energy-rich countries in good times of high energy prices and even now, could have created important common funds for common institutions like OPEC, GECF, EAEU, that could have increased their ability to meet the new energy order challenges in a much better way. The financial part of the Kyoto Protocol mechanism was also an important attempt to provide financial backing for the envisaged transformation.

The decisions of the above mentioned G20 Global Summit on Financing Energy Efficiency, Innovation and Clean Technology in Tokyo, June 2019 are a clear recent example of moving in a proper direction. Access to more finance for energy transformation in a sustainable way, is not only a problem for specialized classical and new energy organizations, but also for the general competence global institutions as well. The banking and the business community are all more involved in the energy transformation activity, as the market realities demonstrate to what extent energy efficiency investments or renewables investment, may happen to be profitable with or even without the support of the regulator, which seemed unlikely only a decade ago.

Transparency is fundamental for consensus building on energy transformation. Practically all global energy organizations do contribute in one way or another to increased transparency of the system by data collection, spreading of information, preparing reports, undertaking analysis, and working on statistics. Some of them do contribute more by preparing the regular comprehensive reports, like IAE or IRENA.

Others make an additional contribution by a joint effort, like in the case of the recent Energy Data Report presented to the G20 to support energy policies that was prepared by IEA in collaboration with IEF and IRENA (IEA 2018). However, some of the data sources are still missing or considered as not sufficiently reliable, like in the case of Iran's nuclear facilities, or like in the case of the actual oil production facilities, or CO_2 country related emissions. On the spot missions, the way they are provided for in the IAEA Treaty, could help to fill in the data gap. However, this mechanism is missing both in specialized organizations as in the case of OPEC, or similarly in the general Treaties, like in the Paris Agreement on climate change. More transparency necessitates more trust and vice versa. On the other hand, digital data platforms, modern systems of space tracking, and other technologies on the rise, as well as intensified governmental cooperation, could help filling in the remaining gaps in reliable data collection and verification.

Extended mandate of bodies that are set within the structures of international organizations (Secretariats, Commissions, Councils) predetermines to a large extent their governing role to build the future energy order. The downgraded role of common bodies remains the weak point for global energy governance. The big problem is that most of these bodies do not have enough competences to push forward the concepts and ideas of the organizations they represent. Unlike in the EU, most of the global energy governance bodies do have a very limited mandate. These energy organizations do have Councils, Commissions, and Secretariats, but their mandate is far from being so comprehensive and extended, compared to the one of the European Commission—an executive body of the EU, for example. Most structures of the energy organizations do not have the power of legal initiative, they can't exercise strict control over the policies of the member-states of the organization and they do not have the power of enforcement or arbitration. Among other things, setting up a kind of specialized *Dispute settlement body* for energy governance, the way it was useful in trade within the WTO for many years before the current deadlock of trade wars, seems to be a helpful move.

One can ask if this is rational or feasible at all for the energy organizations to have a structure more or less similar to the EU structure? However, if we consider the efficiency criteria and if we accept the fact that the EU energy policies could serve as a model for other institutions, then the mandate of the other organizations should be further extended.

Decision-making procedures represent yet another focal element for the success of any governance system. Most of the energy governance organizations do have consensus ruled systems. Reaching consensus for many countries is extremely difficult, especially when you have as members, countries from different continents, with different traditions and policy priorities with developed or developing economies and not least—energy abundant or energy-dependent economies. Thus, their interests differ to a large extent, and finding consensus is difficult. It makes the governance system slow to advance new decisions needed for the world. Consensus is easier to achieve for ecology and climate change, as awareness of urgency and utility of action is higher in these respective areas. Though, we know how difficult it was to reach consensus at the time of Kyoto protocol, and more so during the Paris agreement. It is still more difficult to reach universal agreement on decarbonisation paths. Some actors will not share the mere idea of decarbonization to the extent other countries will consider that necessary.

Simple majority or qualified majority decision-making prevails in the EU, and that is the reason that member countries are able to make so many important decisions for economic cooperation between them. Moving away from consensus and organizing the redistribution of votes between countries within the organization according to, and in proportion with certain criteria, is not only the reality of the European Union. For instance, in the IMF the voting power depends upon the financial contribution to the Fund. It is true that certain decisions in the energy organizations are also not made by the consensus rule. For example, the decision to accept a new member into OPEC could be passed with a three-fourth majority vote. That is a way forward to increasing the efficiency of decision-making in many global energy organizations (and probably not only in the energy sector). Not all (that is not realistic), but at least some decisions on specific issues relevant for the daily operation of an organization should be passed on a majority vote procedure. That is becoming more of a necessity, taking into consideration that more energy-ecology-climate related problems require faster, if not immediate solutions, and waiting a long time to reach a consensus would be unacceptable where a decision is desperately needed.

Technology transfer lies at the heart of the global energy sector transformation. Many organizations and conventions aim to support the increase in energy sustainability by means of technological exchange. That is clearly a distinctive feature of the emerging system of global energy governance. The Kyoto Protocol, the GECF, the IEA, the IEF, the IAEA, the IRENA, and even OPEC, all of the existing energy-related conventions and energy organizations do support, in an open and clearly defined way, the transfer of advanced technologies as an important element to arrive to a sustainable economy on the basis of sustainable energy. October, 2017 at the 2nd General Conference of the IEA on technological cooperation, the main focus was to identify opportunities for cooperation between the states in the exchange. The World Energy Council's annual report on Innovation insights is an important toolkit for countries wanting to navigate through the fast-changing innovation landscape. From a tech point of view, it becomes feasible and clear, how to make the transition. For instance, the IEA 2019 Report with guidance for policymakers on how to accelerate the decarbonization of the power sector, shows a clear way forward to sustainable energy. An additional point for consideration is the technical assistance procedures that will facilitate the transfer.

If there is a political will, the financial ability (see above) and the technical feasibility for the transfer, the one major remaining issue largely concerns the appropriate protection of intellectual property rights. If the matter relates to trade, there are mechanisms within the WTO for the protection of intellectual property rights. However, the multilateral trade governance system is currently in a mess. Does that mean that the global energy governance should depend upon the system of global trade governance? There should be a link, for sure. On the other hand, there should be some institutions within the energy governance system to deal with the specific issues of protection of intellectual property rights. Alternatively, an intensified cooperation with the WIPO with appropriate legal arrangements could be an important element of the secure transfer of advanced energy technology. The ambition could be not dealing only with the one-way transfer of technologies from the West toward the East, but increasingly on deals the other way around. Organizing the transfer of technologies and safeguarding the protection of the related intellectual property rights in a sound way, could be considered as a fundamental element of the emerging system of sustainable energy governance. The usage of the best practices and an exchange of information in this regard is already strong and supported by many energy organizations (IRENA, IEA).

Technical progress will make it possible to advance new secure and sustainable standards that humanity never dreamed of. Hydro-power and nuclear power stations became that much more advanced that a small-scale production becomes cost-efficient. Will we live to see the time when new global standards will set the maximum size for the hydro or the nuclear power plants, in order to minimize the environmental damage in event of the natural disasters, for instance? Bringing different security concepts together in an integrated global sustainable vision for the technically advanced new energy order becomes a realistic scenario under certain assumptions of the globalization of energy innovations.

Compliance and enforcement are at the heart of another critical issue for governance. Enforcement procedures, which are strong in the EU, are practically non-existent in the global energy governance institutions, and agreements (OPEC, GECF, G20, Paris climate). The system relies on nationally driven systems. Though the legal enforcement procedures and instruments of common control are weak in most global energy organizations with rare exceptions (IAEA), stakes for non-compliance are too high. Be it the consequences of non-compliance for climate change, or for nuclear safety, or for security of energy supplies, or for stability of energy prices. In the case of non-compliance with a tariff line in trading metals, or in another case of not abiding fully with the WTO rules of origin for a product, a country or a company can miss million or even billion-dollar deals. In the case of not-respecting the emissions' quotas under the Paris agreement, or the OECD or IAEA nuclear safety standards, humanity risks not only money, but people's lives. This situation of high stakes and high risks of energy governance non-compliance puts additional pressure on governments and companies, and puts forward the importance of a responsible attitude to energy usages major decisions. Thus, the chances for higher compliance on critical energy issues depend not only upon procedures and mechanisms within the energy organizations, but also to a large extent on social, moral, and ethical aspects of high-level officials in charge of the energy agenda. Conclusions on this point are left for the readers.

Looking at the present system of energy governance one can suggest many more things to come. According to the IRENA report "Global energy transformation—roadmap to 2050", "*The challenge that policy makers around the world face, is how to accelerate the transition. Transitioning from a global fossil-fuel powered energy system, built-up over several hundred years, to one that is sustainable, will require a much greater transformation than current policies envisage*" (IRENA 2018). We fully share this opinion and will continue the efforts to suggest ways and means for the transformation, to the benefit of all people concerned.

References

Barnes J, Jaffe AM (2006) The Persian Gulf and the geopolitics of oil. Survival 48(2006):143–162. https://doi.org/10.1080/00396330600594348

Blumer Y, Moser C, Patt A, Seidl R (2015) The precarious consensus on the importance of energy security: Contrasting views between Swiss energy users and experts. Renew Sustain Energy Rev 52:927

European Commission (2019a) A European Green Deal. https://ec.europa.eu/info/strategy/priorities-2019-2024/european-green-deal_en

European Commission (2019b) Energy charter treaty modernization: European Commission presents draft negotiating directives, Brussels. https://trade.ec.europa.eu/doclib/press/index.cfm?id=2017

FMWG (2013) Fissile Materials Working Group: Consensus Policy Recommendations. https://pgstest.files.wordpress.com/2013/04/fmwg-consensus-recommendations-june-2013.pdf

G20 (2016) G20 Energy efficiency leading programme. https://ec.europa.eu/energy/sites/ener/files/documents/G20%20Energy%20Efficiency%20Leading%20Programme.pdf

G20 (2019a) G20 Energy efficiency finance and investment 2019 Stock-take Report. https://www.meti.go.jp/press/2019/06/20190618008/20190618008_02.pdf

G20 (2019b) G20 Karuizawa innovation action plan on energy transitions and global environment for sustainable growth. http://www.g20.utoronto.ca/2019/2019-G20-Karuizawa-Innovation-Action-Plan.pdf

Hassanzadeh E, Henderson J, Honore A, El-Katiri L, Pirani S, Dickel R (2014) Reducing European dependence on Russian gas: distinguishing natural gas security from geopolitics. The Oxford Institute For Energy Studies, vol 92, p 3. https://www.oxfordenergy.org/wpcms/wp-content/uploads/2014/10/NG92.pdf

IAEA (2008) 20/20 vision for future, background report. https://www.iaea.org/sites/default/files/18/10/20-20vision_220208.pdf

IEA (2018) Energy Transitions in G20 Countries—Energy data transparency and market digitalization. IEA-IRENA-IEF-OLADE. https://webstore.iea.org/energy-transitions-in-g20-countries-energy-data-transparency-and-markets-digitalisation

IEA (2019) World Energy Outlook 2019. https://www-oecd-ilibrary-org.libproxy.ucl.ac.uk/docserver/caf32f3b-en.pdf?=157589697820196E1B520562

IEA-IEF-OPEC (2018) Joint Report on the eighth symposium on energy outlooks. Riyadh. https://www.ief.org/_resources/files/events/the-eighth-iea-ief-opec-symposium-on-energy-outlooks/agreed-summary-record-eighth-iea-ief-opec-symposium-on-energy-outlooks-clean1.pdf

UNEP Finance Initiative (2019) Senior financiers, investors and policy-makers agree importance of boosting energy efficiency investment at official G20 side event, Tokyo. https://www.unepfi.org/news/industries/investment/senior-financiers-investors-and-policy-makers-agree-methods-to-boost-investment-into-energy-efficiency-at-official-g20-side-event/

IRENA (2018) Global energy transformation: a roadmap to 2050. International Renewable Energy Agency, Abu Dhabi. https://www.irena.org/-/media/Files/IRENA/Agency/Publication/2018/Apr/IRENA_Report_GET_2018.pdf

Kelanic RA (2016) The petroleum paradox: oil, coercive vulnerability, and great power behavior. Secur Stud 25(2016):181–213. https://doi.org/10.1080/09636412.2016.1171966

OPEC (2010) OPEC Statement to the UN climate change conference. (COP 25/ CMP 15/CMA 2), Madrid. https://www.opec.org/opec_web/en/press_room/5809.htm

OPEC (2019) OPEC long-term strategy, Vienna, p 22. https://www.opec.org/opec_web/static_files_project/media/downloads/publications/OPECLTS.pdf
Overland I (2019) The geopolitics of renewable energy: debunking four emerging myths. Energy Res Soc Sci 49:36–40. https://doi.org/10.1016/j.erss.2018.10.018
Sharples JD (2016) The shifting geopolitics of Russia's natural gas exports and their impact on EU-Russia gas relations. Geopolitics 21:880–912. https://doi.org/10.1080/14650045.2016.1148690
UN (2015) Sustainable development goal 7. https://sustainabledevelopment.un.org/sdg7

Printed by Printforce, the Netherlands